INTERNATIONAL SERIES OF MONOGRAPHS ON PHYSICS

Quantum Gravity

CLAUS KIEFER

Institute for Theoretical Physics, University of Cologne

CLARENDON PRESS · OXFORD
2004

OXFORD

UNIVERSITY PRESS

Great Clarendon Street, Oxford OX2 6DP

Oxford University Press is a department of the University of Oxford.
It furthers the University's objective of excellence in research, scholarship,
and education by publishing worldwide in

Oxford New York

Auckland Bangkok Buenos Aires Cape Town Chennai
Dar es Salaam Delhi Hong Kong Istanbul Karachi Kolkata
Kuala Lumpur Madrid Melbourne Mexico City Mumbai
Nairobi São Paulo Shanghai Taipei Tokyo Toronto

Published in the United States
by Oxford University Press Inc., New York

© Oxford University Press 2004

First published 2004

A catalogue record for this title is available from the British Library

Library of Congress Cataloging in Publication Data

(Data available)

ISBN 0 19 850687 2

10 9 8 7 6 5 4 3 2 1

Typeset by the Author
Printed and bound in India
by Thomson Press (India) Ltd.

PREFACE

The unification of quantum theory with Einstein's theory of general relativity is perhaps *the* biggest open problem of theoretical physics. Such a theory is not only needed for conceptual reasons, but also for the understanding of fundamental issues such as the origin of the Universe, the final evaporation of black holes, and the structure of space and time.

Historically, the oldest approach is the direct quantization of Einstein's theory of general relativity, an approach which is still being actively pursued. This includes covariant methods such as path-integral quantization as well as canonical methods like the Wheeler–DeWitt approach or the more recent loop quantization. Although one arrives at a perturbatively non-renormalizable theory, quantum general relativity can yield physically interesting results in both the perturbative and the non-perturbative regime. It casts light, in particular, on the fundamental nature of space and time.

The second main approach is string theory. It encapsulates the idea that the problem of constructing a viable quantum theory of gravity can only be solved within a unification of all interactions. In this respect, it goes far beyond quantum general relativity. From a methodological point of view, however, string theory does not stand much apart from it. It is a natural extension of perturbative quantum gravity (from which it inherits the concept of a graviton), and methods of constrained quantization, which are crucial for canonical quantum gravity, appear at key stages in the theory.

Whereas there exist excellent textbooks that discuss string theory at great depth, the present monograph is the first that, to my knowledge, covers quantum gravity in this broad sense. The main part of the book is devoted to general concepts, the quantization of general relativity, and applications to cosmology and black holes. String theory is discussed from the point of view of its quantum gravitational aspects and its connection to other approaches. The edifice of theoretical physics cannot be completed without the conceptual unification that will be provided by quantum gravity. I hope that my book will convince my readers of this outstanding problem and encourage them to work on its solution.

This book has grown out of lectures that I gave at the Universities of Zürich, Freiburg, and Cologne between 1990 and 2003. My main intention is to discuss the general features that a quantum theory of gravity is expected to show and to give an up-to-date overview of the main approaches. The reader is assumed to have some familiarity with general relativity and quantum field theory. Comments can be sent to my e-mail address `kiefer@thp.uni-koeln.de` and are highly welcome.

It is clear that my book could not have been written in this form without the influence of many people over the past 20 years. I am in particular indebted to H.-Dieter Zeh for encouraging me to enter this field of research and for many stimulating and inspiring interactions. I also thank Norbert Straumann in Zürich and Hartmann Römer in Freiburg for providing me with the excellent working conditions that gave me the freedom to follow the research I wanted to do. Many people have read through drafts of this book, and their critical comments have helped me to improve various parts. I owe thanks to Julian Barbour, Andrei Barvinsky, Domenico Giulini, Alexander Kamenshchik, Thomas Mohaupt, Paulo Moniz, Andreas Rathke, Thomas Thiemann, and H.-Dieter Zeh. I am also deeply indebted to them and also, for discussions over many years, to Mariusz Dąbrowski, Lajos Diósi, Petr Hájíček, Erich Joos, Jorma Louko, David Polarski, T. P. Singh, Alexei Starobinsky, and Andreas Wipf. With most of the mentioned theoreticians I have actively collaborated and I want to take this opportunity to thank them for the pleasure that I could experience during these collaborations.

Cologne
January 2004

CONTENTS

1

WHY QUANTUM GRAVITY?

1.1 Quantum theory and the gravitational field

1.1.1 Introduction

Quantum theory seems to be a universal theory of Nature. More precisely, it provides a general framework for all theories describing particular interactions. Quantum theory has passed a plethora of experimental tests and is considered a well-established theory, except for the ongoing discussion about its interpretational foundations.

The only interaction that has not been fully accomodated within quantum theory is the gravitational field, the oldest known interaction. It is described very successfully by a classical (i.e. non-quantum) theory, Einstein's *general theory of relativity* (GR), also called *geometrodynamics*. From a theoretical, or even aesthetical, point of view, it is highly appealing, since the fundamental equations can be formulated in simple geometrical terms. Moreover, there exist by now plenty of experimental tests that have been passed by this theory without problems. One particular impressive example is the case of the binary pulsar PSR 1913+16: the decrease of its orbital period can be fully explained due to the emission of gravitational waves as predicted by GR. The accuracy of this test is only limited by the accuracy of clocks on Earth, which according to recent proposals for rubidium fountain clocks (Fertig and Gibble 2000) should approach an accuracy of about 10^{-16} (such a clock would go wrong by less than 1 s during a time as long as the age of the Universe). The precision is so high that one even needs to model the gravitational influence of the Milky Way on the binary pulsar in order to find agreement with the theoretical prediction (Damour and Taylor 1991).

The formalism of general relativity is discussed in many textbooks, see for example, Hawking and Ellis (1973), Misner *et al.* (1973), Straumann (1984), or Wald (1984). It can be defined by the Einstein–Hilbert action,

$$S_{\text{EH}} = \frac{c^4}{16\pi G} \int_{\mathcal{M}} \mathrm{d}^4 x \, \sqrt{-g} \, (R - 2\Lambda) \pm \frac{c^4}{8\pi G} \int_{\partial\mathcal{M}} \mathrm{d}^3 x \, \sqrt{h} K \ . \qquad (1.1)$$

Note that $c^4/16\pi G \approx 2.41 \times 10^{47}$ g cm s$^{-2} \approx 2.29 \times 10^{74}$(cm s)$^{-1}$ $\hbar$. The integration in the first integral of (1.1) goes over a region $\mathcal{M}$ of the space–time manifold, and the second integral over its boundary $\partial\mathcal{M}$ (with the positive sign referring to a space-like, the negative sign referring to a time-like boundary). The integrand of the latter contains the determinant, h, of the three-dimensional metric on the boundary, and K is the trace of the second fundamental form (see Section 4.2.1).

That a surface term is needed in order to obtain a consistent variational principle had been already noted by Einstein (1916a).

In addition to the action (1.1) one considers actions for non-gravitational fields, in the following called S_m ('matter action'). They give rise to the energy–momentum tensor

$$T_{\mu\nu} = \frac{2}{\sqrt{-g}} \frac{\delta S_m}{\delta g^{\mu\nu}} , \tag{1.2}$$

which acts as a 'source' for the gravitational field. In general, it does not coincide with the canonical energy–momentum tensor. From the variation of $S_{EH} + S_m$, the Einstein field equations are obtained,

$$G_{\mu\nu} \equiv R_{\mu\nu} - \frac{1}{2} g_{\mu\nu} R + \Lambda g_{\mu\nu} = \frac{8\pi G}{c^4} T_{\mu\nu} . \tag{1.3}$$

(Our convention is $R_{\mu\nu} = R^\rho_{\mu\rho\nu}$.)

A natural generalization of general relativity is the Einstein–Cartan theory, see for example, Hehl (1985) and Hehl *et al.* (1976) for details. It is found by gauging translations and Lorentz transformations (i.e. the Poincaré group), leading to the tetrad e^μ_α and the connection $\omega^{\alpha\beta}_\mu$ as gauge potentials, respectively. The corresponding gravitational field strengths are torsion and curvature. Torsion vanishes outside matter and does not propagate, but it is straightforward to formulate extensions with a propagating torsion. The occurrence of torsion is a natural consequence of the presence of spin currents. Its effects are tiny on macroscopic scales (which is why it has not been seen experimentally), but it should be of high relevance in the microscopic realm, for example, on the scale of the electronic Compton wavelength and in the very early universe. In fact, the Einstein–Cartan theory is naturally embedded in theories of supergravity (see Section 2.3), where a spin-3/2 particle (the 'gravitino') plays a central role.

In Chapter 2, we shall discuss some 'uniqueness theorems', which state that every theory of the gravitational field must contain GR (or the Einstein–Cartan theory) in an appropriate limit. Generalizations of GR such as the Jordan–Brans–Dicke theory, which contains an additional scalar field in the gravitational sector, are therefore mainly of interest as effective theories arising from fundamental theories such as string theory (see Chapter 9). They are usually not meant as classical alternatives to GR, except for the parametrization of experimental tests. That GR cannot be true at the most fundamental level is clear from the *singularity theorems* (cf. Hawking and Penrose 1996): under very general conditions, singularities in space–time are unavoidable, signalling the breakdown of GR.

The theme of this book is to investigate the possibilities of unifying the gravitational field with the quantum framework in a consistent way. This may lead to a general avoidance of space–time singularities.

1.1.2 *Main motivations for quantizing gravity*

The first motivation is **unification**. The history of science shows that a reductionist viewpoint has been very fruitful in physics (Weinberg 1993). The standard model of particle physics is a *quantum* field theory that has united in a certain sense all non-gravitational interactions. It has been very successful experimentally, but one should be aware that its concepts are poorly understood beyond the perturbative level; in this sense, the classical theory of GR is in a much better condition.

The universal coupling of gravity to all forms of energy would make it plausible that gravity has to be implemented in a quantum framework too. Moreover, attempts to construct an exact semiclassical theory, where gravity stays classical but all other fields are quantum, have failed up to now, see Section 1.2. This demonstrates, in particular, that classical and quantum *concepts* (phase space versus Hilbert space, etc.) are most likely incompatible.

Physicists have also entertained the hope that unification entails a solution to the notorious divergence problem of quantum field theory; as is shown in Chapter 2, perturbative quantum GR leads to even worse divergences, due to its non-renormalizability, but a full non-perturbative framework without any divergences may exist.

The second motivation comes from **cosmology** and **black holes**. As the singularity theorems and the ensuing breakdown of GR demonstrate, a fundamental understanding of the early universe—in particular, its initial conditions near the 'big bang'—and of the final stages of black-hole evolution requires an encompassing theory. From the historical analogy of quantum mechanics (which due to its stationary states rescued the atoms from collapse), the general expectation is that this encompassing theory is a *quantum* theory. Classically, the generic behaviour of a solution to Einstein's equations near a big-bang singularity is assumed to consist of 'BKL oscillations', cf. Belinskii *at al.* (1982) and the references therein. A key feature of this scenario is the decoupling of different spatial points. A central demand on a quantum theory of gravity is to provide a consistent quantum description of BKL oscillations.

The concept of an 'inflationary universe'[1] is often invoked to claim that the present universe can have emerged from generic initial conditions. This is only partly true, since one can of course trace back *any* present conditions to the past to find the 'correct' initial conditions. In fact, the crucial point lies in the assumptions that enter the *no-hair conjecture*, see for example, Frieman *et al.* (1997). This conjecture states that space–time approaches locally a de Sitter form for large times if a (probably effective) cosmological constant is present. It can be proved, provided some assumptions are made. In particular, it must be assumed that modes on very small scales (smaller than the Planck length, see below) are not amplified to cosmological scales. This assumption thus refers to the unknown regime of quantum gravity.

It must be emphasized that *if* gravity is quantized, the kinematical non-separability of quantum theory demands that the whole universe must be de-

---

[1] Following Harrison (2000), we shall write 'universe' instead of 'Universe' to emphasize that we talk about a *model* of the Universe, in contrast to 'Universe' which refers to 'everything'.

scribed in quantum terms. This leads to the concepts of quantum cosmology and the wave function of the universe, see Chapters 8 and 10.

A third motivation is the **problem of time**. Quantum theory and GR (in fact, every general covariant theory) contain a drastically different concept of time (and space–time). Strictly speaking, they are incompatible. In quantum theory, time is an external (absolute) element, *not* described by an operator (in special relativistic quantum field theory, the role of time is played by the external Minkowski space–time). In contrast, in GR, space–time is a dynamical object. It is clear that a unification of quantum theory with GR must lead to modifications of the concept of time. One might expect that the metric has to be turned into an operator. In fact, as a detailed analysis will show (Chapters 5 and 9), this will lead to novel features. Related problems concern the role of background structures in quantum gravity, the role of the diffeomorphism group (Poincaré invariance, as used in ordinary quantum field theory, is no longer a symmetry group), and the notion of 'observables'. That a crucial point lies in the presence of a more general invariance group was already noted by Pauli (1955)[2]:

It seems to me ... that it is not so much the linearity or non-linearity which forms the heart of the matter, but the very fact that here a more general group than the Lorentz group is present ... .

1.1.3 *Relevant scales*

In a universally valid quantum theory, genuine quantum effects can occur on any scale, while classical properties are an emergent phenomenon only (see Chapter 10). This is a consequence of the superposition principle. Independent of this, there exist scales where quantum effects of a particular interaction should definitely be non-negligible.

It was already noted by Planck (1899) that the fundamental constants, speed of light (c), gravitational constant (G), and quantum of action $(\hbar)$, can be combined in a unique way to yield units of length, time, and mass. In Planck's honour they are called Planck length, l_P, Planck time, t_P, and Planck mass, m_P, respectively. They are given by the expressions

$$l_\mathrm{P} = \sqrt{\frac{\hbar G}{c^3}} \approx 1.62 \times 10^{-33} \text{ cm} , \tag{1.4}$$

$$t_\mathrm{P} = \frac{l_\mathrm{P}}{c} = \sqrt{\frac{\hbar G}{c^5}} \approx 5.40 \times 10^{-44} \text{ s} , \tag{1.5}$$

$$m_\mathrm{P} = \frac{\hbar}{l_\mathrm{P} c} = \sqrt{\frac{\hbar c}{G}} \approx 2.17 \times 10^{-5} \text{ g} \approx 1.22 \times 10^{19} \text{ GeV} . \tag{1.6}$$

The Planck mass seems to be a rather large quantity on microscopic standards. One has to keep in mind, however, that this mass (energy) must be concentrated

---

[2]'Es scheint mir ... , daß nicht so sehr die Linearität oder Nichtlinearität Kern der Sache ist, sondern eben der Umstand, daß hier eine allgemeinere Gruppe als die Lorentzgruppe vorhanden ist ... .'

in a region of linear dimension l_P in order to see direct quantum-gravity effects. In fact, the Planck scales are attained for an elementary particle whose Compton wavelength is (apart from a factor of 2) equal to its Schwarzschild radius,

$$\frac{\hbar}{m_P c} \approx R_S \equiv \frac{2G m_P}{c^2} ,$$

which means that the space–time curvature of an elementary particle is not negligible. Sometimes (e.g. in cosmology), one also uses the Planck temperature,

$$T_P = \frac{m_P c^2}{k_B} \approx 1.41 \times 10^{32} \text{ K} . \tag{1.7}$$

It is interesting to observe that Planck had introduced his units one year before he wrote the famous paper containing the quantum of action, see Planck (1899). How had this been possible? The constant $\hbar$ appears in Wien's law, $\hbar\omega_{max} \approx 2.82 k_B T$, which was phenomenologically known at that time. Planck learnt from this that a new constant of nature is contained in this law, and he called it b. Planck concludes his article by writing[3]:

These quantities retain their natural meaning as long as the laws of gravitation, of light propagation in vacuum, and the two laws of the theory of heat remain valid; they must therefore, if measured by all kinds of intelligent beings, always turn out to be the same.

It is also interesting that similar units had already been introduced by the Irish physicist Johnstone Stoney (1881). Of course, $\hbar$ was not known at that time, but one could (in principle) get the elementary electric charge e from Avogadro's number L and Faraday's number $F = eL$. With e, G, and c, one can construct the same fundamental units as with $\hbar, G$, and c (since the fine structure constant is $\alpha = e^2/\hbar c \approx 1/137$; therefore, Stoney's units differ from Planck's units by factors of $\sqrt{\alpha}$. Quite generally one can argue that there are three fundamental dimensional quantities (cf. Okun 1992).

The Planck length is indeed very small. If one imagines an atom to be of the size of the Moon's orbit, l_P would only be as small as about a tenth of the size of a nucleus. Still, physicists have already for a while entertained the idea that something dramatically happens at the Planck length, from the breakdown of the continuum to the emergence of non-trivial topology ('space–time foam'), see for example, Misner *et al.* (1973). We shall see in the course of this book how such ideas can be made more precise in quantum gravity. Unified theories may contain an intrinsic length scale from which l_P may be deduced. In string theory, for example, this is the string length l_s. A generalized uncertainty relation shows that scales smaller than l_s have no operational significance, see Chapter 9.

---

[3]'Diese Grössen behalten ihre natürliche Bedeutung so lange bei, als die Gesetze der Gravitation, der Lichtfortpflanzung im Vacuum und die beiden Hauptsätze der Wärmetheorie in Gültigkeit bleiben, sie müssen also, von den verschiedensten Intelligenzen nach den verschiedensten Methoden gemessen, sich immer wieder als die nämlichen ergeben.'

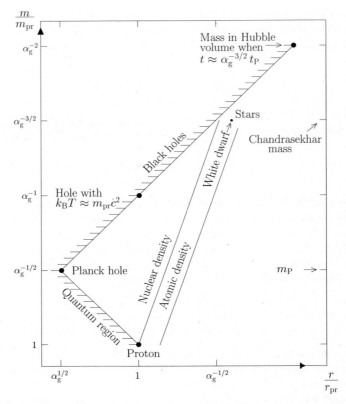

FIG. 1.1. Structures in the Universe (adapted from Rees (1995)).

Figure 1.1 presents some of the important structures in our universe in a mass-versus-length diagram. A central role is played by the 'fine structure constant of gravity',

$$\alpha_g = \frac{G m_{pr}^2}{\hbar c} = \left(\frac{m_{pr}}{m_P} \right)^2 \approx 5.91 \times 10^{-39} \, , \tag{1.8}$$

where m_{pr} denotes the proton mass. Its smallness is responsible for the unimportance of quantum-gravitational effects on laboratory and astrophysical scales, and for the separation between micro- and macrophysics. As can be seen from the diagram, important features occur for masses that contain simple powers of α_g (in terms of m_{pr}), cf. Rees (1995). For example, the Chandrasekhar mass M_C is given by

$$M_C \approx \alpha_g^{-3/2} m_{pr} \approx 1.4 M_\odot \, . \tag{1.9}$$

It gives the upper limit for the mass of a white dwarf and sets the scale for stellar masses. The minimum stellar life-times contain $\alpha_g^{-3/2} t_P$ as the important factor. It is also interesting to note that the size of human beings is roughly the

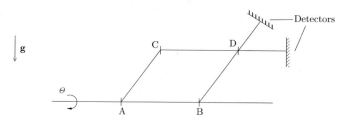

FIG. 1.2. Schematic description of the 'COW'-experiment for neutron interferometry in the gravitational field of the Earth.

geometric mean of Planck length and size of the observable part of the universe. It is an open question whether fundamental theories such as quantum gravity can provide an explanation for such values, for example, for the ratio m_{pr}/m_{P}, or not. We shall come back to this in Chapter 10.

As far as the relationship between quantum theory and the gravitational field is concerned, one can distinguish between different levels. The first, lowest level deals with quantum *mechanics* in *external* gravitational fields (either described by GR or its Newtonian limit). No back reaction on the gravitational field is taken into account. This is the only level where experiments exist so far. The next level concerns quantum *field* theory in *external* gravitational fields described by GR. Back reaction can be taken into account in a perturbative sense. These two levels will be dealt with in the next two subsections. The highest level, *full* quantum gravity, will be discussed in the rest of this book.

1.1.4 *Quantum mechanics and Newtonian gravity*

Consider first the level of Newtonian gravity. There exist experiments that test the classical trajectories of elementary particles, such as thermal neutrons that fall like mass points, see for example, Hehl *et al.* (1991). This is not so much of interest here. We are more interested in quantum-mechanical *interference* experiments concerning the motion of neutrons and atoms in external gravitational fields.

Historically, two experiments have been of significance. The experiment by Colella, Overhauser, and Werner ('COW') in 1975 was concerned with neutron interferometry in the gravitational field of the Earth. According to the equivalence principle, an analogous experiment should be possible with neutrons in accelerated frames. Such an experiment was performed by Bonse and Wroblewski in 1983. Details and references can be found in the reviews by Hehl *et al.* (1991) and Werner and Kaiser (1990).

In the following, we shall briefly describe the 'COW' experiment, see Fig. 1.2. A beam of neutrons is split into two parts, such that they can travel on different heights in the terrestrial gravitational field. They are then recombined and sent to detectors. The whole apparatus can be rotated with a varying angle θ around the horizontal axis. The interferences are then measured in dependence on θ.

The theoretical description makes use of the Schrödinger equation for neutrons (Hehl *et al.* 1991). The Hamiltonian in the system of the rotating Earth is given by

$$H = \frac{\mathbf{p}^2}{2m_i} + m_g \mathbf{g}\mathbf{r} - \omega \mathbf{L} \ . \tag{1.10}$$

We have distinguished here between the inertial mass, m_i, of the neutron and its (passive) gravitational mass, m_g, because 'COW' have also used this experiment as a test of the equivalence principle. In the last term, ω and $\mathbf{L}$ denote the angular velocity of the Earth and the angular momentum of the neutron with respect to the centre of the Earth (given by $\mathbf{r} = 0$), respectively. This term describes centrifugal and Coriolis forces. Note that the canonical momentum is given by

$$\mathbf{p} = m_i \dot{\mathbf{r}} + m_i \omega \times \mathbf{r} \ . \tag{1.11}$$

The phase shift in the interferometer experiment is given by

$$\Delta\beta = \frac{1}{\hbar} \oint \mathbf{p}\mathrm{d}\mathbf{r} \ , \tag{1.12}$$

where the integration runs over the parallelogram ABDC of Fig. 1.2. According to (1.11), there are two contributions to the phase shift. The term containing ω describes the influence of the terrestrial rotation on the interference pattern ('neutron Sagnac effect'). It yields

$$\Delta\beta_{\text{Sagnac}} = \frac{m_i}{\hbar} \oint (\omega \times \mathbf{r})\mathrm{d}\mathbf{r} = \frac{2m_i}{\hbar}\omega \mathbf{A} \ , \tag{1.13}$$

where $\mathbf{A}$ denotes the normal area vector of the loop ABDC.

Of main interest here is the gravitational part of the phase shift. Since the contributions of the sides $\overline{AC}$ and $\overline{DB}$ cancel, one has

$$\Delta\beta_g = \frac{m_i}{\hbar} \oint \mathbf{v}\mathrm{d}\mathbf{r} \approx \frac{m_i(v_0 - v_1)}{\hbar}\overline{AB} \ , \tag{1.14}$$

where v_0 and v_1 denote the absolute values of the velocities along $\overline{AB}$ and $\overline{CD}$, respectively. From energy conservation one gets

$$v_1 = v_0 \sqrt{1 - \frac{2\Delta V}{m_i v_0^2}} \approx v_0 - \frac{m_g g h_0 \sin\theta}{m_i v_0} \ ,$$

where $\Delta V = m_g g h_0 \sin\theta$ is the potential difference, h_0 denotes the perpendicular distance between $\overline{AB}$ and $\overline{CD}$, and the limit $2\Delta V/m_i v_0^2 \ll 1$ (about 10^{-8} in the experiment) has been used. The neutrons are prepared with a de Broglie wavelength $\lambda = 2\pi\hbar/p \approx 2\pi\hbar/m_i v_0$ (neglecting the ω part, since the Sagnac effect contributes only 2 per cent of the effect), attaining a value of about 1.4

Å in the experiment. One then gets for the gravitational phase shift, the final result

$$\Delta\beta_{\mathrm{g}} \approx \frac{m_{\mathrm{i}} m_{\mathrm{g}} g \lambda A \sin\theta}{2\pi\hbar^2} , \tag{1.15}$$

where A denotes the area of the parallelogram ABDC. This result has been confirmed by 'COW' with 1 per cent accuracy. The phase shift (1.15) can be rewritten in an alternative form such that only those quantities appear that are directly observable in the experiment (Lämmerzahl 1996). It then reads

$$\Delta\beta_{\mathrm{g}} \approx \frac{m_{\mathrm{g}}}{m_{\mathrm{i}}} \mathbf{g G} T T' , \tag{1.16}$$

where T (T') denotes the flight time of the neutron from A to B (from A to C), and $\mathbf{G}$ is the reciprocal lattice vector of the crystal layers (from which the neutrons are scattered in the beam splitter). Now m_{g} and m_{i} appear like in the classical theory as a ratio, not as a product. The COW experiment has also confirmed the validity of the (weak) equivalence principle in the quantum domain. Modern tests prefer to use atom interferometry because atoms are easier to handle and the experiments allow tests of higher precision (Lämmerzahl 1996, 1998). There the flight time is just the time between laser pulses, that is, the interaction time with the gravitational field; T is chosen by the experimentalist. Still, neutrons are useful to study quantum systems in the gravitational field. An experiment with ultracold neutrons has shown that their vertical motion in the gravitational field has discrete energy states, as predicted by the Schrödinger equation (Nesvizhevsky *et al.* 2002). The minimum energy is 1.4×10^{-12} eV, which is much smaller than the ground-state energy of the hydrogen atom.

It is also of interest to discuss the Dirac equation instead of the Schrödinger equation because this may give rise to additional effects. In Minkowski space (and cartesian coordinates), it reads

$$\left(\mathrm{i}\hbar\gamma^\mu \partial_\mu + \frac{mc}{\hbar} \right) \psi(x) = 0 , \tag{1.17}$$

where $\psi(x)$ is a Dirac spinor, and

$$[\gamma^\mu, \gamma^\nu]_+ \equiv \gamma^\mu\gamma^\nu + \gamma^\nu\gamma^\mu = 2\eta^{\mu\nu} . \tag{1.18}$$

The transformation into an accelerated frame is achieved by replacing partial derivatives with covariant derivatives, see for example, Hehl *et al.* (1991),

$$\partial_n \longrightarrow D_n \equiv \partial_n + \frac{\mathrm{i}}{2}\sigma^{mk}\omega_{nmk} , \tag{1.19}$$

where $\sigma^{mk} = \mathrm{i}[\gamma^m, \gamma^k]$ is the generator of the Lorentz group, and ω_{nmk} denotes the anholonomic components of the connection. From the equivalence principle, one would expect that this gives also the appropriate form in curved space–time, where

$$[\gamma^\mu, \gamma^\nu]_+ = 2g^{\mu\nu} \ . \tag{1.20}$$

For the formulation of the Dirac equation in curved space–time, one has to use the tetrad ('vierbein') formalism, in which a basis $e_n = \{e_0, e_1, e_2, e_3\}$ is chosen at each space–time point. One can expand the tetrads with respect to the tangent vectors along coordinate lines ('holonomic basis') according to

$$e_n = e_n^\mu \partial_\mu \ . \tag{1.21}$$

Usually one chooses the tetrad to be orthonormal,

$$e_n \cdot e_m \equiv g_{\mu\nu} e_n^\mu e_m^\nu = \eta_{nm} \equiv \mathrm{diag}(-1, 1, 1, 1) \ . \tag{1.22}$$

The reason why one has to go beyond the pure metric formalism is the fact that spinors (describing fermions) are objects whose wave components transform with respect to a two-valued representation of the Lorentz group. One therefore needs a local Lorentz group and local orthonormal frames.

One can define anholonomic Dirac matrices according to

$$\gamma^n \equiv e_\mu^n \gamma^\mu \ , \tag{1.23}$$

where $e_n^\mu e_\mu^m = \delta_n^m$. This leads to

$$[\gamma^n, \gamma^m]_+ = 2\eta^{nm} \ . \tag{1.24}$$

The Dirac equation in curved space–time or accelerated frames then reads

$$\left(\mathrm{i}\hbar\gamma^n D_n + \frac{mc}{\hbar} \right) \psi(x) = 0 \ . \tag{1.25}$$

In order to study quantum effects of fermions in the gravitational field of the Earth, one specializes this equation to the non-inertial frame of an accelerated and rotating observer, with acceleration $\mathbf{a}$ and angular velocity ω, respectively (see e.g. Hehl et $al.$ 1991). A non-relativistic approximation with relativistic corrections is then obtained by the standard Foldy–Wouthuysen transformation. This leads to (writing $\beta \equiv \gamma^0$)

$$\mathrm{i}\hbar\frac{\partial\psi}{\partial t} = H_{\mathrm{FW}}\psi \ , \tag{1.26}$$

with

$$H_{\mathrm{FW}} = \beta mc^2 + \frac{\beta}{2m}\mathbf{p}^2 - \frac{\beta}{8m^3c^2}\mathbf{p}^4 + \beta m(\mathbf{a}\,\mathbf{x})$$
$$-\omega(\mathbf{L} + \mathbf{S}) + \frac{\beta}{2m}\mathbf{p}\frac{\mathbf{a}\,\mathbf{x}}{c^2}\mathbf{p} + \frac{\beta\hbar}{4mc^2}\vec{\sigma}(\mathbf{a} \times \mathbf{p}) + \mathcal{O}\left(\frac{1}{c^3}\right) \tag{1.27}$$

($\vec{\sigma}$ denotes the Pauli matrices). The interpretation of the various terms in (1.27) is straightforward. The first four terms correspond to the rest mass, the usual

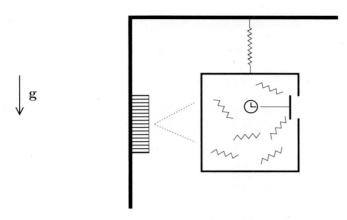

FIG. 1.3. Setting in the gedanken experiment of the Einstein–Bohr debate on the time–energy uncertainty relation.

non-relativistic kinetic term, the first relativistic correction to the kinetic term, and the 'COW' effect (or its analogue for pure acceleration), respectively. The term $\omega\mathbf{L}$ describes the Sagnac effect, while $\omega\mathbf{S}$ corresponds to a new spin-rotation effect ('Mashhoon effect') that cannot be found from the Schrödinger equation. One can estimate that for typical values of a neutron interferometer experiment, the Mashhoon effect contributes only 10^{-9} of the Sagnac effect. This is very small, but it has been indirectly observed (Mashhoon 1995). We mention that this framework is also of use in the study of a 'generalized' Dirac equation to parametrize quantum tests of general relativity (Lämmerzahl 1998) and to construct an axiomatic approach to space–time geometry, yielding a Riemann–Cartan geometry (see Audretsch *et al.* 1992).

In concluding this subsection, we want to discuss briefly one important occasion where GR seems to play a role in the foundations of quantum mechanics. This is the discussion of the time–energy uncertainty relations by Bohr and Einstein at the sixth Solvay conference which took place in Brussels in 1930 (cf. Bohr 1949).

Einstein came up with the following counter-argument against the validity of this uncertainty relation. Consider a box filled with radiation. A clock controls the opening of a shutter for a short time interval such that a single photon can escape at a fixed time t. The energy E of the photon is, however, also fixed because it can be determined by weighing the box before and after the escape of the photon. It thus seems as if the time–energy uncertainty relation were violated.

In his response to Einstein's attack, Bohr came up with the following arguments. Consider the details of the weighing process in which a spring is attached to the box, see Fig. 1.3. The null position of the balance is known with an accuracy Δq. This leads to an uncertainty in the momentum of the box, $\Delta p \sim \hbar/\Delta q$.

Bohr then makes the assumption that Δp must be smaller than the total momentum imposed by the gravitational field during the time T of the weighing process on the mass uncertainty Δm of the box. This leads to

$$\Delta p < v\Delta m = gT\Delta m \ , \tag{1.28}$$

where g is the gravitational acceleration. Now GR enters the game: the tick rate of clocks depends on the gravitational potential according to the 'redshift formula'

$$\frac{\Delta T}{T} = \frac{g\Delta q}{c^2} \ , \tag{1.29}$$

so that, using (1.28), the uncertainty in ΔT after the weighing process is

$$\Delta T = \frac{g\Delta q}{c^2}T > \frac{\hbar}{\Delta mc^2} = \frac{\hbar}{\Delta E} \ , \tag{1.30}$$

in accordance with the time–energy uncertainty relation. (After this, Einstein gave up to find an inconsistency in quantum mechanics, but focused instead on its possible incompleteness.) But are Bohr's arguments really consistent? There are, in fact, some possible loopholes (cf. Shi 2000). First, it is unclear whether (1.28) must really hold, since Δp is an intrinsic property of the apparatus. Second, the relation (1.29) cannot hold in this form, because T is *not* an operator, and therefore ΔT cannot have the same interpretation as Δq. In fact, if T is considered as a classical quantity, it would be more consistent to relate Δq to an uncertainty in g, which in fact would suggest to consider the quantization of the gravitational field. One can also change the gedanken experiment by using an electrostatic field instead of the gravitational field, where a relation of the form (1.29) no longer holds (see von Borzeszkowski and Treder 1988). It should also be emphasized that not much of GR is needed, in fact, the relation (1.29) follows from energy conservation and $E = h\nu$ alone. A general criticism of all these early gedanken experiments deals with their inconsistent interpretation of measurements as being related to uncontrollable interactions; see Shi (2000) and Chapter 10. The important feature is, however, entanglement between quantum systems.

It thus seems as if Bohr's analysis was mainly based on dimensional arguments. In fact, the usual application of the time–energy uncertainty relation relates linewidths of spectra for unstable systems with the corresponding half-life time. In quantum gravity, no time parameter appears on the fundamental level (see Chapter 5). A time–energy uncertainty relation can only be derived in the semiclassical limit.

1.1.5 *Quantum field theory in curved space–time*

Some interesting new aspects appear when quantum *fields* play a role. They mainly concern the notions of *vacuum* and *particles*. A vacuum is only invariant with respect to Poincaré transformations, so that observers that are not related by inertial motion refer in general to a different type of vacuum (Fulling 1973). 'Particle creation' can occur in the presence of external fields or for the state of non-inertial motion. An external electric field, for example, can lead to the creation of electron–positron pairs ('Schwinger effect'), see for example, Grib *et al.* (1994). We shall be mainly concerned with particle creation in the presence of external gravitational fields (Birrell and Davies 1982). This was first discussed by Schrödinger (1939).

One example of particular interest is particle creation from black holes (Hawking 1975), see for example, Frolov and Novikov (1998), Fré *et al.* (1999), and Hehl *et al.* (1998) for a detailed review. This is not only of fundamental theoretical interest, but could also lead to observational consequences. A black hole radiates with a universal temperature ('Hawking temperature') according to

$$T_{\mathrm{BH}} = \frac{\hbar \kappa}{2\pi k_{\mathrm{B}} c} \; , \tag{1.31}$$

where κ is the surface gravity of a stationary black hole which by the no-hair theorem is uniquely characterized by its mass M, its angular momentum J, and (if present) its electric charge Q. In the particular case of the spherically symmetric Schwarzschild black hole, one has $\kappa = c^4/4GM = GM/R_{\mathrm{S}}^2$ and therefore

$$T_{\mathrm{BH}} = \frac{\hbar c^3}{8\pi k_{\mathrm{B}} GM} \approx 6.17 \times 10^{-8} \left(\frac{M_{\odot}}{M} \right) \; \mathrm{K} \; . \tag{1.32}$$

This temperature is unobservationally small for solar-mass (and bigger) black holes, but may be observable for primordial black holes. It must be emphasized that the expression for T_{BH} contains all fundamental constants of Nature. One may speculate that this expression—relating the macroscopic parameters of a black hole with thermodynamic quantities—plays a similar role for quantum gravity as de Broglie's relations $E = \hbar\omega$ and $p = \hbar k$ once played for the development of quantum theory (Zeh 2001).

Hawking radiation was derived in the semiclassical limit in which the gravitational field can be treated classically. According to (1.32), the black hole loses mass through its radiation and becomes hotter. After it has reached a mass of the size of the Planck mass (1.6), the semiclassical approximation breaks down and the full theory of quantum gravity should be needed. Black-hole evaporation thus plays a crucial role in any approach to quantum gravity; cf. Chapter 7.

There exists a related effect to (1.31) in flat Minkowski space. An observer in uniform acceleration experiences the standard Minkowski vacuum not as empty, but as filled with *thermal* radiation with temperature

$$T_{\mathrm{DU}} = \frac{\hbar a}{2\pi k_{\mathrm{B}} c} \approx 4.05 \times 10^{-23} \, a \left[\frac{\mathrm{cm}}{\mathrm{s}^2} \right] \; \mathrm{K} \; . \tag{1.33}$$

This temperature is often called the 'Davies–Unruh temperature' after the work by Davies (1975) and Unruh (1976), with important contributions also by Fulling

(1973). Formally, it arises from (1.31) through the substitution of κ by a. This can be understood from the fact that *horizons* are present in both the black-hole case and the acceleration case, see for example, Kiefer (1999) for a detailed review. Although (1.33) seems to be a small effect, people have suggested to look for it in accelerators (Leinaas 2002) or in experiments with ultraintense lasers (Chen and Tajima 1999), without definite success up to now.

A central role in the theory of quantum fields on an external space–time is played by the semiclassical Einstein equations. These equations are obtained by replacing the energy–momentum tensor in (1.1) by the expectation value of the energy–momentum operator with respect to some quantum state Ψ,

$$R_{\mu\nu} - \frac{1}{2}g_{\mu\nu}R + \Lambda g_{\mu\nu} = \frac{8\pi G}{c^4}\langle\Psi|\hat{T}_{\mu\nu}|\Psi\rangle \ . \tag{1.34}$$

A particular issue is the regularization and renormalization of the object on the right-hand side (Birrell and Davies 1982). This leads, for example, to a flux of negative energy into the black hole, which can be interpreted as the origin of Hawking radiation. As we shall discuss in the next section, (1.34) is of limited value if seen from the viewpoint of the full quantum theory. We shall see in Section 5.4 that (1.34) can be derived approximately from canonical quantum gravity as a kind of mean-field equation.

1.2   Problems of a fundamental semiclassical theory

When dealing with approaches to quantum gravity, the question is sometimes asked whether it is really necessary to quantize the gravitational field. And even if it is, doubts have occasionally been put forward whether such a theory can operationally be distinguished from an 'exact' semiclassical theory.[4] As a candidate for the latter, the semiclassical Einstein equations (1.34) are often presented, cf. Møller (1962). The more general question behind this issue concerns the possibility of a consistent *hybrid dynamics* through which a quantum and a classical system is being coupled.

Eppley and Hannah (1977) argued that the coupling between a classical gravitational wave with a quantum system leads to inconsistencies. In fact, the gravitational nature of this wave is not crucial—it can be any classical wave. Without going into details, their arguments demonstrate that the quantum nature of the measuring apparatus has to be taken into account, in order to avoid inconsistencies. This resembles the old debate between Bohr and Einstein, in which Bohr had to impose the uncertainty relations also for a macroscopic object (the screen in the double-slit experiment), in order to save them from Einstein's attacks. In this sense the arguments by Eppley and Hannah give a general hint for the quantum nature of the gravitational field.

---

[4]'Semiclassical' here means a theory that couples exactly quantum degrees of freedom to classical degrees of freedom. It therefore has nothing to do with the WKB approximation which usually is referred to as the semiclassical approximation. The latter is discussed in Section 5.4.

It is often argued that the famous gedanken experiments by Bohr and Rosenfeld (1933) imply that a coupling between a classical and a quantum system is inconsistent. For this reason, here we shall briefly review their arguments (see also Heitler (1984) for a lucid discussion). Historically, Landau and Peierls (1931) had claimed that the quantum nature of the *electromagnetic field* cannot be tested, since there exists a fundamental minimal uncertainty for single field amplitudes, not only for conjugate pairs. Bohr and Rosenfeld have then shown that this is not true. Their line of thought runs as follows. Consider a charged body with mass M and charge Q, acting as a measuring device for the electric field $\mathcal{E}$ being present in a volume $V \equiv l^3$. Momentum measurements of the body are being made at the beginning and the end of the measurement time interval. In order to qualify the body as a measurement device, the following assumptions are made (in order to avoid back reaction, etc.); cf. also von Borzeszkowski and Treder (1988),

$$Q^2 \gtrsim \hbar c \approx 137 e^2 \ , \tag{1.35}$$

$$l > \frac{Q^2}{Mc^2} \ . \tag{1.36}$$

The latter condition expresses the fact that the electrostatic energy should be smaller than the rest mass. Bohr and Rosenfeld then found from their detailed analysis, the following conditions with $\mathcal{E}$ denoting the field average over the volume V,

$$\Delta\mathcal{E} \, l^2 \gtrsim \frac{\hbar c}{Q} \ , \tag{1.37}$$

$$\Delta\mathcal{E} \, l^3 \gtrsim \frac{\hbar Q}{Mc} \ . \tag{1.38}$$

One can now always choose a measurement device such that the ratio Q/M in the last expression can be made arbitrarily small. Therefore, $\mathcal{E}$ can be measured with arbitrary precision, contrary to the arguments of Landau and Peierls (1931). Bohr and Rosenfeld then show that Eqns (1.37) and (1.38) are in agreement with the uncertainty relations as being derived from the quantum commutators of quantum electrodynamics (QED). Their discussion, therefore, shows the *consistency* of the formalism with the measurement analysis. It does *not* provide a logical proof that the electromagnetic field must be quantized; cf. Rosenfeld (1963).[5]

Although the final formalism for quantum gravity is not yet at hand, the Bohr–Rosenfeld analysis can at least formally be extended to the gravitational field, cf. Bronstein (1936), DeWitt (1962) and von Borzeszkowski and Treder (1988). One can substitute the electric field $\mathcal{E}$ by the Christoffel symbols Γ

---

[5]From *empirical* arguments, we know, of course, that the electromagnetic field is of quantum nature.

('gravitational force'). Since one can then perform in the Newtonian approximation $\Gamma \sim GM_g/r^2c^2$ (M_g denoting the gravitational mass) the substitutions

$$\Delta\mathcal{E} \rightarrow \frac{\Delta\Gamma c^2}{G}, \quad Q \rightarrow M_g , \tag{1.39}$$

one gets from (1.38), the relation (writing M_i instead of M to emphasize that it is the inertial mass)

$$\Delta\Gamma \, l^3 \gtrsim \frac{\hbar G}{c^3} \frac{M_g}{M_i} . \tag{1.40}$$

Using the (weak) equivalence principle, $M_g = M_i$, and recalling the definition (1.4) of the Planck length, one can write

$$\Delta\Gamma \gtrsim \frac{l_P^2}{l^3} . \tag{1.41}$$

The analogous relation for the metric g would then read

$$\Delta g \gtrsim \left(\frac{l_P}{l}\right)^2 . \tag{1.42}$$

Thus, the measurement of a single quantity (the metric) is operationally restricted.[6] This is of course possible because, unlike QED, the fundamental length scale l_P is available. On the other hand, a gedanken experiment by Smith and Bergmann (1979) shows that the magnetic-type components of the Weyl tensor in linearized quantum gravity can be measured, provided a suitable average over space–time domains is performed.

Does (1.42) imply that the quantum nature of the gravitational field cannot be tested? We want to answer this question in the negative, for the following reasons. First, there might be other measurement devices which do not necessarily obey the above relations. Second, this analysis does not say anything about global situations (black holes, cosmology) and about non-trivial applications of the superposition principle. And third, in fact, quantum gravity seems to predict the existence of a smallest scale with operational meaning, see Chapters 6 and 9. Then, (1.42) could be interpreted as a confirmation of quantum gravity. It might of course be possible, as argued in von Borzeszkowski and Treder (1988), that quantum-gravitational analogues of effects such as Compton scattering or Lamb shift are unobservable in the laboratory. This has only little bearing on the above discussion, since one would expect that such quantum-gravitational effects are anyway not seen directly in laboratory experiments.

Returning to the specific equations (1.34) for a semiclassical theory, there are a number of problems attached with them. First, the expectation value of

---

[6]Equation (1.42) is similar to the heuristic relation $\Delta g \gtrsim l_P/l$ of Misner *et al.* (1973), although the exponent is different.

the energy–momentum tensor that occurs on the right-hand side is usually divergent and needs some regularization and renormalization.[7] In this process, however, counter-terms arise that invoke higher powers of the curvature such as R^2, which may alter the semiclassical equations at a fundamental level. Second, (1.34) introduces the following element of non-linearity. The space–time metric g depends on the quantum state in a complicated way, since in (1.34) $|\Psi\rangle$ depends on g also through the (functional) Schrödinger equation (an equivalent statement holds in the Heisenberg picture). Consequently, if g_1 and g_2 correspond to states $|\Psi_1\rangle$ and $|\Psi_2\rangle$, respectively, there is no obvious relation between a superposition $A|\Psi_1\rangle + B|\Psi_2\rangle$ (which still satisfies the Schrödinger equation) and the metrics g_1 and g_2. This was already remarked by Anderson in Møller (1962) and by Belinfante in a discussion with Rosenfeld (see Infeld 1964). It was also the reason why Dirac strongly objected to (1.34); cf. von Borzeszkowski and Treder (1988).

Rosenfeld insisted on (1.34) because he strongly followed Bohr's interpretation of the measurement process for which classical concepts should be indispensable. This holds in particular for the structure of space–time, so he wished to have a c-number representation for the metric. He rejects a quantum description for the total system and answers to Belinfante in Infeld (1964) that Einstein's equations may merely be thermodynamical equations of state that break down for large fluctuations, that is, the gravitational field may only be an effective, not a fundamental, field, cf. also Jacobson (1995).

The problem with the superposition principle can be demonstrated by the following argument that has even been put to an experimental test (Page and Geilker 1981). One assumes that there is no explicit collapse of $|\Psi\rangle$, because otherwise one would expect the covariant conservation law $\langle \hat{T}_{\mu\nu}\rangle;^{\nu} = 0$ to be violated, in contradiction to (1.34). If the gravitational field were quantized, one would expect that each component of the superposition in $|\Psi\rangle$ would act as a source for the gravitational field. This is of course the Everett interpretation for quantum theory; cf. Chapter 10. On the other hand, eqn (1.34) depends on *all* components of $|\Psi\rangle$ simultaneously. Page and Geilker (1981) envisaged the following gedanken experiment, reminiscent of Schrödinger's cat, to distinguish between these options.

In a box, there is a radioactive source together with two masses that are connected by a spring. Initially, the masses are rigidly connected, so that they cannot move. If a radioactive decay happens, the rigid connection will be broken and the masses can swing towards each other. Outside the box, there is a Cavendish balance that is sensitive to the location of the masses and therefore acts as a device to 'measure' their position. Following Unruh (1984), the situation can be described by the following simple model. We denote with $|0\rangle$, the quantum state of the masses with rigid connection, and with $|1\rangle$, the corresponding state in which they can move towards each other. For the purpose of this

---

[7]This procedure leads to an essentially unique result for $\langle \hat{T}_{\mu\nu}\rangle$ if certain physical requirements are imposed, cf. Birrell and Davies (1982). The ambiguities can then be absorbed by a redefinition of constants appearing in the action.

experiment, it is sufficient to go to the Newtonian approximation of GR and to use the Hamilton operator $\hat{H}$ instead of the full energy–momentum tensor $\hat{T}_{\mu\nu}$. For initial time $t = 0$, it is assumed that the state is given by $|0\rangle$. For $t > 0$, the state then evolves into a superposition of $|0\rangle$ and $|1\rangle$,

$$|\Psi\rangle(t) = \alpha(t)|0\rangle + \beta(t)|1\rangle ,$$

with the coefficients $|\alpha(t)|^2 \approx \mathrm{e}^{-\lambda t}$, $|\beta|^2 \approx 1 - \mathrm{e}^{-\lambda t}$, according to the law of radioactive decay, with a decay constant λ. From this, one finds for the evolution of the expectation value,

$$\langle\Psi|\hat{H}|\Psi\rangle(t) = |\alpha(t)|^2\langle 0|\hat{H}|0\rangle + |\beta(t)|^2\langle 1|\hat{H}|1\rangle + 2\mathrm{Re}\left[\alpha^*\beta\langle 0|\hat{H}|1\rangle\right] .$$

If one makes the realistic assumption that the states are approximate eigenstates of the Hamiltonian, the last term, which describes interferences, vanishes. Anyway, this is not devised as an interference experiment (in contrast to Schrödinger's cat), and interferences would become small due to decoherence (Chapter 10). One is thus left with

$$\langle\Psi|\hat{H}|\Psi\rangle(t) \approx \mathrm{e}^{-\lambda t}\langle 0|\hat{H}|0\rangle + \left(1 - \mathrm{e}^{-\lambda t}\right)\langle 1|\hat{H}|1\rangle . \tag{1.43}$$

According to semiclassical gravity as described by (1.34), therefore, the Cavendish balance would follow the dynamics of the expectation value and slightly swing in the course of time. This is in sharp contrast to the prediction of linear quantum gravity, where in each component the balance reacts to the mass configuration and would thus be observed to swing instantaneously at a certain time. This is, in fact, what has been observed in the actual experiment (Page and Geilker 1981). This experiment, albeit simple, demonstrates convincingly that (1.34) cannot fundamentally be true.

In the above experiment, the reason for the deviation between the predictions of the semiclassical theory and the 'full' theory lies in the large fluctuation for the Hamiltonian. In fact, the experiment was devised to generate such a case. Large fluctuations also occur in another interesting situation—the gravitational radiation emitted by quantum systems (Ford 1982). The calculations are performed for linearized gravity, that is, for a small metric pertubation around flat space–time with metric $\eta_{\mu\nu}$, see for example, Misner *et al.* (1973) and Chapter 2. Denoting by $G_r(x, x')$ the retarded Green function, one finds for the *integrated* energy–momentum tensor $S_{\mu\nu}$ in the semiclassical theory described by (1.34), the expression,[8]

$$S_{\mathrm{sc}}^{\mu\nu} = -8\pi G \int \mathrm{d}^3x\mathrm{d}^4x'\mathrm{d}^4x'' \; \partial^\mu G_r(x, x')\partial^\nu G_r(x, x'')$$

---

[8]Hats on operators are avoided for simplicity; from now on we set $c = 1$ in most cases.

$$\times \left[\langle T_{\alpha\beta}(x') \rangle \langle T^{\alpha\beta}(x'') \rangle - 1/2 \; \langle T(x') \rangle \langle T(x'') \rangle \right] , \qquad (1.44)$$

where $T \equiv T^{\mu\nu}\eta_{\mu\nu}$ denotes the trace of the energy–momentum tensor. On the other hand, quantization of the linear theory (see Chapter 2) yields

$$S_{\mathrm{q}}^{\mu\nu} = -8\pi G \int \mathrm{d}^3 x \mathrm{d}^4 x' \mathrm{d}^4 x'' \; \partial^\mu G_r(x, x') \partial^\nu G_r(x, x'')$$

$$\times \langle T_{\alpha\beta}(x') T^{\alpha\beta}(x'') - 1/2 \; T(x') T(x'') \rangle . \qquad (1.45)$$

The difference in these results can be easily interpreted: in the semiclassical theory, $\langle T_{\mu\nu} \rangle$ acts as a source, and so no two-point functions $\langle T \ldots T \rangle$ can appear, in contrast to linear quantum theory.

It is obvious that the above two expressions strongly differ, once the fluctuation of the energy–momentum tensor is large. As a concrete example, Ford (1982) takes a massless real scalar field as matter source. For coherent states there is no difference between (1.44) and (1.45). This is not unexpected, since coherent states are as 'classical' as possible, and so the semiclassical and the full theory give identical results. For a superposition of coherent states, however, this is no longer true, and the energies emitted by the quantum system via gravitational waves can differ by macroscopic amounts. For a number eigenstate of the scalar field, the semiclassical theory does not predict any radiation at all ($\langle T_{\mu\nu} \rangle$ is time-independent), whereas there is radiation in quantum gravity ($\langle T_{\mu\nu} T_{\rho\lambda} \rangle$ is time-dependent).[9] Therefore, one can in principle have macroscopic quantum-gravity effects even far away from the Planck scale!

Kuo and Ford (1993) have extended this analysis to situations where the expectation value of the energy density can be negative. They show that in such cases the fluctuations in the energy–momentum tensor are large and that the semiclassical theory gives different predictions than the quantum theory. This is true, in particular, for a squeezed vacuum state describing particle creation—a case that is relevant, for example, for structure formation in the universe, see the remarks in Section 10.1.3. Another example is the Casimir effect. Kuo and Ford (1993) show that the gravitational field produced by the Casimir energy is *not* described by a fixed classical metric.

It will be discussed in Section 5.4 to what extent the semiclassical equations (1.34) can be derived as approximations from full quantum gravity. Modern developments in quantum mechanics discuss the possibility of a consistent formulation of 'hybrid dynamics', coupling a quantum to a classical system, see for example, Diósi *et al.* (2000). This leads to equations that generalize mean-field equations such as (1.34), although no one has applied this formalism to the gravitational case. It seems that such a coupling can be formulated consistently if the 'classical' system is, in fact, a decohered quantum system (Halliwell 1998). However, this already refers to an effective and not to a fundamental level of description. It seems that DeWitt is right, who wrote (DeWitt 1962):

---

[9] Analogous results hold for electrodynamics, with the current j_μ instead of $T_{\mu\nu}$.

It is shown in a quite general manner that the quantization of a given system implies also the quantization of any other system to which it can be coupled.

1.3   Approaches to quantum gravity

As we have seen in the last sections, there exist strong arguments supporting the idea that the gravitational field is of *quantum* nature at the fundamental level. The major task, then, is the construction of a consistent quantum theory of gravity that can be subject to experimental tests.

Can one get hints how to construct such a theory from observation? A direct probe of the Planck scale (1.6) in high-energy experiments would be illusory. In fact, an accelerator of current technology would have to be of the size of several thousand lightyears in order to probe the Planck energy $m_\mathrm{P}c^2 \approx 10^{19}$ GeV. However, we have seen in Section 1.2 that macroscopic effects of quantum gravity could in principle occur at lower energy scales, and we will encounter some other examples in the course of this book. Among these are effects of the full theory such as non-trivial applications of the superposition principle for the quantized gravitational field or the existence of discrete quantum states in black-hole physics or the early universe. But one might also be able to observe quantum-gravitational correction terms to established theories, such as correction terms to the functional Schrödinger equation in an external space–time, or effective terms violating the weak equivalence principle. Such effects could potentially be measured in the anisotropy spectrum of the cosmic microwave background radiation or in the forthcoming satellite tests of the equivalence principle such as the mission STEP.

One should also keep in mind that the final theory (which is not yet available) will make its own predictions, some perhaps in a totally unexpected direction. As Heisenberg recalls from a conversation with Einstein[10]:

From a fundamental point of view it is totally wrong to aim at basing a theory only on observable quantities. For in reality it is just the other way around. Only the theory decides about what can be observed.

A really fundamental theory should have such a rigid structure that all phenomena in the low-energy regime, such as particle masses or coupling constants, can be predicted in an unique way. As there is no direct experimental hint yet, most work in quantum gravity focuses on the attempt to construct a mathematically and conceptually consistent (and appealing) framework.

There is, of course, no a priori given starting point in the methodological sense. In this context, Isham (1987) makes a distinction between a 'primary theory of quantum gravity' and a 'secondary theory'. In the primary approach, one starts with a given classical theory and applies heuristic quantization rules. This

---

[10]'Aber vom prinzipiellen Standpunkt aus ist es ganz falsch, eine Theorie nur auf beobachtbare Größen gründen zu wollen. Denn es ist ja in Wirklichkeit genau umgekehrt. Erst die Theorie entscheidet darüber, was man beobachten kann.' (Einstein according to Heisenberg (1979))

is the approach usually adopted, which was successful, for example, in QED. Often the starting point is general relativity, leading to 'quantum general relativity' or 'quantum geometrodynamics', but one could also start from another classical theory such as the Brans–Dicke theory. One usually distinguishes between canonical and covariant approaches. The former employ at the classical level a split of space–time into space and time, whereas the latter aim at preserving four-dimensional covariance at each step. They will be discussed in Chapters 5, 6, and Chapter 2, respectively. The main advantage of these approaches is that the starting point is given. The main disadvantage is that one does not arrive immediately at a unified theory of all interactions.

The opposite holds for a 'secondary theory'. One starts with a fundamental quantum framework of all interactions and tries to derive (quantum) general relativity in certain limiting situations, for example, through an energy expansion. The most important example here is string theory (see Chapter 9). The main advantage is that the fundamental quantum theory automatically yields a unification, a 'theory of everything'; cf. Weinberg (1993). The main disadvantage is that the starting point is entirely speculative. A short review of the main approaches to quantum gravity is given by Carlip (2001).

In this book, we shall mainly focus on quantum GR because it is closer to established theories and because it exhibits many general aspects clearer. In any case, even if quantum GR is superseded by a more fundamental theory such as string theory, it should be valid as an *effective theory* in some appropriate limit. The reason is that far away from the Planck scale, classical general relativity is the appropriate theory, which in turn must be the classical limit of an underlying quantum theory. Except perhaps close to the Planck scale itself, quantum GR should be a viable framework (such as QED, which is also supposed to be only an effective theory). It should also be emphasized that string theory automatically implements many of the methods used in the primary approach, such as quantization of constrained systems and covariant perturbation theory.

An important question in the heuristic quantization of a given classical theory is which of the classical structures should be subjected to the superposition principle and which should remain classical (or absolute, non-dynamical) structures. Isham (1994) distinguishes the following hierarchy of structures; see also Butterfield and Isham (1999),

Point set of events $\longrightarrow$ topological structure $\longrightarrow$ differentiable manifold $\longrightarrow$ causal structure $\longrightarrow$ Lorentzian structure.

Most approaches subject the Lorentzian and the causal structure to quantization, but keep the manifold structure fixed. This is, however, not clear. More general approaches include attempts to quantize topological structure, see for example, Isham (1989), or to quantize causal sets, see for example, Sorkin (2003) and the references therein. We shall not discuss such approaches in this volume. According to the Copenhagen interpretation of quantum theory, all structures related to space–time would probably have to stay classical because they are thought to be necessary ingredients for the measurement process, cf. Chapter 10. For the purpose of quantum gravity, such a viewpoint is, however, insufficient and probably inconsistent.

Historically, the first remark on the necessity to deal with quantum gravity was made by Einstein (1916b). This was, of course, in the framework of the 'old' quantum theory and does not yet reflect his critical attitude against quantum theory, which he adopted later. He writes[11]:

In the same way the atoms would have to emit, because of the inneratomic electronic motion, not only electromagnetic, but also gravitational energy, although in tiny amounts. Since this does hardly hold true in nature, it seems that quantum theory will have to modify not only Maxwell's electrodynamics, but also the new theory of gravitation.

---

[11] 'Gleichwohl müßten die Atome zufolge der inneratomischen Elektronenbewegung nicht nur elektromagnetische, sondern auch Gravitationsenergie ausstrahlen, wenn auch in winzigem Betrage. Da dies in Wahrheit in der Natur nicht zutreffen dürfte, so scheint es, daß die Quantentheorie nicht nur die Maxwellsche Elektrodynamik, sondern auch die neue Gravitationstheorie wird modifizieren müssen.'

COVARIANT APPROACHES TO QUANTUM GRAVITY

2.1 The concept of a graviton

A central role in the quantization of the gravitational field is played by the *graviton*—a massless particle of spin-2, which is the mediator of the gravitational interaction. It is analogous to the photon in quantum electrodynamics. Its definition requires, however, the presence of a background structure, at least in an approximate sense. We shall, therefore, first review weak gravitational waves in Minkowski space–time and the concept of helicity. It will then be explained how gravitons are defined as spin-2 particles from representations of the Poincaré group. Finally, the gravitational field in its linear approximation is quantized. It is shown, in particular, how Poincaré invariance ensues the equivalence principle and therefore the *full* theory of general relativity (GR) in the classical limit.

2.1.1 *Weak gravitational waves*

Starting point is the decomposition of a space–time metric $g_{\mu\nu}$ into a *fixed* (i.e. non-dynamical) background and a 'perturbation'; see for example, Weinberg (1972) and Misner *et al.* (1973). In the following, we take for the background the flat Minkowski space–time with the standard metric $\eta_{\mu\nu} = \mathrm{diag}(-1, 1, 1, 1)$ and call the perturbation $f_{\mu\nu}$. Thus,

$$g_{\mu\nu} = \eta_{\mu\nu} + f_{\mu\nu} \ . \tag{2.1}$$

We assume that the perturbation is small, that is, that the components of $f_{\mu\nu}$ are small in the standard cartesian coordinates. Using (2.1), the Einstein equations (1.3) read in the linear approximation

$$\Box f_{\mu\nu} = -16\pi G \left(T_{\mu\nu} - 1/2 \ \eta_{\mu\nu} T \right) \ , \tag{2.2}$$

where $T \equiv \eta^{\mu\nu} T_{\mu\nu}$, and the 'harmonic condition' (also called 'de Donder gauge')

$$f_{\mu\nu,}{}^{\nu} = \frac{1}{2} f^{\nu}{}_{\nu,\mu} \tag{2.3}$$

has been used.[1] This condition is analogous to the Lorenz[2] gauge condition in electrodynamics and is used here to partially fix the coordinates. Namely, the invariance of the full theory under coordinate transformations

---

[1] Indices are raised and lowered by $\eta^{\mu\nu}$ and $\eta_{\mu\nu}$, respectively.

[2] This is not a misprint. The Lorenz condition is named after the Danish physicist Ludwig Lorenz (1829–91).

$$x^\mu \to x'^\mu = x^\mu + \epsilon^\mu(x) \tag{2.4}$$

leads to the invariance of the linear theory under

$$f_{\mu\nu} \to f_{\mu\nu} - \epsilon_{\mu,\nu} - \epsilon_{\nu,\mu} \ . \tag{2.5}$$

It is often useful to employ instead of $f_{\mu\nu}$, the combination

$$\bar{f}_{\mu\nu} \equiv f_{\mu\nu} - 1/2 \ \eta_{\mu\nu} f^\rho{}_\rho \ , \tag{2.6}$$

so that (2.2) assumes the simple form

$$\Box \bar{f}_{\mu\nu} = -16\pi G T_{\mu\nu} \ . \tag{2.7}$$

The harmonic gauge condition (2.3) then reads $\partial_\nu \bar{f}_\mu{}^\nu = 0$, in direct analogy to the Lorenz gauge condition $\partial_\nu A^\nu = 0$. Since (2.7) is analogous to the wave equation $\Box A^\mu = -4\pi j^\mu$, the usual solutions (retarded waves, etc.) can be found. Note that the harmonic gauge condition is consistent with $\partial_\nu T^{\mu\nu} = 0$ (which is analogous to $\partial_\nu j^\nu = 0$), but *not* with $\nabla_\nu T^{\mu\nu} = 0$ (vanishing of covariant derivative). Therefore, although $T_{\mu\nu}$ acts as a source for $f_{\mu\nu}$, there is in the linear approximation no exchange of energy between matter and the gravitational field.

In the vacuum case ($T_{\mu\nu} = 0$), the simplest solutions to (2.2) are plane waves,

$$f_{\mu\nu} = e_{\mu\nu} e^{ikx} + e^*_{\mu\nu} e^{-ikx} \ , \tag{2.8}$$

where $e_{\mu\nu}$ is the polarization tensor. One has $k_\mu k^\mu = 0$ and, from (2.3), $k^\nu e_{\mu\nu} = (1/2) k_\mu e^\nu{}_\nu$. With $f_{\mu\nu}$ obeying (2.3), one can perform a new coordinate transformation of the type (2.4) to get

$$f'_{\mu\nu,}{}^\nu - 1/2 \ f'^\nu{}_{\nu,\mu} = -\Box \epsilon_\mu \ .$$

Without leaving the harmonic condition (2.3), one can thus *fix* the coordinates by choosing the four functions $\epsilon_\mu(x)$ to satisfy $\Box \epsilon_\mu = 0$ (this equation has plane-wave solutions and is therefore not in conflict with (2.8)). In total, one thus finds 10–4–4=2 independent degrees of freedom for the gravitational field in the linear approximation. The question of how many degrees of freedom the *full* field possesses will be dealt with in Chapter 4. With the ϵ_μ chosen as plane waves, $\epsilon_\mu(x) = 2\mathrm{Re}[if_\mu e^{ikx}]$ (f_μ being real numbers), the transformed plane wave reads the same as (2.8), with

$$e_{\mu\nu} \to e_{\mu\nu} + k_\mu f_\nu + k_\nu f_\mu \ . \tag{2.9}$$

It is most convenient for plane waves to choose the 'transverse-traceless (TT)' gauge, in which the wave is purely spatial and transverse to its own direction of propagation and where $e^\nu{}_\nu = 0$. This turns out to have a gauge-invariant meaning, so that the gravitational waves are really transversal. The two independent linear polarization states are usually called the + polarization and the × polarization (Misner *et al.* 1973).

Consider, for example, a plane wave moving in $x^1 \equiv x$ direction. In the transversal (y and z) directions, a ring of test particles will be deformed into a pulsating ellipse, with the axis of the $+$ polarization being rotated by $45°$ compared to the $\times$ polarization. One has explicitly

$$f_{\mu\nu} = 2\mathrm{Re}\left(e_{\mu\nu}\mathrm{e}^{-\mathrm{i}\omega(t-x)}\right) , \tag{2.10}$$

with $x^0 \equiv t$, $k^0 = k^1 \equiv \omega > 0$, $k^2 = k^3 = 0$. Denoting with $\mathbf{e}_y$ and $\mathbf{e}_z$ the unit vectors in y and z direction, respectively, one has for the $+$ and the $\times$ polarization, the expressions

$$e_{22}\mathbf{e}_+ = e_{22}(\mathbf{e}_y \otimes \mathbf{e}_y - \mathbf{e}_z \otimes \mathbf{e}_z) \tag{2.11}$$

and

$$e_{23}\mathbf{e}_\times = e_{23}(\mathbf{e}_y \otimes \mathbf{e}_z + \mathbf{e}_z \otimes \mathbf{e}_y) \tag{2.12}$$

for the polarization tensor, respectively (e_{22} and e_{23} are numbers giving the amplitude of the wave.) General solutions of the wave equation can be found by performing *superpositions* of the linear polarization states. In particular,

$$\mathbf{e}_\mathrm{R} = \frac{1}{\sqrt{2}}(\mathbf{e}_+ + \mathrm{i}\mathbf{e}_\times), \quad \mathbf{e}_\mathrm{L} = \frac{1}{\sqrt{2}}(\mathbf{e}_+ - \mathrm{i}\mathbf{e}_\times) \tag{2.13}$$

are the right and the left circular polarization states, respectively. The general case of an elliptic polarization also changes the shape of the ellipse.

Of special interest is the behaviour of the waves with respect to a rotation around the axis of propagation (here: the x-axis). Rotating counterclockwise with an angle θ, the polarization states transform according to

$$\mathbf{e}'_+ = \mathbf{e}_+ \cos 2\theta + \mathbf{e}_\times \sin 2\theta ,$$
$$\mathbf{e}'_\times = \mathbf{e}_\times \cos 2\theta - \mathbf{e}_+ \sin 2\theta . \tag{2.14}$$

For (2.13), this corresponds to

$$\mathbf{e}'_\mathrm{R} = \mathrm{e}^{-2\mathrm{i}\theta}\mathbf{e}_\mathrm{R} , \quad \mathbf{e}'_\mathrm{L} = \mathrm{e}^{2\mathrm{i}\theta}\mathbf{e}_\mathrm{L} . \tag{2.15}$$

The polarization tensors thus rotate with an angle 2θ. This corresponds to a symmetry with respect to a rotation by $180°$.

If a plane wave φ transforms as $\varphi \to \mathrm{e}^{\mathrm{i}h\theta}$ under a rotation around the direction of propagation, one calls h its *helicity*. The left (right) circular polarized gravitational wave thus has helicity 2 (-2). In the quantum theory, these states will become the states of the 'graviton', see Section 2.1.2. For plane waves with helicity h, the axes of linear polarization are inclined towards each other by an angle $90°/h$. For a spin-1/2 particle, for example, this is $180°$, which is why invariance for them is only reached after a rotation by $720°$.

For electrodynamics, right and left polarized states are given by the *vectors*

$$\mathbf{e}_R = \frac{1}{\sqrt{2}}(\mathbf{e}_y + i\mathbf{e}_z) , \quad \mathbf{e}_L = \frac{1}{\sqrt{2}}(\mathbf{e}_y - i\mathbf{e}_z) . \tag{2.16}$$

Under the above rotation, they transform as

$$\mathbf{e}'_R = e^{-i\theta}\mathbf{e}_R , \quad \mathbf{e}'_L = e^{i\theta}\mathbf{e}_L . \tag{2.17}$$

The left (right) circular polarized electromagnetic wave thus has helicity 1 (-1). Instead of (2.8), one has here

$$A_\mu = e_\mu e^{ikx} + e^*_\mu e^{-ikx} \tag{2.18}$$

with $k_\mu k^\mu = 0$ and $k_\nu e^\nu = 0$. It is possible to perform a gauge transformation without leaving the Lorenz gauge, $A_\mu \to A'_\mu = A_\mu + \partial_\mu \Lambda$ with $\Box\Lambda = 0$. With Λ being a plane-wave solution, $\Lambda = 2\mathrm{Re}[i\lambda e^{ikx}]$, one has instead of (2.9),

$$e_\mu \to e_\mu - \lambda k_\mu . \tag{2.19}$$

The field equations of linearized gravity can be obtained from the Lagrangian (Fierz and Pauli 1939),

$$\mathcal{L} = \frac{1}{64\pi G} \ (f^{\mu\nu,\sigma} f_{\mu\nu,\sigma} - f^{\mu\nu,\sigma} f_{\sigma\nu,\mu} - f^{\nu\mu,\sigma} f_{\sigma\mu,\nu}$$

$$-f^\mu{}_{\mu,\nu} f^\rho{}_{\rho,}{}^\nu + 2f^{\rho\nu}{}_{,\nu} f^\sigma{}_{\sigma,\rho}) + \frac{1}{2}T_{\mu\nu}f^{\mu\nu} . \tag{2.20}$$

The Euler–Lagrange field equations yield (writing $f \equiv f^\mu{}_\mu$)

$$f_{\mu\nu,\sigma}{}^\sigma - f_{\sigma\mu,\nu}{}^\sigma - f_{\sigma\nu,\mu}{}^\sigma + f_{,\mu\nu} + \eta_{\mu\nu}\left(f^{\alpha\beta}{}_{,\alpha\beta} - f_{,\sigma}{}^\sigma\right) = -16\pi G T_{\mu\nu} . \tag{2.21}$$

Performing the trace and substituting the $\eta_{\mu\nu}$-term yields

$$\Box f_{\mu\nu} - f_{\sigma\mu,\nu}{}^\sigma - f_{\sigma\nu,\mu}{}^\sigma + f_{,\mu\nu} = -16\pi G \left(T_{\mu\nu} - 1/2\,\eta_{\mu\nu}T\right) . \tag{2.22}$$

Using the harmonic condition (2.3), one finds the linearized Einstein equations (2.2).[3] From (2.22) one also gets the linearized Bianchi identity $\partial_\nu G^{\mu\nu} = 0$, consistent with $\partial_\nu T^{\mu\nu} = 0$. The Bianchi identity is a consequence of the gauge invariance (modulo a total divergence) of the Lagrangian (2.20) with respect to (2.4).[4]

---

[3]It is often useful to make a redefinition $f_{\mu\nu} \to \sqrt{32\pi G}f_{\mu\nu}$, cf. Section 2.2.2.
[4]The Bianchi identity in electrodynamics reads $\partial_\mu(\partial_\nu F^{\mu\nu}) = 0$, consistent with the charge-conservation law $\partial_\nu j^\nu = 0$.

Exploiting the Poincaré invariance of the flat background, one can calculate from the Fierz–Pauli Lagrangian (2.20) without the $T_{\mu\nu}$-term the *canonical* energy–momentum tensor of the linearized gravitational field,

$$t_{\mu\nu} = \frac{\partial \mathcal{L}}{\partial f_{\alpha\beta,}{}^{\nu}} f_{\alpha\beta,\mu} - \eta_{\mu\nu}\mathcal{L} \; . \tag{2.23}$$

The resulting expression is lengthy, but can be considerably simplified in the TT gauge where $f_{\mu\nu}$ assumes the form

$$f_{\mu\nu} = \begin{pmatrix} 0 & 0 & 0 & 0 \\ 0 & 0 & 0 & 0 \\ 0 & 0 & f_{22} & f_{23} \\ 0 & 0 & f_{23} & -f_{22} \end{pmatrix} \; .$$

Then,

$$t_{00} \stackrel{\mathrm{TT}}{=} \frac{1}{16\pi G} \left(\dot{f}_{22}^{2} + \dot{f}_{23}^{2} \right) = -t_{01} \; , \tag{2.24}$$

which can be written covariantly as

$$t_{\mu\nu} \stackrel{\mathrm{TT}}{=} \frac{1}{32\pi G} f_{\alpha\beta,\mu} f^{\alpha\beta}{}_{,\nu} \; . \tag{2.25}$$

It is sometimes appropriate to average this expression over a region of space–time much larger than ω^{-1}, where ω is the frequency of the weak gravitational wave, so that terms such as $\exp(-2i\omega(t-x))$ drop out. In the TT gauge, this leads to

$$\bar{t}_{\mu\nu} = \frac{k_{\mu}k_{\nu}}{16\pi G} e^{\alpha\beta*} e_{\alpha\beta} \; . \tag{2.26}$$

In the general harmonic gauge, without necessarily specifying to the TT gauge, the term $e^{\alpha\beta*}e_{\alpha\beta}$ is replaced by $e^{\alpha\beta*}e_{\alpha\beta} - \frac{1}{2}|e^{\alpha}{}_{\alpha}|^{2}$. This expression remains invariant under the gauge transformations (2.9), as it should.

Regarding the Fierz–Pauli Lagrangian (2.20), the question arises whether it could serve as a candidate for a fundamental helicity-2 theory of the gravitational field *in* a flat background. As it stands, this is certainly not possible because, as already mentioned, one has $\partial_{\nu}T^{\mu\nu} = 0$ and there is therefore no back reaction of the gravitational field onto matter. One might therefore wish to add the canonical energy–momentum tensor $t_{\mu\nu}$, Eqn (2.23), to the right-hand side of the linearized Einstein equations,

$$\Box \bar{f}_{\mu\nu} = -16\pi G(T_{\mu\nu} + t_{\mu\nu}) \; .$$

This modified equation would, however, lead to a Lagrangian *cubic* in the fields which in turn would give a new contribution to $t_{\mu\nu}$, and so on. Deser (1970, 1987) was able to show that this infinite process can actually be performed in one single step. The result is that the original metric $\eta_{\mu\nu}$ is unobservable and that all matter couples to the metric $g_{\mu\nu} = \eta_{\mu\nu} + f_{\mu\nu}$; the resulting action is the Einstein–Hilbert action and the theory therefore GR. Minkowski space–time as a background structure has completely disappeared. Boulanger *et al.* (2001) have shown that, starting from a finite number of Fierz–Pauli Lagrangians, no consistent coupling between the various helicity-2 fields is possible if the fields occur at most with second derivatives—leading only to a sum of uncoupled Einstein–Hilbert actions.

Since GR follows uniquely from (2.20), the question arises whether one would be able to construct a pure scalar or vector theory of gravity, cf. Straumann (2000). As has already been known to Maxwell, a vector theory is excluded because it would lead to repulsing forces. A scalar theory, on the other hand, would only lead to attraction. In fact, even before the advent of GR, Nordström had tried to describe gravity by a scalar theory, which can be defined by the Lagrangian

$$\mathcal{L}_n = 1/2 \; \eta^{\mu\nu} \partial_\mu \phi \partial_\nu \phi - 4\pi G T \phi + \mathcal{L}_{\text{matter}} \; , \tag{2.27}$$

where $T = \eta_{\mu\nu} T^{\mu\nu}$. This leads to the field equation

$$\Box \phi = -4\pi G T \; . \tag{2.28}$$

The physical metric (as measured by rods and clocks) turns out to be $g_{\mu\nu} \equiv \phi^2 \eta_{\mu\nu}$. A non-linear generalization of the Nordström theory was given by Einstein and Fokker (1914); their field equations read

$$R = 24\pi G T \; . \tag{2.29}$$

However, this theory is in contradiction with observation, since it contains no light deflection (the electromagnetic field has $T = 0$) and the perihelion motion of Mercury comes out incorrectly. Moreover, this theory contains an absolute structure, cf. Section 1.3: the conformal structure (the 'lightcone') is given from the outset and the theory thus possesses an invariance group (the conformal group), which in four dimensions is a finite-dimensional Lie group and which must be conceptually distinguished from the diffeomorphism group of GR. While pure scalar fields are thus unsuitable for a theory of the gravitational field, they can nevertheless occur *in addition* to the metric of GR. In fact, this happens quite frequently in unified theories, cf. Chapter 9.

2.1.2 *Gravitons from representations of the Poincaré group*

We shall now turn to the quantum theory of the linear gravitational field. The discussion of the previous subsection suggests that it is described by the behaviour of a massless spin-2 particle. Why massless? From the long-range nature of the gravitational interaction, it is clear that the graviton must have a small mass. Since the presence of a non-vanishing mass, however small, affects the deflection of light discontinuously, one may conclude that the graviton mass is strictly zero, see van Dam and Veltman (1970) and Carrera and Giulini (2001). Such an argument cannot be put forward for the photon.

In the following, we shall give a brief derivation of the spin-2 nature in the framework of representation theory (see e.g. Weinberg 1995 or Sexl and Urbantke 2001). In this subsection, we shall only deal with one-particle states ('quantum mechanics'), while field-theoretic aspects will be discussed in Section 2.1.3. The important ingredient is the presence of flat Minkowski space–time with metric $\eta_{\mu\nu}$ as an absolute background structure and the ensuing Poincaré symmetry. The use of the *Poincaré group*, not available beyond the linearized level, already indicates the approximate nature of the graviton concept.

According to Wigner, 'particles' are classified by irreducible representations of the Poincaré group. We describe a Poincaré transformation as

$$x'^{\mu} = \Lambda^{\mu}{}_{\nu}x^{\nu} + a^{\mu} \ , \tag{2.30}$$

where the Λ denotes Lorentz transformations and a denote space–time translations. According to Wigner's theorem, (2.30) induces a unitary transformation[5] in the *Hilbert space* of the theory,

$$\psi \to U(\Lambda, a)\psi \ . \tag{2.31}$$

This ensures that probabilities remain unchanged under the Poincaré group. Since this group is a Lie group, it is of advantage to study group elements close to the identity,

$$\Lambda^{\mu}{}_{\nu} = \delta^{\mu}{}_{\nu} + \omega^{\mu}{}_{\nu} \ , \quad a^{\mu} = \epsilon^{\mu} \ , \tag{2.32}$$

where $\omega_{\mu\nu} = -\omega_{\nu\mu}$. This corresponds to the unitary transformation[6]

$$U(1 + \omega, \epsilon) = 1 + 1/2 \ i\omega_{\mu\nu}J^{\mu\nu} - i\epsilon_{\mu}P^{\mu} + \dots \ , \tag{2.33}$$

where $J^{\mu\nu}$ and P^{μ} denote the 10 Hermitian generators of the Poincaré group, which are the boost generators, the angular momentum and the four-momentum, respectively. They obey the following Lie-algebra relations,

$$[P^{\mu}, P^{\rho}] = 0 \ , \tag{2.34}$$

$$i[J^{\mu\nu}, J^{\lambda\rho}] = \eta^{\nu\lambda}J^{\mu\rho} - \eta^{\mu\lambda}J^{\nu\rho} - \eta^{\rho\mu}J^{\lambda\nu} + \eta^{\rho\nu}J^{\lambda\mu} \ , \tag{2.35}$$

$$i[P^{\mu}, J^{\lambda\rho}] = \eta^{\mu\lambda}P^{\rho} - \eta^{\mu\rho}P^{\lambda} \ . \tag{2.36}$$

One-particle states are classified according to their behaviour with respect to Poincaré transformations. Since the components P^{μ} of the four-momentum commute with each other, we shall choose their eigenstates,

$$P^{\mu}\psi_{p,\sigma} = p^{\mu}\psi_{p,\sigma} \ , \tag{2.37}$$

where σ stands symbolically for all other variables. Application of the unitary operator then yields

$$U(1, a)\psi_{p,\sigma} = e^{-ip^{\mu}a_{\mu}}\psi_{p,\sigma} \ . \tag{2.38}$$

How do these states transform under Lorentz transformations (we only consider orthochronous proper transformations)? According to the method of induced

---

[5]The theorem also allows anti-unitary transformations, but these are relevant only for discrete symmetries.

[6]The plus sign on the right-hand side is enforced by the commutation relations $[J_i, J_k] = i\epsilon_{ikl}J_l$ (where $J_3 \equiv J_{12}$, etc.), whereas the minus sign in front of the second term is pure convention.

representations, it is sufficient to find the representations of the *little group*. This group is characterized by the fact that it leaves a 'standard' vector k^μ invariant (within each class of given $p^2 \leq 0$[7] and given sign of p^0). For positive p^0, one can distinguish between the following two cases. The first possibility is $p^2 = -m^2 < 0$. Here one can choose $k^\mu = (m, 0, 0, 0)$, and the little group is SO(3), since these are the only Lorentz transformations that leave a particle with $\mathbf{k} = 0$ at rest. The second possibility is $p^2 = 0$. One chooses $k^\mu = (1, 0, 0, 1)$, and the little group is ISO(2), the invariance group of Euclidean geometry (rotations and translations in two dimensions). Any p^μ within a given class can be obtained from the corresponding k^μ by a Lorentz transformation. The normalization is chosen such that

$$\langle \psi_{p', \sigma'}, \psi_{p, \sigma} \rangle = \delta_{\sigma\sigma'} \delta(\mathbf{p} - \mathbf{p}') \ . \tag{2.39}$$

Consider first the case $m \neq 0$ where the little group is SO(3). As is well known from quantum mechanics, its unitary representations are a direct sum of irreducible unitary representations $\mathcal{D}^{(j)}_{\sigma\sigma'}$, with dimensions $2j + 1$ ($j = 0, \frac{1}{2}, 1 \ldots$). Denoting the angular momentum with respect to the z-axis by $J_3^{(j)} \equiv J_{12}^{(j)}$, one has $(J_3^{(j)})_{\sigma\sigma'} = \sigma\delta_{\sigma\sigma'}$ with $\sigma = -j, \ldots, +j$.

On the other hand, for $m = 0$, the little group is ISO(2). It turns out that the quantum-mechanical states are only distinguished by the eigenvalue of J_3, the component of the angular momentum in the direction of motion (recall $k^\mu = (1, 0, 0, 1)$ from above),

$$J_3 \psi_{k, \sigma} = \sigma \psi_{k, \sigma} \tag{2.40}$$

($\psi_{k, \sigma}$ is annihilated by J_1 and J_2). The eigenvalue σ is called the *helicity*. One then gets

$$U(\Lambda, 0)\psi_{p, \sigma} = N e^{i\sigma\theta(\Lambda, p)} \psi_{\Lambda p, \sigma} \ , \tag{2.41}$$

where θ denotes the angle contained in the rotation part of Λ. Since massless particles are not at rest in any inertial system, helicity is a Lorentz-invariant property and may be used to characterize a particle with $m = 0$.

Comparison with (2.17) exhibits that $\sigma = \pm 1$ characterizes the *photon*.[8] Because of the helicity-2 nature of weak gravitational waves in a flat background, see (2.15), we attribute the particle with $\sigma = \pm 2$ with the gravitational interaction and call it the *graviton*. Since for a massless particle, $|\sigma|$ is called its *spin*, we recognize that the graviton has spin 2. The helicity eigenstates (2.41) correspond to circular polarization (see Section 2.1.1), while their superpositions correspond in the generic case to elliptic polarization or (for equal absolute values of the amplitudes) to linear polarization.

---

[7]This restriction is imposed in order to avoid tachyons (particles with $m^2 < 0$).

[8]Because of space inversion symmetry, $\sigma = 1$ and $\sigma = -1$ describe the same particle. This holds also for the graviton, but due to parity violation not, for example, for neutrinos.

2.1.3 Quantization of the linear field theory

We now turn to field theory. One starts from a superposition of plane-wave solutions (2.8) and formally turns this into an operator,

$$f_{\mu\nu}(x) = \sum_\sigma \int \frac{\mathrm{d}^3 k}{\sqrt{2|\mathbf{k}|}} \left[a(\mathbf{k}, \sigma) e_{\mu\nu}(\mathbf{k}, \sigma) \mathrm{e}^{ikx} + a^\dagger(\mathbf{k}, \sigma) e_{\mu\nu}^*(\mathbf{k}, \sigma) \mathrm{e}^{-ikx} \right] . \quad (2.42)$$

As in the usual interpretation of free quantum field theory, $a(\mathbf{k}, \sigma)$ $(a^\dagger(\mathbf{k}, \sigma))$ is interpreted as the annihilation (creation) operator for a graviton of momentum $\hbar\mathbf{k}$ and helicity σ (see e.g. Weinberg 1995). They obey

$$[a(\mathbf{k}, \sigma), a^\dagger(\mathbf{k}', \sigma')] = \delta_{\sigma\sigma'}\delta(\mathbf{k} - \mathbf{k}') , \quad (2.43)$$

with all other commutators vanishing. The quantization of the linearized gravitational field was already discussed by Bronstein (1936).

Since we only want the presence of helicities ± 2, $f_{\mu\nu}$ cannot be a true tensor with respect to Lorentz transformations (note that the TT-gauge condition is not Lorentz invariant). As a consequence, one is *forced* to introduce gauge invariance and demand that $f_{\mu\nu}$ transform under a Lorentz transformation according to

$$f_{\mu\nu} \to \Lambda_\mu{}^\lambda \Lambda_\nu{}^\rho f_{\lambda\rho} - \partial_\nu \epsilon_\mu - \partial_\mu \epsilon_\nu , \quad (2.44)$$

in order to stay within the TT-gauge, cf. (2.5). Therefore, the coupling in the Lagrangian (2.20) must be to a *conserved* source, $\partial_\nu T^{\mu\nu} = 0$, because otherwise the coupling is not gauge invariant.

The occurrence of gauge invariance can also be understood in a group-theoretic way. We start with a symmetric tensor field $f_{\mu\nu}$ with vanishing trace (nine degrees of freedom). This field transforms according to the irreducible $\mathcal{D}^{(1,1)}$ representation of the Lorentz group, see for example, section 5.6 in Weinberg (1995). Its restriction to the subgroup of rotations yields

$$\mathcal{D}^{(1,1)} = \mathcal{D}^{(1)} \otimes \mathcal{D}^{(1)} , \quad (2.45)$$

where $\mathcal{D}^{(1)}$ denotes the $j = 1$ representation of the rotation group. (Since one has three 'angles' in each of them, this yields the $3 \times 3 = 9$ degrees of freedom of the trace-free $f_{\mu\nu}$.) The representation (2.45) is reducible with 'Clebsch–Gordon decomposition'

$$\mathcal{D}^{(1)} \otimes \mathcal{D}^{(1)} = \mathcal{D}^{(2)} \oplus \mathcal{D}^{(1)} \oplus \mathcal{D}^{(0)} , \quad (2.46)$$

corresponding to the five degrees of freedom for a massive spin-2 particle, three degrees of freedom for a spin-1 particle, and one degree of freedom for a spin-0 particle, respectively. The latter $3 + 1$ degrees of freedom can be eliminated by the four conditions $\partial_\nu f^{\mu\nu} = 0$ (transversality). To obtain only two degrees of freedom one needs, however, to impose the gauge freedom (2.5). This yields three additional conditions (the four ϵ_μ in (2.5) have to satisfy $\partial_\mu \epsilon^\mu = 0$—one condition—to preserve tracelessness; the condition $\Box \epsilon_\mu = 0$—needed to preserve transversality—is automatically fulfilled for plane waves). In total, one arrives at $(10-1)-4-3=2$ degrees of freedom, corresponding to the two helicity states of the graviton.[9]

The same arguments also apply of course to electrodynamics: A^μ cannot transform as a Lorentz vector, since, for example, the temporal gauge $A^0 = 0$ can be chosen. Instead, one is forced to introduce gauge invariance, and the transformation law is

$$A_\mu \to \Lambda_\mu{}^\nu A_\nu + \partial_\mu \epsilon \;, \tag{2.47}$$

in analogy to (2.44). Therefore, a Lagrangian is needed that couples to a *conserved* source, $\partial_\mu j^\mu = 0$.[10] The group-theoretic argument for QED goes as follows. A vector field transforms according to the $\mathcal{D}^{(1/2,1/2)}$ representation of the Lorentz group which, if restricted to rotations, can be decomposed as

$$\mathcal{D}^{(1/2,1/2)} = \mathcal{D}^{(1/2)} \otimes \mathcal{D}^{(1/2)} = \mathcal{D}^{(1)} \oplus \mathcal{D}^{(0)} \;. \tag{2.48}$$

The $\mathcal{D}^{(0)}$ describes spin-0, which is eliminated by the Lorenz condition $\partial_\nu A^\nu = 0$. The $\mathcal{D}^{(1)}$ corresponds to the three degrees of freedom of a massive spin-1 particle. One of these is eliminated by the gauge transformation $A_\mu \to A_\mu + \partial_\mu \epsilon$ (with $\Box \epsilon = 0$ to preserve the Lorenz condition) to arrive at the two degrees of freedom for the massless photon.

Weinberg (1964) concluded (see also p. 537 of Weinberg 1995) that one can derive the equivalence principle (and thus GR if no other fields are present) from the *Lorentz invariance* of the spin-2 theory (plus the pole structure of the S-matrix). Similar arguments can be put forward in the electromagnetic case to show that electric charge must be conserved. No arguments of gauge invariance are needed, at least not explicitly. The gravitational mass m_g is defined in this approach by the strength of interaction with a soft graviton, that is, a graviton with four-momentum $k \to 0$. The amplitude for the emission (see Fig. 2.1) of a single soft graviton is given by the expression

$$M^{\mu\nu}_{\beta\alpha}(k) = M_{\beta\alpha} \cdot \sum_n \frac{\eta_n g_n p_n^\mu p_n^\nu}{p_n^\mu k_\mu - i\eta_n \epsilon} \;, \tag{2.49}$$

where $M_{\beta\alpha}$ denotes the amplitude for the process without soft-graviton emission, and the sum runs over all ingoing and outgoing particles; α refers to the ingoing, β to the outgoing particles (Fig. 2.1), g_n denotes the coupling of the graviton to particle n ($\eta_n = 1$ for outgoing, $\eta_n = -1$ for ingoing particles), and p_n is the four-momentum of the nth particle (in the initial and final state, respectively).

---

[9]The counting in the canonical version of the theory leads of course to the same result and is presented in Section 4.2.3.

[10]This is also connected with the fact that massless spins ≥ 3 are usually excluded, because no conserved tensor is available. Massless spins ≥ 3 cannot generate long-range forces, cf. Weinberg (1995, p. 252).

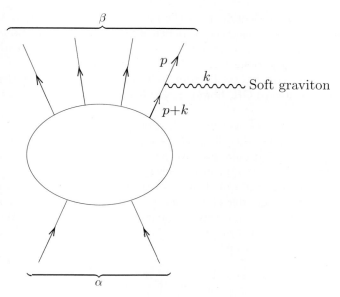

FIG. 2.1. Emission of a soft graviton.

To calculate the amplitude for graviton emission with definite helicity, one has to contract (2.49) with the polarization tensor $e_{\mu\nu}(\mathbf{k}, \sigma)$. As argued above, however, the latter is not a Lorentz tensor, but transforms according to

$$e_{\mu\nu} \rightarrow \Lambda_\mu{}^\lambda \Lambda_\nu{}^\rho e_{\lambda\rho} + k_\nu f_\mu + k_\mu f_\nu \ .$$

To obtain a Lorentz-invariant amplitude, one must therefore demand that

$$k_\mu M^{\mu\nu}_{\beta\alpha}(k) = 0 \ .$$

From (2.49), one then finds

$$\sum_n \eta_n g_n p_n^\nu = 0 \ ,$$

which is equivalent to the statement that $\sum_n g_n p_n^\nu$ is conserved. But for non-trivial processes, the only linear combination of momenta that is conserved is the total momentum $\sum_n \eta_n p_n^\nu$. Consequently, the couplings g_n must all be *equal*, and one can set $g_n \equiv \sqrt{8\pi G}$. Therefore, all low-energy particles with spin 2 and $m = 0$ couple to all forms of energy in an equal way. As in Section 2.1.1, this shows the 'equivalence' of a spin-2 theory with GR. From this point of view, GR is a consequence of quantum theory. Weinberg (1964) also showed that the effective gravitational mass m_{g} is given by

$$m_{\mathrm{g}} = 2E - \frac{m_{\mathrm{i}}^2}{E} \ ,$$

where m_i denotes as in Section 1.1.4 the inertial mass, and E is the energy. For $E \to m_i$, this leads to the usual equivalence of inertial and gravitational mass. On the other hand, one has $m_g = 2E$ for $m_i \to 0$. How can this be interpreted? In his 1911-calculation of the deflection of light, Einstein found from the equivalence principle alone (setting $m_g = E$), the Newtonian expression for the deflection angle. The full theory of GR, however, yields twice this value, corresponding to $m_g = 2E$.

Weinberg's arguments, as well as the approaches presented in Section 2.1.1, are important for unified theories such as string theory (see Chapter 9) in which a massless spin-2 particle emerges with necessity, leading to the claim that such theories contain GR in an appropriate limit.

Upon discussing linear quantum gravity, the question arises whether this framework leads to observable effects in the laboratory. A brief estimate suggests that such effects would be too tiny: comparing in atomic physics, the quantum-gravitational decay rate Γ_g with its electromagnetic counterpart Γ_e, one would expect for dimensional reasons

$$\frac{\Gamma_g}{\Gamma_e} \sim \alpha^n \left(\frac{m_e}{m_P}\right)^2 \qquad (2.50)$$

with some power n of the fine-structure constant α, and m_e being the electron mass. The square of the mass ratio already yields the tiny number 10^{-45}. Still, it is instructive to discuss some example in detail. Following Weinberg (1972), the transition rate from the $3d$ level to the $1s$ level in the hydrogen atom due to the emission of a graviton will be calculated. One needs at least the $3d$ level, since $\Delta l = 2$ is needed for the emission of a spin-2 particle.

One starts from the classical formula for gravitational radiation and interprets it as the emission rate of gravitons with energy $\hbar\omega$,

$$\Gamma_g = \frac{P}{\hbar\omega} , \qquad (2.51)$$

where P denotes the classical expression for the emitted power. This has already been suggested by Bronstein (1936). In the quadrupole approximation one then finds for the transition rate from an initial state i to a final state f (restoring c),

$$\Gamma_g(i \to f) = \frac{2G\omega^5}{5\hbar c^5} \left(\sum_{kl} Q_{kl}^*(i \to f) Q_{kl}(i \to f) - \frac{1}{3}\sum_k |Q_{kk}(i \to f)|^2\right) , \qquad (2.52)$$

where

$$Q_{kl}(i \to f) \equiv m_e \int \mathrm{d}^3 x \; \psi_f^*(\mathbf{x}) x_k x_l \psi_i(\mathbf{x}) \qquad (2.53)$$

is the quantum-mechanical analogue of the classical quadrupole moment,

$$Q_{kl} = \frac{1}{c^2} \int \mathrm{d}^3 x \; x_k x_l T^{00} . \qquad (2.54)$$

Inserting the hydrogen eigenfunctions ψ_{100} and ψ_{32m} ($m = -2, \ldots, 2$) for ψ_i and ψ_f, respectively, and averaging over m, one finds after some calculations,[11]

$$\Gamma_g = \frac{Gm_e^3 c\alpha^6}{360\hbar^2} \approx 5.7 \times 10^{-40} \text{ s}^{-1} \ . \tag{2.55}$$

This corresponds to a life-time of

$$\tau_g \approx 5.6 \times 10^{31} \text{ years} \ , \tag{2.56}$$

which is too large to be observable (although it is of the same order as the value for the proton life-time predicted by some unified theories). Since electromagnetic dipole transitions are of the order of $\Gamma_e \sim m_e \alpha^5 c^2 / \hbar$, one has

$$\frac{\Gamma_g}{\Gamma_e} \sim \alpha \left(\frac{m_e}{m_P} \right)^2 \ , \tag{2.57}$$

cf. (2.50).

Through similar heuristic considerations, one can also calculate the number N of gravitons per unit volume associated with a gravitational wave (Weinberg 1972). We recall from (2.26) and the accompanying remarks that the space–time average of the energy–momentum tensor for the weak gravitational waves is in the general harmonic gauge given by the expression

$$\bar{t}_{\mu\nu} = \frac{k_\mu k_\nu c^4}{16\pi G} \left(e^{\alpha\beta*} e_{\alpha\beta} - \frac{1}{2} |e^\alpha{}_\alpha|^2 \right) \ . \tag{2.58}$$

Since the energy–momentum tensor of a collection of particles reads (cf. Weinberg 1972, section 2.8)

$$T^{\mu\nu} = c^2 \sum_n \frac{p_n^\mu p_n^\nu}{E_n} \delta(\mathbf{x} - \mathbf{x}_n(t)) \ , \tag{2.59}$$

a collection of gravitons with momenta $p^\mu = \hbar k^\mu$ possesses the energy–momentum tensor

$$T_g^{\mu\nu} = \frac{\hbar c^2 k^\mu k^\nu N}{\omega} \ . \tag{2.60}$$

Comparison with (2.58) then yields

$$N = \frac{\omega c^2}{16\pi\hbar G} \left(e^{\alpha\beta*} e_{\alpha\beta} - \frac{1}{2} |e^\alpha{}_\alpha|^2 \right) \ , \tag{2.61}$$

where the expression in parentheses is the amplitude squared, $|A|^2$. Taking, for example, $\omega \approx 1$ kHz, $|A| \approx 10^{-21}$ (typical values for gravitational waves arriving

---

[11] The result is different from Weinberg (1972) because it seems that Weinberg has used the eigenfunctions from Messiah's textbook on quantum mechanics, which contain a misprint. I am grateful to N. Straumann for discussions on this point.

from a supernova), one finds $N \approx 3 \times 10^{14}$ cm^{-3}. For the stochastic background of gravitons from the early universe (cf. Section 10.1), one would expect typical values of $\omega \approx 10^{-3}$ Hz, $|A| \approx 10^{-22}$, leading to $N \approx 10^{6}$ cm^{-3}.

The ground-state wave function of the quantum-mechanical oscillator is given by a Gaussian, $\psi_0(x) \propto \exp(-\omega x^2/2)$. Similarly, one can express the ground state of a free quantum field as a Gaussian wave *functional*. We shall discuss such functionals in detail in Section 5.3, but mention here that there exists an explicit form for the ground-state functional of the linear graviton field discussed above (Kuchař 1970). This has a form similar to the ground-state functional of free QED, which reads

$$\Psi_0[\mathbf{A}] \propto \exp\left(-\frac{1}{2} \int d^3x \, d^3y \, A^a(\mathbf{x})\omega_{ab}(\mathbf{x},\mathbf{y})A^b(\mathbf{y}) \right) \tag{2.62}$$

with

$$\omega_{ab}(\mathbf{x},y) = (-\nabla^2 \delta_{ab} + \partial_a \partial_b) \int d^3p \, \frac{e^{-i\mathbf{p}(\mathbf{x}-\mathbf{y})}}{|\mathbf{p}|} . \tag{2.63}$$

Note that a and b are only spatial indices, since the wave functional is defined on space-like hypersurfaces. This state can also be written in a manifest gauge-invariant form,

$$\Psi_0 \propto \exp\left(-\frac{1}{4\pi^2\hbar} \int d^3x \, d^3y \, \frac{\mathbf{B}(\mathbf{x})\mathbf{B}(\mathbf{y})}{|\mathbf{x}-\mathbf{y}|^2} \right) , \tag{2.64}$$

where $\mathbf{B}$ is the magnetic field and $\hbar$ has been re-inserted. In the functional picture, it is evident that the ground state is a highly non-local state. The ground state of the linearized gravitational field can be written in a manifest gauge-invariant way similar to (2.64) as

$$\Psi_0 \propto \exp\left(-\frac{1}{8\pi^2 G\hbar} \int d^3x \, d^3y \, \frac{f_{ab,c}^{\mathrm{TT}}(\mathbf{x})f_{ab,c}^{\mathrm{TT}}(\mathbf{y})}{|\mathbf{x}-\mathbf{y}|^2} \right) , \tag{2.65}$$

where f_{ab} are the spatial components of the metric perturbation introduced in (2.1).

2.2 Path integrals and the background-field method

2.2.1 General properties of path integrals

A popular method in both quantum mechanics and quantum field theory is path-integral quantization. In the case of quantum mechanics, the propagator for a particle to go from position $\mathbf{x}'$ at time t' to position $\mathbf{x}''$ at time t'' can be expressed as a formal sum over all possible paths connecting these positions,

$$\langle \mathbf{x}'', t''|\mathbf{x}', t' \rangle = \int \mathcal{D}\mathbf{x}(t) \, e^{iS[\mathbf{x}(t)]/\hbar} . \tag{2.66}$$

It is important to remember that most 'paths' in this sum are continuous, but nowhere differentiable, and that $\mathcal{D}\mathbf{x}(t)$ is, in fact, a formal notation for the following limiting process (spelled out for simplicity for a particle in one spatial dimension),

$$\langle \mathbf{x}'', t'' | \mathbf{x}', t' \rangle = \lim_{N \to \infty} \int dx_1 \cdots dx_{N-1} \left(\frac{mN}{2\pi it\hbar} \right)^{N/2}$$

$$\times \prod_{j=0}^{N-1} \exp \left(-\frac{m(x_{j+1} - x_j)^2 N}{2it\hbar} - \frac{itV(x_j)}{\hbar N} \right) , \quad (2.67)$$

where m denotes the mass, V the potential, and $t \equiv t'' - t'$. The path integral is especially suited for performing a semiclassical approximation (expansion of the action around classical solutions).

The quantum-mechanical path integral, which can be put on a firm mathematical footing, can be formally generalized to quantum field theory where such a footing is lacking (there is no measure-theoretical foundation). Still, the field-theoretical path integral is of great heuristic value and plays a key role especially in gauge theories (see e.g. Böhm *et al.* (2001) among many other references). Consider, for example, a real scalar field $\phi(x)$.[12] Then one has instead of (2.66), the expression (setting $\hbar = 1$ again)

$$Z[\phi] = \int \mathcal{D}\phi(x) \, e^{iS[\phi(x)]} , \quad (2.68)$$

where $Z[\phi]$ is the usual abbreviation for the path integral in the field-theoretical context (often referring to in-out transition amplitudes or to partition sums, see below). The path integral is very useful in perturbation theory and gives a concise possibility to derive Feynman rules (via the notion of the generating functional, see below). Using the methods of Grassmann integration, a path integral such as (2.68) can also be defined for fermions. For systems with constraints (such as gauge theories and gravity), however, the path-integral formulation has to be generalized, as will be discussed in the course of this section. It must also be noted that the usual operator ordering ambiguities of quantum theory are also present in the path-integral approach, in spite of integrating over classical configurations: the ambiguities are here reflected in the ambiguities for the integration measure.

Instead of the original formulation in space–time, it is often appropriate to perform a rotation to four-dimensional Euclidean space via the *Wick rotation* $t \to -i\tau$. In the case of the scalar field, this leads to

$$iS[\phi] = i \int dt d^3x \left(\frac{1}{2} \left(\frac{\partial \phi}{\partial t} \right)^2 - \frac{1}{2} (\nabla \phi)^2 - V(\phi) + g\mathcal{L}_{\text{int}} \right) ,$$

[12] The notation x is a shorthand for x^μ, $\mu = 0, 1, 2, 3$.

$$\overset{t \to -i\tau}{\longrightarrow} -\int \mathrm{d}\tau \mathrm{d}^3 x \; \left(\frac{1}{2} \left(\frac{\partial \phi}{\partial \tau} \right)^2 + \frac{1}{2}(\nabla \phi)^2 + V(\phi) - g\mathcal{L}_{\mathrm{int}} \right) ,$$
$$\equiv -S_{\mathrm{E}}[\phi] , \tag{2.69}$$

where $V(\phi)$ is the potential and an interaction $g\mathcal{L}_{\mathrm{int}}$ to other fields has been taken into account. This formal rotation to Euclidean space has some advantages. First, since S_{E} is bounded from below, it improves the convergence properties of the path integral: instead of an oscillating integrand one has an exponentially damped integrand (remember, however, that e.g. the Fresnel integrals used in optics are convergent in spite of the e^{ix^2}-integrand). Second, for the extremization procedure, one has to now deal with elliptic instead of hyperbolic equations, which are more suitable for the boundary problem of specifying configurations at initial and final instants of time. Third, in the Euclidean formulation, the path integral can be directly related to the partition sum in statistical mechanics (e.g. for the canonical ensemble one has $Z = \mathrm{tr}\,\mathrm{e}^{-\beta H}$). Fourth, the Euclidean formulation is convenient for lattice gauge theory where one considers

$$Z[U] = \int \mathcal{D}U \; \mathrm{e}^{-S_{\mathrm{W}}[U]} , \tag{2.70}$$

with U denoting the lattice gauge fields defined on the links and S_{W} the Wilson action, see also Chapter 6. The justification of performing Wick rotations in quantum field theory relies on the fact that Euclidean Green functions can be analytically continued back to real time while preserving their pole structure; cf. Osterwalder and Schrader (1975).

The quantum-gravitational path integral, first formulated by Misner (1957), would be of the form

$$Z[g] = \int \mathcal{D}g_{\mu\nu}(x) \; \mathrm{e}^{iS[g_{\mu\nu}(x)]} , \tag{2.71}$$

where the sum runs over all metrics on a four-dimensional manifold $\mathcal{M}$ divided by the diffeomorphism group $\mathrm{Diff}\mathcal{M}$ (see below). One might expect that an additional sum has to be performed over all topologies, but this is a contentious issue. As we shall see in Section 5.3.4, the path integral (2.71) behaves more like an energy Green function instead of a propagator.

Needless to say that (2.71) is of a tremendous complicated nature, both technically and conceptually. One might therefore try, for the reasons stated above, to perform a Wick rotation to the Euclidean regime. This leads, however, to problems which are not present in ordinary quantum field theory. First, not every Euclidean metric (in fact, only very few) possesses a Lorentzian section, that is, leads to a signature $(-,+,+,+)$ upon $\tau \to it$. Such a section exists only for metrics with special symmetries. (The Wick rotation is definitely not a diffeomorphism-invariant procedure.) Second, a sum over topologies cannot be performed even in principle because four-manifolds are not classifiable (Geroch and Hartle 1986).

The third, and perhaps most severe, problem is the fact that the Euclidean gravitational action is not bounded from below. Performing the same Wick rotation
as above (in order to be consistent with the matter part), one finds from (1.1)
for the Euclidean action, the expression (using a space-like boundary)

$$S_E[g] = -\frac{1}{16\pi G} \int_{\mathcal{M}} \mathrm{d}^4x \ \sqrt{g}\,(R - 2\Lambda) - \frac{1}{8\pi G} \int_{\partial\mathcal{M}} \mathrm{d}^3x \ \sqrt{h}K \ . \tag{2.72}$$

To see the unboundedness of this action, consider a conformal transformation of
the metric, $g_{\mu\nu} \to \tilde{g}_{\mu\nu} = \Omega^2 g_{\mu\nu}$. This yields (Gibbons *et al.* 1978; Hawking 1979)

$$S_E[\tilde{g}] = -\frac{1}{16\pi G} \int_{\mathcal{M}} \mathrm{d}^4x \ \sqrt{g}(\Omega^2 R + 6\Omega_{;\mu}\Omega_{;\nu}g^{\mu\nu} - 2\Lambda\Omega^4) - \frac{1}{8\pi G} \int_{\partial\mathcal{M}} \mathrm{d}^3x \ \sqrt{h}\Omega^2 K \ .$$
$$\tag{2.73}$$

One recognizes that the action can be made arbitrarily negative by choosing a
highly varying conformal factor Ω. The presence of such metrics in the path integral then leads to its divergence. This is known as the *conformal-factor problem*.
There are, however, strong indications for a solution of this problem. As Dasgupta and Loll (2001) have argued, the conformal divergence can cancel with a
similar term of opposite sign arising from the measure in the path integral (cf.
Section 2.2.3 for a discussion of the measure), see Hartle and Schleich (1987) for
a similar result in the context of linearized gravity. Euclidean path integrals are
often used in quantum cosmology, being related to boundary conditions of the
universe (see Section 8.3), so a clarification of these issues is of central interest.

Since the gravitational path integral is of a highly complicated nature, the
question arises whether it can be evaluated by discretization and performing the
continuum limit. In fact, among others, the two following methods have been
employed,

1. *Regge calculus* (see e.g. Williams (1997) for a review): Originally conceived
   by Tullio Regge as a method for classical numerical relativity, it was applied to the Euclidean path integral from the 1980s on. The central idea
   is to decompose four-dimensional space into a set of simplices and treat
   the edge lengths as dynamical entities. An important technical feature in
   the evaluation of the path integral is the implementation of the triangle
   inequalities for the lengths—one must tell the formalism that the sum of
   the length of two sides in a triangle is bigger than the third side. Therefore,
   the path integral can only be evaluated numerically.

2. *Dynamical triangulation* (see e.g. Loll (2003) for a review): In contrast to
   Regge calculus, all edge lengths are kept fixed, and the sum in the path
   integral is instead taken over all possible manifold-gluings of equilateral
   simplices. The evaluation is thus reduced to a combinatorical problem. In
   contrast to Regge calculus, this method is applied to Lorentzian geometries,
   emphasizing the importance of the lightcone structure already at the level
   of geometries in the path integral. Due to this well-defined causal structure,
   a topology change of the spatial slice does not occur.

The discussion of path integrals will be continued in Section 2.2.3, where emphasis is put on the integration measure and the derivation of Feynman rules for gravity. In the next subsection, we shall give an introduction into the use of perturbation theory in quantum gravity.

2.2.2   *Background-field method*

In Section 2.1, we have treated the concept of a graviton similar to the photon—within the representation theory of the Poincaré group. We have, in particular, discussed the Fierz–Pauli Lagrangian (2.20) which is (up to a total derivative) gauge invariant and which at the classical level inevitably leads to GR. The question thus arises whether this Lagrangian can be quantized in a way similar to electrodynamics where one arrives at the very successful theory of QED. More generally then, why should one not perform a quantum *perturbation theory* of the Einstein–Hilbert action (1.1)?

The typical situation for applications of perturbation theory in quantum field theory addresses 'scattering' situations in which asymptotically free quantum states (representing ingoing and outgoing particles) are connected by a region of interaction. This is the standard situation in accelerators. In fact, most quantum field theories are only understood in the perturbative regime.

Perturbation theory in quantum gravity belongs to the class of covariant quantization schemes to which also the path-integral methods belong. These methods intend to maintain four-dimensional (space–time) covariance. They are distinguished from the canonical methods to be discussed in Chapters 4–6. Can perturbation theory be useful in quantum gravity? One might think that the gravitational interaction is intrinsically non-perturbative, and that objects such as black holes or the early universe cannot be described in perturbation theory. On the other hand, as has been discussed in Chapter 1, it is hopeless to probe Planck-scale effects in accelerators. While this is true, it is not excluded *per se* that perturbative quantum gravity effects are unobservable. For example, such effects could in principle show up in the anisotropy spectrum of the cosmic microwave background, cf. Section 5.4.

A major obstacle to the viability of perturbation theory is the *non-renorma-lizability* of quantum GR. What does this mean? Quantum field theory uses *local* field operators $\phi(x)$. This leads to the occurrence of arbitrarily small distances and, therefore, to arbitrarily large momenta. As a consequence, *divergences* show up usually in calculations of cross-sections coming from integrals in momentum space. The theory is called renormalizable if these divergences can all be removed by a redefinition of a *finite* number of physical constants (masses, charges, etc.) and fields; see e.g. Weinberg (1995) for details. It turns out that the mass dimensionality (in units where $\hbar = c = 1$) of the coupling constant for a certain interaction decides about renormalizability. This dimensionality is given by a coefficient Δ which is called the superficial degree of divergence and which must not be negative. It can be calculated by the formula

$$\Delta = 4 - d - \sum_f n_f(s_f + 1) \, , \qquad (2.74)$$

where d is the number of derivatives, n_f the number of fields of type f, and $s_f = 0, 1/2, 1, 0$ for scalars, fermions, massive vector fields, and photons and gravitons, respectively. Considering, for example, the standard QED interaction $-ie\bar{\psi}A^\mu\gamma_\mu\psi$, one obtains $\Delta = 4 - 0 - 3/2 - 3/2 - 1 = 0$; the electric charge e is thus dimensionless and the coupling is renormalizable. On the other hand, the presence of a 'Pauli term' $\bar{\psi}[\gamma_\mu, \gamma_\nu]\psi F^{\mu\nu}$, for example, would lead to $\Delta = -1$ and thus to a non-renormalizable interaction. One can show, for example, that Yang–Mills theories are renormalizable. A most important consequence of this is the renormalizability of the standard model of particle physics.

Why are renormalizable theories successful? Consider a non-renormalizable interaction $g \sim M^{-|\Delta|}$, where M is a typical mass scale. For momenta $k \ll M$, therefore, g must be accompanied by a factor $k^{|\Delta|}$; as a consequence, this non-renormalizable interaction is suppressed by a factor $(k/M)^{|\Delta|} \ll 1$ and not seen at low momenta. The success of the renormalizable standard model thus indicates that any such mass scale must be much higher than currently accessible energies.

Whereas non-renormalizable theories had been fully discarded originally, a more modern viewpoint attributes to them a possible use as *effective theories* (Weinberg 1995). If all possible terms allowed by symmetries are included in the Lagrangian, then there is a counterterm present for any ultraviolet (UV) divergence (see the discussion below for gravity). For energies much smaller than M, effective theories might therefore lead to useful predictions. Incidentally, also the standard model of particle physics, albeit renormalizable, is today interpreted as an effective theory. The only truly fundamental theories seem to be those which unify all interactions at the Planck scale.

An early example of an effective theory is given by the Euler–Heisenberg Lagrangian,

$$\mathcal{L}_{\text{E–H}} = \frac{1}{8\pi}(\mathbf{E}^2 - \mathbf{B}^2) + \frac{e^4\hbar}{360\pi^2 m_{\text{e}}^2}\left[(\mathbf{E}^2 - \mathbf{B}^2)^2 + 7(\mathbf{EB})^2\right] \, , \qquad (2.75)$$

where m_{e} is the electron mass, see for example, section 12.3 in Weinberg (1995). The second term in (2.75) arises after the electrons are integrated out and terms with order $\propto \hbar^2$ and higher are neglected. Already at this effective level one can calculate observable physical effects such as Delbrück scattering (scattering of a photon at an external field).

In the *background-field method* to quantize gravity (DeWitt 1967b, c), one expands the metric about an arbitrary curved background solution to the Einstein equations,[13]

$$g_{\mu\nu} = \bar{g}_{\mu\nu} + \sqrt{32\pi G}f_{\mu\nu} \, . \qquad (2.76)$$

Here, $\bar{g}_{\mu\nu}$ denotes the background field with respect to which (four-dimensional) covariance will be implemented in the formalism; $f_{\mu\nu}$ denotes the quantized

---

[13]Sometimes the factor $\sqrt{8\pi G}$ is chosen instead of $\sqrt{32\pi G}$.

field and has the dimension of a mass. If one chooses a flat background space–time, $\bar{g}_{\mu\nu} = \eta_{\mu\nu}$ (cf. (2.1)), one finds from the Einstein–Hilbert Lagrangian the Fierz–Pauli Lagrangian (2.20) plus higher order terms having the symbolic form (omitting indices)

$$\sqrt{32\pi G} f(\partial f)(\partial f) + \cdots + (\sqrt{32\pi G} f)^n (\partial f)(\partial f) + \cdots \qquad (2.77)$$

These are infinitely many terms because the inverse of the metric, $g^{\mu\nu}$, enters the Einstein–Hilbert Lagrangian $\propto \sqrt{-g} R_{\mu\nu} g^{\mu\nu}$. Each term contains two f-derivatives because the Ricci scalar has two derivatives; each vertex has a factor $\sqrt{G}$.

Before going into the details, it will be shown that the gravitational interaction is indeed non-renormalizable. Considering the first interaction term $\sqrt{G} f(\partial f)(\partial f)$ one finds for (2.74) a negative mass dimension, $\Delta = 4 - 2 - 3(0 + 1) = -1$, consistent with the fact that $\sqrt{G} \propto m_{\mathrm{P}}^{-1}$. Recalling the discussion after (2.49), one could say that the non-renormalizability for gravity is a consequence of the equivalence principle.

To see this more explicitly, let us first consider the kinetic term of (2.20): after an appropriate gauge fixing (see Section 2.2.3), it leads to a propagator of the usual form with momentum dependence $D \propto k^{-2}$. Since the interaction terms in (2.77) all contain two derivatives, one finds for the vertex $V \propto k^2$ unlike, for example, QED where the vertices are momentum-independent. Therefore, $DV \propto 1$ (k-independent). Consider now a Feynman diagram with one loop, r propagators and r vertices (Fig. 2.2); cf. Deser (1989). It involves momentum integrals that together lead to the following integral (assuming for the moment n instead of four space–time dimensions)

$$\int^{p_{\mathrm{c}}} \mathrm{d}^n k \, (DV)^r \propto \int^{p_{\mathrm{c}}} \mathrm{d}^n k \propto p_{\mathrm{c}}^n ,$$

where p_{c} is some cutoff momentum. The addition of a new internal line (Fig. 2.2) then gives the new factor

$$\int^{p_{\mathrm{c}}} \mathrm{d}^n k \, D(DV)^2 \propto \int^{p_{\mathrm{c}}} \mathrm{d}^n k \, D \propto p_{\mathrm{c}}^{n-2} .$$

Therefore, if L loops are present, the degree of divergence is

$$p_{\mathrm{c}}^n \cdot p_{\mathrm{c}}^{(L-1)(n-2)} = p_{\mathrm{c}}^{(L-1)(n-2)+n} .$$

For $n = 4$, for example, this yields $p_{\mathrm{c}}^{2(L+1)}$ and thus an unbounded increase with increasing order of the diagram. In other words, since $\sqrt{G}$ has inverse mass dimension, $\sqrt{G} p_{\mathrm{c}}$ is dimensionless and can appear at any order. Consequently, an infinite number of divergences emerges, making the perturbation theory non-renormalizable. An exception is $n = 2$, where G is a pure number. GR is, however, trivial in two space–time dimensions, so that one can construct a sensible theory only if additional fields are added to the gravitational sector; cf. Section 5.3.5.

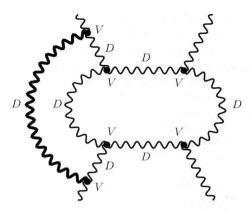

FIG. 2.2. Adding a new internal line to a Feynman diagram.

It must be emphasized that the counting of the degree of divergences only reflects the expectations. This degree might well be lower due to the presence of symmetries and the ensuing cancellations of divergences. In QED, for example, divergences are at worst logarithmic due to gauge invariance. The situation in the gravitational case will be discussed more explicitly in the next subsection.

The situation with divergences would be improved if the propagator behaved as $D \propto k^{-4}$ instead of $D \propto k^{-2}$, for then the factor corresponding to the new internal line in Fig. 2.2 would be

$$\int^{p_{\mathrm{c}}} \mathrm{d}^n k \ D \propto p_{\mathrm{c}}^{n-4}$$

and would therefore be independent of the cutoff in $n = 4$ dimensions, that is, higher loops would not lead to new divergences. This can be achieved, for example, by adding terms with the curvature squared to the Einstein–Hilbert action because this would involve *fourth-order* derivatives. Such a theory would indeed be renormalizable, but with a high price; as Stelle (1977) has shown (and as has already been noted by DeWitt (1967b)), the ensuing quantum theory is *not unitary*. The reason is that the propagator D can then be written in the form

$$D \propto \frac{1}{k^4 + Ak^2} = \frac{1}{A}\left(\frac{1}{k^2} - \frac{1}{k^2 + A}\right) \ ,$$

and the negative sign in front of the second term spoils unitarity (for $A < 0$ a tachyon—a particle with a negative mass squared—can also appear). For this reason, 'exact' R^2-theories are abandoned. However, Lagrangians with R^2-terms can, and indeed do, appear as correction terms for not-too-big curvatures (see below). (For high curvatures, R is comparable to R^2-terms, and therefore also R^3 and higher orders are needed.) On this effective level, R^2-terms only lead to a modification of the vertices, not the propagator.

Since the loop expansion of Feynman diagrams is also an expansion in $\hbar$, giving $\hbar^L$ for L loops, one finds in the gravitational case a matching with the G-expansion: since one recognizes from the above that there must be for $n = 4$ a factor G^{L+1} for L loops in order to compensate for the power of p_c, it is clear that $G\hbar$ is the relevant expansion parameter for pure gravity.

The occurrence of divergences in the quantization of gravity was first noticed by Rosenfeld (1930). He calculated the gravitational energy generated by an electromagnetic field to see whether an infinite self-energy occurs in this case (the infinite self-energy for an electron had already been recognized) and found in fact a linear divergence.

2.2.3 Effective action and Feynman rules

In order to apply path-integral methods to derive the effective action and Feynman diagrams, the formalism of Section 2.2.1 must be generalized to include 'gauge fixing'. There exists a general procedure which can be found in many references (e.g. in Weinberg 1996 or Böhm et al. 2001) and which will be briefly outlined before being applied to quantum gravity.

If we apply the general path integral in (2.71) to the expansion (2.76), we find an integral over the quantum field $f_{\mu\nu}$,

$$Z = \int \mathcal{D}f_{\mu\nu} \; e^{iS[f_{\mu\nu}, \bar{g}_{\mu\nu}]} \; , \qquad (2.78)$$

where in the following, we shall frequently use f as an abbreviation for $f_{\mu\nu}$ (and $\bar{g}$ for the background field). The point is now that (2.78) is formally infinite because $f_{\mu\nu}$ is invariant under the gauge transformations (2.5), $f_{\mu\nu} \to f_{\mu\nu} - \partial_\nu \epsilon_\mu - \partial_\mu \epsilon_\nu \equiv f_{\mu\nu}^\epsilon$, the infinity arising from integrations over gauge directions. Faddeev and Popov (1967) gave a general prescription on how to deal with this problem. This prescription has become especially popular in Yang–Mills theories. It consists of the following steps.

In a first step, a gauge constraint is chosen in order to fix the gauge. In the case of gravity, this would be four conditions, $G_\alpha[f, \bar{g}] = 0$. One desires to choose them such that the gauge is uniquely fixed, that is, such that each 'group orbit' f^ϵ is hit exactly once. It is known, however, that this cannot always be achieved ('Gribov ambiguities'), but this problem is usually not relevant in perturbation theory. In the path integral one then integrates over the subspace given by $G_\alpha[f, \bar{g}] = 0$. To implement this, one defines in a second step, a functional $\Delta_G[f, \bar{g}]$ through (neglecting in the following for notational convenience the dependence on the background field $\bar{g}$)

$$\Delta_G[f] \cdot \int \mathcal{D}\epsilon \; \prod_\alpha \delta(G_\alpha[f^\epsilon]) = 1 \; . \qquad (2.79)$$

The integration measure is a formal integration over the gauge group and is 'left invariant', that is, $\mathcal{D}\epsilon = \mathcal{D}(\epsilon'\epsilon)$. Using this invariance of the measure one can show that $\Delta_G[f]$ is gauge invariant, that is, $\Delta_G[f] = \Delta_G[f^\epsilon]$.

In a third step, one introduces the '1' of (2.79) into the naive path integral in (2.78). Making the substitution $f^\epsilon \to f$ and using the gauge invariance of Δ, one gets the expression

$$\int \mathcal{D}\epsilon \int \mathcal{D}f \prod_\alpha \delta(G_\alpha[f]) \Delta_G[f] e^{iS[f]} .$$

The infinite term coming from the $\mathcal{D}\epsilon$ integration can now be omitted (it just corresponds to the volume of the gauge orbit). One then arrives at the following definition for the path integral (again called Z for simplicity),

$$Z \equiv \int \mathcal{D}f \prod_\alpha \delta(G_\alpha[f]) \Delta_G[f] e^{iS[f]} . \tag{2.80}$$

It depends formally on the gauge G but is in fact gauge invariant; $\Delta_G[f]$ is called 'Faddeev–Popov determinant'.

Since the delta function appears in (2.80), we can in (2.79) expand $G_\alpha[f^\epsilon]$ around $\epsilon = 0$ to evaluate Δ_G,

$$G_\alpha[f^\epsilon] = G_\alpha[f^0] + (\hat{A}\epsilon)^\alpha ,$$

where the first term on the right-hand side is zero (it is just the gauge condition), and $\hat{A}$ is the matrix (with elements $A_{\alpha\beta}$) containing the derivatives of the G_α with respect to the ϵ^μ. Therefore, $\Delta_G[f] = \det\hat{A}$. For the derivation of Feynman rules it is convenient to use this expression and to rewrite the determinant as a Grassmann path integral over anticommuting fields $\eta^\alpha(x)$,

$$\det\hat{A} = \int \prod_\alpha \mathcal{D}\eta^{*\alpha}(x) \mathcal{D}\eta^\alpha(x) \ \exp\left(i \int d^4x \ \eta^{*\alpha}(x) A_{\alpha\beta}(x) \eta^\beta(x)\right) . \tag{2.81}$$

The fields $\eta^\alpha(x)$ and $\eta^{*\alpha}(x)$ are called 'vector ghosts' or 'Faddeev–Popov ghosts' because they are fermions with spin 1; they cannot appear as physical particles (external lines in a Feynman diagram) but are only introduced for mathematical convenience (and only occur inside loops in Feynman diagrams).

Apart from Δ_G, the gauge-fixing part $\delta(G_\alpha)$ also can be rewritten as an effective term contributing to the action. Choosing instead of $G_\alpha = 0$ a condition of the form $G_\alpha(x) = c_\alpha(x)$, the corresponding path integral Z^c is in fact independent of the c_α (shortly called c). Therefore, if one integrates Z over c with an arbitrary weight function, only the (irrelevant) normalization of Z will be changed. Using a Gaussian weight function, one obtains

$$Z \propto \int \mathcal{D}c \ \exp\left(-\frac{i}{4\xi} \int d^4x \ c_\alpha c^\alpha\right) \int \mathcal{D}f \prod_\alpha \Delta_G[f] \delta(G_\alpha - c_\alpha) e^{iS} . \tag{2.82}$$

Performing the c-integration then yields

$$Z \propto \int \mathcal{D}f \; \Delta_G[f] \exp\left(\mathrm{i}S[f] - \frac{\mathrm{i}}{4\xi} \int \mathrm{d}^4x \; G_\alpha G^\alpha\right) . \tag{2.83}$$

The second term in the exponential is called 'gauge-fixing term'. Taking all contributions together, the final path integral can be written in the form (disregarding all normalization terms and re-inserting the dependence on the background field)

$$Z = \int \mathcal{D}f \mathcal{D}\eta^\alpha \mathcal{D}\eta^{*\alpha} \; \mathrm{e}^{\mathrm{i}S_{\mathrm{tot}}[f,\eta,\bar{g}]} , \tag{2.84}$$

where

$$\begin{aligned} S_{\mathrm{tot}}[f,\eta,\bar{g}] &= S[f,\bar{g}] - \frac{1}{4\xi} \int \mathrm{d}^4x \; G_\alpha[f,\bar{g}]G^\alpha[f,\bar{g}] \\ &\quad + \int \mathrm{d}^4x \; \eta^{*\alpha}(x)A_{\alpha\beta}[f,\bar{g}](x)\eta^\beta(x) \\ &\equiv \int \mathrm{d}^4x \; (\mathcal{L}_{\mathrm{g}} + \mathcal{L}_{\mathrm{gf}} + \mathcal{L}_{\mathrm{ghost}}) . \end{aligned} \tag{2.85}$$

In this form, the path integral is suitable for the derivation of Feynman rules.

As it stands, (2.85) is also the starting point for the 'conventional' perturbation theory (i.e. without an expansion around a background field). In that case, the gauge-fixed Lagrangian is no longer gauge invariant but instead invariant under more general transformations including the ghost fields. This is called BRST symmetry (after the names Becchi, Rouet, Stora, Tyutin, see e.g. DeWitt (2003) for the original references) and encodes the information about the original gauge invariance at the gauge-fixed level. We shall give a brief introduction into BRST quantization in the context of string theory; see Chapter 9. For gravitational systems one needs a generalization of BRST quantization known as BFV quantization; see Batalin and Vilkovisky (1977), and Batalin and Fradkin (1983). This generalization is, for example, responsible for the occurrence of a four-ghost vertex in perturbation theory.

In the background-field method, on the other hand, the gauge symmetry (here: symmetry with respect to coordinate transformations) is preserved for the background field $\bar{g}$. For this reason, one can call this method a *covariant* approach to quantum gravity. The formal difference between 'conventional' and background-field method is the form of the gauge-fixing terms (Böhm *et al.* 2001), leading to different diagrams in this sector. It turns out that, once the Feynman rules are obtained, calculations are often simpler in the background-field method. For non-gauge theories, both methods are identical. Background-field method and BRST method are alternative procedures to arrive at the same physical results.

For the gravitational field, we have the following Lagrangian (for vanishing cosmological constant), see 't Hooft and Veltman (1974),

$$\mathcal{L}_{\mathrm{g}} = \frac{\sqrt{-g}R}{16\pi G} = \sqrt{-\bar{g}}\left(\frac{\bar{R}}{16\pi G} + \mathcal{L}_{\mathrm{g}}^{(1)} + \mathcal{L}_{\mathrm{g}}^{(2)} + \dots\right) , \tag{2.86}$$

where the 'barred' quantities refer to the background metric, see (2.76). We have for $\mathcal{L}_g^{(1)}$ the expression

$$\mathcal{L}_g^{(1)} = \frac{f_{\mu\nu}}{\sqrt{32\pi G}} \left(\bar{g}^{\mu\nu} \bar{R} - 2\bar{R}^{\mu\nu} \right) . \tag{2.87}$$

This vanishes if the background is a solution of the (vacuum) Einstein equations. The expression for $\mathcal{L}_g^{(2)}$ reads

$$\mathcal{L}_g^{(2)} = 1/2 \, f_{\mu\nu;\alpha} f^{\mu\nu;\alpha} - 1/2 \, f_{;\alpha} f^{;\alpha} + f_{;\alpha} f^{\alpha\beta}_{\;\;;\beta} - f_{\mu\beta;\alpha} f^{\mu\alpha;\beta}$$
$$+ \bar{R}(1/2 \, f_{\mu\nu} f^{\mu\nu} - 1/4 \, f^2) + \bar{R}^{\mu\nu} \left(f f_{\mu\nu} - 2 f_\mu^{\;\alpha} f_{\nu\alpha} \right) . \tag{2.88}$$

The first line corresponds to the Fierz–Pauli Lagrangian (2.20), while the second line describes the interaction with the background (not present in (2.20) because there the background was flat); we recall that $f \equiv f_\mu^{\;\mu}$. For the gauge-fixing part $\mathcal{L}_{gf}$, one chooses

$$\mathcal{L}_{gf} = \sqrt{-\bar{g}}((f_{\mu\nu}^{\;\;;\nu} - 1/2 \, f_{;\mu})(f^{\mu\rho}_{\;\;;\rho} - 1/2 \, f^{;\mu})) . \tag{2.89}$$

This condition corresponds to the 'harmonic gauge condition' (2.3) and turns out to be a convenient choice. For the ghost part, one finds

$$\mathcal{L}_{ghost} = \sqrt{-\bar{g}} \, \eta^{*\mu} \left(\eta_{\mu;\sigma;}^{\;\;\;\;\sigma} - \bar{R}_{\mu\nu} \eta^\nu \right) . \tag{2.90}$$

The first term in the brackets is the covariant d'Alembertian, $\Box \eta_\mu$. The full action in the path integral (2.84) then reads, with the background metric obeying Einstein's equations,

$$S_{tot} = \int d^4x \, \sqrt{-\bar{g}} \Big(\frac{\bar{R}}{16\pi G} - \frac{1}{2} f_{\mu\nu} D^{\mu\nu\alpha\beta} f_{\alpha\beta} + \frac{f_{\mu\nu}}{\sqrt{32\pi G}} \left[\bar{g}^{\mu\nu} \bar{R} - 2\bar{R}^{\mu\nu} \right]$$
$$+ \eta^{*\mu}(\bar{g}_{\mu\nu}\Box - \bar{R}_{\mu\nu})\eta^\nu + \mathcal{O}(f^3) \Big) , \tag{2.91}$$

where $D^{\mu\nu\alpha\beta}$ is a shorthand for the terms occurring in (2.88) and (2.89). The desired Feynman diagrams can then be obtained from this action. The operator $D^{\mu\nu\alpha\beta}$ is—in contrast to the original action without gauge fixing—invertible and defines both propagator and vertices (interaction with the background field). The explicit expressions are complicated, see 't Hooft and Veltman (1974) and Donoghue (1994). They simplify considerably in the case of a flat background where one has, for example, for the propagator (in harmonic gauge), the expression

$$\frac{1}{2(k^2 - i\epsilon)} (\eta_{\mu\alpha}\eta_{\nu\beta} + \eta_{\mu\beta}\eta_{\nu\alpha} - \eta_{\mu\nu}\eta_{\alpha\beta}) . \tag{2.92}$$

(The structure of the term in brackets is reminiscent of the DeWitt metric discussed in Section 4.1.2.) The action (2.91) leads to diagrams with at most one

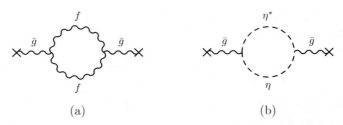

FIG. 2.3. (a) Graviton loop in interaction with the background field. (b) Ghost loop in interaction with the background field.

loop such as those depicted in Fig. 2.3 ('one-loop approximation'). Figure 2.3 (a) describes a graviton loop in interaction with the background field. It is a virtue of the background-field method that an arbitrary number of external lines can be considered. Figure 2.3 (b) describes a ghost loop interacting with the background field. Ghosts are needed to guarantee the unitarity of the S-matrix, as has already been noted by Feynman (1963) and DeWitt (1967b).

The use of the background-field method guarantees covariance with respect to the background field. This is achieved in particular by the implementation of dimensional regularization to treat the arising divergences.[14] The divergences are local and give rise to covariant terms (such as R, R^2, etc.) which must be absorbed by terms of the same form already present in the Lagrangian. Since curvature terms different from R are absent in the original Einstein–Hilbert Lagrangian,[15] one must start instead of $\mathcal{L}_{\mathrm{g}}$ with a Lagrangian of the form

$$\tilde{\mathcal{L}}_{\mathrm{g}} = \mathcal{L}_{\mathrm{g}} + \sqrt{-\bar{g}}(\mathcal{L}_{(2)} + \mathcal{L}_{(3)} + \ldots) + \mathcal{L}_{\mathrm{m}} , \qquad (2.93)$$

where

$$\mathcal{L}_{(2)} = c_1 \bar{R}^2 + c_2 \bar{R}_{\mu\nu} \bar{R}^{\mu\nu},$$
$$\mathcal{L}_{(3)} = \mathcal{O}(\bar{R}^3) , \quad \text{etc.} \qquad (2.94)$$

Note that because of the Gauß–Bonnet theorem, the expression

$$\int \mathrm{d}^4 x \ \sqrt{-g}(R_{\mu\nu\sigma\tau}R^{\mu\nu\sigma\tau} - 4R_{\mu\nu}R^{\mu\nu} + R^2)$$

is a topological invariant, which is why a term proportional to $R_{\mu\nu\sigma\tau}R^{\mu\nu\sigma\tau}$ does not have to be considered.

Can one justify the use of the modified gravitational Lagrangian (2.93)? Higher powers in curvature emerge generically from fundamental theories as

---

[14]Dimensional regularization employs an auxiliary extension of the number n of space–time dimensions in a continuous way such that no divergences appear for $n \neq 4$. The divergences that occur for $n \to 4$ are then subtracted.

[15]This is different from QED where only terms occur with a form already present in the Maxwell Lagrangian.

effective theories. This is the case, for example, in string theory (Chapter 9). Experimental limits on the parameters c_1, c_2, etc., are very weak because curvatures are usually very small (e.g. in the solar system). Stelle (1978) has estimated that from the perihelion motion of Mercury one only gets $c_1, c_2 \lesssim 10^{88}$. As one would expect c_1 and c_2 to be of order one, there is no hope to measure them in solar system experiments.

Using (2.91) in the path integral, integration over the $f_{\mu\nu}$ gives a term proportional to $(\det D)^{-1/2} = \exp(-\frac{1}{2}\operatorname{tr}\ln D)$, where D is a shorthand notation for $D^{\mu\nu\alpha\beta}$ ('t Hooft and Veltman 1974). In the one-loop approximation used here, this yields a divergent contribution to the Lagrangian of the form

$$\mathcal{L}_{1-\text{loop}}^{(\text{div})} = \frac{\hbar}{8\pi^2\epsilon}\left(\frac{\bar{R}^2}{120} + \frac{7}{20}\bar{R}_{\mu\nu}\bar{R}^{\mu\nu}\right)\,,\qquad (2.95)$$

where $\epsilon = 4 - n$ is the parameter occurring in dimensional regularization, which diverges in the limit of space–time dimension $n \to 4$. For pure gravity $\mathcal{L}_{1-\text{loop}}^{(\text{div})}$ is zero if the background field is 'on-shell', that is, a solution of the (vacuum) Einstein equations (this is a feature of the background-field method), but this is no longer true in the presence of matter. Adding (2.95) to (2.93), one obtains the 'renormalized' values for c_1 and c_2,

$$c_1^{(\text{ren})} = c_1 + \frac{\hbar}{960\pi^2\epsilon}\,,\qquad c_2^{(\text{ren})} = c_2 + \frac{7\hbar}{160\pi^2\epsilon}\,.\qquad (2.96)$$

These constants can only be 'measured' (or determined from a fundamental theory); here their use is just to absorb the divergences present in (2.95).

The divergences at two-loop order were first calculated by Goroff and Sagnotti (1985) using computer methods. Their result was later confirmed by van de Ven (1992). An analytic calculation of two-loop divergences was performed by Barvinsky and Vilkovisky (1987). The divergent part in the two-loop Lagrangian was found to read

$$\mathcal{L}_{2-\text{loop}}^{(\text{div})} = \frac{209\hbar^2}{2880}\frac{32\pi G}{(16\pi^2)^2\epsilon}\bar{R}^{\alpha\beta}{}_{\gamma\delta}\bar{R}^{\gamma\delta}{}_{\mu\nu}\bar{R}^{\mu\nu}{}_{\alpha\beta}\,.\qquad (2.97)$$

The divergence for $\epsilon \to 0$ can be absorbed by a corresponding term in $\mathcal{L}_{(3)}$ of (2.94). In contrast to (2.95), the gravitational constant G occurs here. For higher loops one finds from dimensional analysis that the divergent part of the Lagrangian is of the form

$$\mathcal{L}_{L-\text{loop}}^{(\text{div})} \sim \hbar^L G^{L-1}\nabla^p\bar{R}^m\frac{1}{\epsilon}\,,\qquad p+m = L+1\qquad (2.98)$$

(recall that the loop expansion is also a WKB expansion), where $\nabla^p\bar{R}^m$ is a shorthand for all curvature terms and their derivatives that can occur at this order.

Divergences remain if other fields (scalars, photon, Yang–Mills fields) are coupled (Deser *et al.* 1975). In the Einstein–Maxwell theory, for example, one obtains instead of (2.95), the expression

$$\mathcal{L}_{1-\text{loop}}^{(\text{div})} = \frac{137\hbar}{60\epsilon} \bar{R}_{\mu\nu} \bar{R}^{\mu\nu} \ . \tag{2.99}$$

We shall see in Section 2.3 how supergravity can improve the situation without, however, avoiding the occurrence of divergences.

The above treatment of divergences concerns the (ill-understood) UV-behaviour of the theory and does not lead to any new prediction.[16] Genuine predictions can, however, be obtained from the action (2.91) in the infrared-limit (Donoghue 1994). These predictions are independent of the unknown coefficients c_1, c_2, etc. Let us consider two examples. The first example is a quantum-gravitational correction term to the Newtonian potential[17]

$$V(r) = -\frac{Gm_1 m_2}{r}$$

between two masses m_1 and m_2. It is appropriate to define the potential through the scattering amplitude. The non-analytic parts of this amplitude give the long-range, low-energy corrections to the Newtonian potential. After a rather long calculation, Bjerrum-Bohr *et al.* (2003*a*) find (restoring c)

$$V(r) = -\frac{Gm_1 m_2}{r} \left(1 + 3\frac{G(m_1 + m_2)}{2rc^2} + \frac{41}{10\pi} \frac{G\hbar}{r^2 c^3} \right) \ . \tag{2.100}$$

All terms are fully determined by the non-analytic parts of the one-loop amplitude; the parameters connected with the higher curvature terms in the action contribute only to the analytic parts. It is for this reason that an unambiguous result can be obtained. Note that (2.100) corresponds to an effective gravitational constant $G_{\text{eff}}(r) > G$.

Although arising from a one-loop amplitude, the first correction term is in fact an effect of classical GR. It can be obtained from the Einstein–Infeld–Hoffmann equations, in which none of the two bodies is treated as a test body. Interestingly, this term had already been derived from quantum-gravitational considerations by Iwasaki (1971).

The second correction is proportional to $\hbar$ and is of genuine quantum-gravitational origin. The sign in front of this term indicates that the strength of the gravitational interaction is increased as compared to the pure Newtonian potential.[18] The result (2.100) demonstrates that a definite prediction from quantum

[16]It has, however, been suggested that in certain situations, theories with infinitely many couplings can be studied perturbatively also at high energies; cf. Anselmi (2003).

[17]The derivation of the Newtonian potential from linear quantum gravity was already performed by Bronstein (1936).

[18]There had been some disagreement about the exact number in (2.100); see the discussion in Bjerrum-Bohr *et al.* (2003*a*).

gravity is in principle possible. Unfortunately, the correction term, being of the order $(l_{\mathrm{P}}/r)^2 \ll 1$, is not measurable in laboratory experiments. We remark that similar techniques are applied successfully in low-energy QCD (in the limit of pion masses $m_\pi \to 0$) and known under the term 'chiral perturbation theory' (which also is a 'non-renormalizable theory' with a dimensionful coupling constant), see for example, Gasser and Leutwyler (1984). Quantum corrections to the Schwarzschild and Kerr metrics are calculated along these lines in Bjerrum-Bohr *et al.* (2003*b*). An analogous calculation within 'M-theory' (see Chapter 9) was performed by Becker and Becker (1999).

The second example is graviton–graviton scattering. This is the simplest low-energy process in quantum gravity. It was originally calculated in tree level by DeWitt (1967*c*). For the scattering of a graviton with helicity $+2$ with a graviton with helicity -2, for example, he found for the cross-section in the centre-of-mass frame, the expression

$$\frac{\mathrm{d}\sigma}{\mathrm{d}\Omega} = 4G^2 E^2 \frac{\cos^{12}\theta/2}{\sin^4\theta/2} \; , \qquad (2.101)$$

where E is the centre-of-mass energy (and similar results for other combinations of helicity). One recognizes in the denominator of (2.101) the term well known from Rutherford scattering. DeWitt (1967*c*) also considered other processes such as gravitational bremsstrahlung.

One-loop calculations can also be carried out. In the background-field method, the quantum fields $f_{\mu\nu}$ occur only in internal lines; external lines contain only the background field $\bar{g}_{\mu\nu}$. It was already mentioned that this makes the whole formalism 'covariant'. Donoghue and Torma (1999) have shown that the one-loop calculations of graviton–graviton scattering yield a finite result in the infrared (IR) limit, independent of any parameters such as c_1 or c_2. The cancellation of IR divergences with the emission of soft gravitons is needed and shown in fact to occur (as e.g. in QED). This yields again a definite result from quantum gravity. There is, in fact, a huge literature about IR-effects from quantum gravity. One example is the dynamical relaxation of the cosmological constant and its possible relevance for the dark-matter problem (Tsamis and Woodard 1993). Another way of addressing this issue is the investigation of renormalization-group equations, which can be applied also to effective theories (Reuter 1998). At least, in principle, one could understand from this method the occurrence of a small positive cosmological constant in agreement with observations because it would arise as a strong IR quantum effect (Reuter and Saueressig 2002).

The idea that non-renormalizable theories can be treated as ordinary physical theories from which phenomenological consequences can be drawn was also discussed in other contexts. Kazakov (1988), for example, generalized the standard formalism of the renormalization group to non-renormalizable theories. Barvinsky *et al.* (1993) developed a version of the renormalization-group formalism for non-renormalizable theories, which is particularly convenient for applications to GR coupled to a scalar field.

2.2.4 *Some general remarks on path integrals in perturbation theory*

Let us first review some important concepts in quantum field theory related to path integrals and the effective action (see e.g. Barvinsky 1990; Buchbinder *et al.* 1992; Böhm *et al.* 2001). For a quantum field φ (with possible components φ^i), the *generating functional* $W[J]$ is defined by the path integral

$$\langle \text{out}, 0 | \text{in}, 0 \rangle_J \equiv Z[J] \equiv \mathrm{e}^{\mathrm{i}W[J]} = \int \mathcal{D}\varphi \; \mathrm{e}^{\mathrm{i}S[\varphi] + \mathrm{i}J_k \varphi^k} \;, \qquad (2.102)$$

where J is an external current and $J_k \varphi^k$ is an abbreviation for $\int \mathrm{d}^4 x J_i(x) \varphi^i(x)$. This is also known as 'DeWitt's condensed notation', cf. DeWitt (1965). If φ is a gauge field, the measure in (2.102) is understood as including gauge-fixing terms and Faddeev–Popov ghosts, see (2.85) above. $W[J]$ is called the generating functional because one can calculate from it Green functions of the theory. More precisely, W generates *connected* Green functions[19] according to

$$\langle \varphi_1 \cdots \varphi_k \rangle_J = \mathrm{e}^{-\mathrm{i}W[J]} \left(\frac{1}{\mathrm{i}} \right)^k \frac{\delta^k}{\delta J_1 \cdots \delta J_k} \mathrm{e}^{\mathrm{i}W[J]} \;, \qquad (2.103)$$

where the expectation value occurring on the left-hand side is defined by the expression

$$\langle A(\varphi) \rangle \equiv \frac{\langle \text{out}, 0 | T(A(\varphi)) | \text{in}, 0 \rangle}{\langle \text{out}, 0 | \text{in}, 0 \rangle} \;, \qquad (2.104)$$

where T is time ordering.[20] Since the expectation value of the field in the presence of the external current (the 'mean field') is given by the expression

$$\langle \varphi \rangle_J = Z^{-1} \int \mathcal{D}\varphi \; \varphi \; \mathrm{e}^{\mathrm{i}S[\varphi] + \mathrm{i}J_k \varphi^k} \;,$$

it can be expressed as

$$\langle \varphi \rangle_J = \frac{\delta W[J]}{\delta J} \;. \qquad (2.105)$$

The two-point function then follows according to

$$\langle \varphi_i \varphi_k \rangle_J = -\mathrm{i} \frac{\delta^2 W}{\delta J_i \delta J_k} + \langle \varphi_i \rangle_J \langle \varphi_k \rangle_J$$

$$\equiv -\mathrm{i} G_J^{ik} + \langle \varphi_i \rangle_J \langle \varphi_k \rangle_J \;. \qquad (2.106)$$

The 'propagator' that yields the complete two-point function of the theory is then given by G_J^{ik} evaluated at $J = 0$ and simply denoted by G^{ik}.

---

[19] A connected Green function is a Green function referring to a connected graph, that is, a graph for which any two of its points are connected by internal lines.

[20] For fermionic fields one must distinguish between functional derivatives from the left and from the right.

One defines an *effective action*, $\Gamma[\langle\varphi\rangle]$ (a functional of the mean field!), in this context by the Legendre transformation

$$\Gamma[\langle\varphi\rangle] = W[J] - J \cdot \langle\varphi\rangle \,, \tag{2.107}$$

where J is expressed through $\langle\varphi\rangle$ (inversion of (2.105)). It therefore follows that

$$\frac{\delta\Gamma}{\delta\langle\varphi\rangle} = -J(\langle\varphi\rangle) \,. \tag{2.108}$$

In the absence of external sources this is

$$\frac{\delta\Gamma}{\delta\langle\varphi\rangle} = 0 \,, \tag{2.109}$$

generalizing the classical equations $\delta S/\delta\varphi_{\mathrm{cl}} = 0$. Equation (2.109) describes the dynamics of the mean field *including all quantum corrections*, since in the absence of external sources one has

$$e^{\mathrm{i}\Gamma[\langle\varphi\rangle]} = \int \mathcal{D}\varphi \, e^{\mathrm{i}S[\varphi]} \,.$$

That Γ contains all information about the full theory can also be understood as follows: solve (2.108) or (2.109) for $\langle\varphi\rangle$ and insert it into (2.106). This yields the two-point function, and a similar procedure leads to all higher correlation functions. Unfortunately, Γ is not invariant under field redefinitions, although the S-matrix is; see Barvinsky (1990).

One can show that Γ generates all one-particle irreducible Feynman diagrams.[21] The irreducible part of the two-point function is given by

$$D_{ik} = \frac{\delta^2\Gamma}{\delta\langle\varphi^i\rangle\delta\langle\varphi^k\rangle} = -\frac{\delta J(\langle\varphi_k\rangle)}{\delta\langle\varphi^i\rangle} \,. \tag{2.110}$$

From (2.106), one has $G_J^{ik} D_{kl} = -\delta_l^i$.

For concrete calculations one usually has to employ a loop expansion of Γ. Re-inserting $\hbar$ for the moment, one writes

$$\Gamma[\langle\varphi\rangle] = \frac{1}{\hbar}S[\langle\varphi\rangle] + \sum_{L=1}^{\infty} \hbar^{L-1}\Gamma^{(L)}[\langle\varphi\rangle] \,.$$

For the one-loop contribution, one can find the following explicit expression

$$\Gamma^{(1)}[\langle\varphi\rangle] = \frac{\mathrm{i}}{2} \ln\det \frac{\delta^2 S}{\delta\langle\varphi^i\rangle\delta\langle\varphi^k\rangle} = \frac{\mathrm{i}}{2} \mathrm{tr}\ln \frac{\delta^2 S}{\delta\langle\varphi^i\rangle\delta\langle\varphi^k\rangle} \,,$$

see for example, section 2.4 in Buchbinder *et al.* (1992). $\Gamma^{(1)}$ is called the 'one-loop effective action'.

---

[21] A diagram is called irreducible if it does not decompose into two separate diagrams by just cutting one internal line.

So far this formalism is completely general. In the background-field method, where one expands around a classical background (see (2.76)), one considers[22]

$$e^{iW[J,\bar{g}]} = \int \mathcal{D}f\mathcal{D}\phi \; e^{iS[\bar{g},f,\phi]+iJ_f f+iJ_\phi \phi} \;, \tag{2.111}$$

where ϕ denotes a general matter field. On the left-hand side, J stands for all external currents. It is again assumed that ghost and gauge-fixing terms are included in the measure. We take for the action S the sum of the action S_{tot}, see (2.85), the matter action $S_{\text{m}}[\bar{g}, f, \phi]$, *and* all possible covariant terms in the sense of effective field theory, that is, not only the Einstein–Hilbert action but also integrals over higher curvature terms such as R^2 with coefficients c_1, c_2, etc. (it is here where the unknown fundamental theory enters). The effective action is then again defined by a Legendre transform similar to (2.107),

$$\Gamma[\langle f \rangle, \langle \phi \rangle, \bar{g}] = W[J, \bar{g}] - J_f \langle f \rangle - J_\phi \langle \phi \rangle \;. \tag{2.112}$$

One can define another action $\Gamma[\bar{g}]$ by

$$\Gamma[\bar{g}] \equiv \Gamma[0, 0, \bar{g}] \;, \tag{2.113}$$

which can again be expanded in the number of loops,

$$\Gamma[\bar{g}] = \sum_{L=0}^{\infty} \Gamma^{(L)}[\bar{g}] \;, \tag{2.114}$$

with

$$\Gamma^{(0)} = S[\bar{g}] \;, \quad \Gamma^{(1)} = \frac{i}{2} \text{tr} \ln D \;, \dots \;, \tag{2.115}$$

where D denotes again the operator in (2.91) and (2.110). The expression for $\Gamma^{(1)}$ stems from the '$fDf/2$'-term in (2.91). For vanishing ϕ-field, this just yields the expansion discussed above (with the divergences of $\Gamma^{(1)}$ being absorbed by curvature-squared terms and those of $\Gamma^{(2)}$ being absorbed by curvature-cubed terms, etc.). The expression $\Gamma^{(1)}$ is again called the 'one-loop effective action'.

A huge literature is devoted to the study of methods for calculating $\Gamma^{(1)}$ (the limit of 'one-loop quantum gravity'), see for example, Barvinsky (1990) and Buchbinder *et al.* (1992). This limit is also referred to as 'quantum field theory in curved space–time' (Birrell and Davies 1982). The reason is that in the absence of self-interactions for matter, the only diagrams without external lines are the diagrams describing 'vacuum polarization'—closed loops corresponding to the matter field and the f-field, respectively; they do not contain G. In fact, one can

---

[22]The formalism is presented here for the case of GR; however, it can be applied to other field theories in the same way.

derive in this limit, the semiclassical Einstein equations (1.34) as *approximate* equations in the following way. From (2.113), one has here

$$\Gamma[\bar{g}] = \Gamma[0, 0, \bar{g}] = W[0, \bar{g}] \equiv W[\bar{g}] \ .$$

This is why $W[\bar{g}]$ can also be called 'effective action'. Recalling (2.111) one gets

$$\frac{\delta W}{\delta \bar{g}} e^{iW[\bar{g}]} = \int \mathcal{D}f \, \frac{\delta S_{\text{tot}}}{\delta \bar{g}} e^{iS_{\text{tot}}} \int \mathcal{D}\phi \, e^{iS_m}$$
$$+ \int \mathcal{D}f \mathcal{D}\phi \, \frac{\delta S_m}{\delta \bar{g}} e^{iS_{\text{tot}} + iS_m} \ .$$

We now expand both S_{tot} and S_m around the background $\bar{g}$ up to first derivatives with respect to $\bar{g}$. For S_{tot} this yields

$$S_{\text{tot}}[\bar{g} + f, \ldots] = S_{\text{EH}}[\bar{g}] + \frac{\delta S_{\text{EH}}}{\delta \bar{g}} f + \cdots \ .$$

Therein, the ghost contributions do not yet play a role and the expansion of S_{tot} in this order corresponds to the expansion of the Einstein–Hilbert action S_{EH}. Recalling from the definition (2.113) that $\langle f \rangle = 0$, one then arrives at

$$\frac{\delta W}{\delta \bar{g}} = \frac{\delta S_{\text{EH}}}{\delta \bar{g}} + \left\langle \frac{\delta S_m}{\delta \bar{g}} \right\rangle + \left\langle \frac{\delta^2 S_m}{\delta \bar{g} \delta \bar{g}} f \right\rangle \ . \tag{2.116}$$

Assuming now that the last terms approximately factorizes into two expectation values and using again $\langle f \rangle = 0$, the stationarity of W with respect to $\bar{g}$ yields

$$\frac{\delta S_{\text{EH}}}{\delta \bar{g}} \approx - \left\langle \frac{\delta S_m}{\delta \bar{g}} \right\rangle \ . \tag{2.117}$$

Multiplication with $2/\sqrt{-g}$ then just leads to the 'semiclassical' Einstein equations (1.34),[23]

$$G_{\mu\nu} + \Lambda g_{\mu\nu} \approx 8\pi G \langle T_{\mu\nu} \rangle \ . \tag{2.118}$$

The expectation value on the right-hand side corresponds to the in-out expectation value (2.102). It can, however, be related to the ordinary expectation value; cf. Barvinsky and Nesterov (2001). This derivation makes a strong case for the approximate nature of the semiclassical Einstein equations which can only be derived if the mean square deviation of $T_{\mu\nu}$ is small; cf. also Section 1.2. As far as graviton effects can be neglected, the action $W_\phi[\bar{g}]$ defined by

$$e^{iW_\phi[\bar{g}]} = \int \mathcal{D}\phi \, e^{iS_m[\bar{g}, \phi]} \tag{2.119}$$

is already the whole effective action.

---

[23]These equations follow also if N matter fields are coupled and an expansion in terms of $32\pi GN$ is performed; cf. Tomboulis (1977).

For an evaluation of the effective action up to the one-loop order, one has to calculate an operator of the form $\operatorname{tr} \ln D$; cf. (2.115). This can be efficiently done by using the 'Schwinger–DeWitt technique', which admits a covariant regularization (Birrell and Davies 1982; Barvinsky 1990; DeWitt 2003). It allows a local expansion of Green functions in powers of dimensional background quantities. Technically, a proper-time representation for the Green functions is used. This leads to a formulation of the effective action in terms of 'DeWitt coefficients' $a_0(x), a_1(x), a_2(x), \ldots$, which are local scalars that are constructed from curvature invariants. The effective action is thereby expanded in powers of the inverse curvature scale of the background. This only works in massive theories, although the divergent part of the action can also be used in the massless case. The Schwinger–DeWitt technique can thus be used to compute UV-divergences for massless theories too. The arising divergences can be absorbed into gravitational constant, cosmological constant, and c_1 and c_2. In this way, one can calculate the renormalized expectation value of the energy–momentum tensor, $\langle T_{\mu\nu} \rangle_{\mathrm{ren}}$, for the right-hand side of (2.118). The various methods of calculation are described in the above references.

Usually one cannot evaluate (2.119) exactly. An example where this can be done is the case of a massless scalar field in two dimensions (see e.g. Vilkovisky 1984). The result is, re-inserting $\hbar$,

$$W_\phi[\bar{g}] = -\frac{\hbar}{96\pi} \int \mathrm{d}^2 x \, \sqrt{-\bar{g}} \; {}^{(2)}\bar{R} \frac{1}{\Box} {}^{(2)}\bar{R} \;, \qquad (2.120)$$

where ${}^{(2)}\bar{R}$ denotes the two-dimensional Ricci scalar. Equation (2.120) arises directly from the so-called 'Weyl anomaly' (also called 'trace anomaly' or 'conformal anomaly'), that is, the breakdown of conformal invariance upon quantization: classically, this invariance leads to the vanishing of the energy–momentum tensor; cf. Section 3.2, while its breakdown in the quantum theory leads to a nonvanishing value. This anomaly plays a central role in string theory (Chapter 9).[24] It is given by

$$\langle T_\mu{}^\mu(x) \rangle_{\mathrm{ren}} = -\frac{a_1(x)\hbar}{4\pi} = -\frac{{}^{(2)}\bar{R}\hbar}{24\pi} \;. \qquad (2.121)$$

In four space–time dimensions the anomaly is proportional to the DeWitt coefficient $a_2(x)$ instead of $a_1(x)$.[25] In the two-dimensional model of (2.120), the flux of Hawking radiation (Section 7.1) is directly proportional to the anomaly.

We have seen that quantum GR is perturbatively non-renormalizable. In spite of this, we have argued that genuine predictions can be obtained through the method of effective action. Still, one can speculate that the occurrence of divergences could automatically be cured by going to a *non-perturbative* framework. We want to conclude this section by reviewing a simple example from Arnowitt *et*

---

[24]The prefactor then reads $\hbar c/96\pi$, where c is the central charge; cf. (3.62).

[25]The anomaly occurs only for an even number n of space–time dimensions and is then proportional to $a_{n/2}$.

al. (1962) showing how non-perturbative gravitational effects could in principle reach this goal. The example is the self-energy of a thin charged shell. Assume that the shell has a 'bare' mass m_0, a charge Q, and a radius ϵ. At the Newtonian level (plus energy-mass equivalence from special relativity), the energy of the shell is given by

$$m(\epsilon) = m_0 + \frac{Q^2}{2\epsilon} \ , \tag{2.122}$$

which diverges in the limit $\epsilon \to 0$ of a point charge. The inclusion of the gravitational self-energy leads to

$$m(\epsilon) = m_0 + \frac{Q^2}{2\epsilon} - \frac{Gm_0^2}{2\epsilon} \ , \tag{2.123}$$

which also diverges (unless one fine-tunes the charge unnaturally). Implementing, however, heuristically the (strong) equivalence principle by noting that gravity contributes itself to gravitational energy, one must substitute in the above expression the term m_0^2 by $m^2(\epsilon)$,

$$m(\epsilon) = m_0 + \frac{Q^2}{2\epsilon} - \frac{Gm^2(\epsilon)}{2\epsilon} \ . \tag{2.124}$$

As $\epsilon \to 0$ this now has a finite limit,

$$m(\epsilon) \xrightarrow{\epsilon \to 0} \frac{|Q|}{\sqrt{G}} \ . \tag{2.125}$$

Since G appears in the denominator, this is a genuine non-perturbative result which cannot be found from any perturbation theory around $G = 0$.[26] The same result can be obtained within GR (Heusler *et al.* 1990). Such simple examples give rise to the hope that a consistent non-perturbative theory of quantum gravity will automatically prevent the occurrence of divergences, see, for example, also the models by DeWitt (1964) and Padmanabhan (1985). One of the main motivations of string theory (Chapter 9) is the avoidance of divergences in the first place. Genuine non-perturbative approaches are also the canonical methods described in Chapters 3–6, where established quantization rules are applied to GR. Before we enter the discussion of these approaches, we shall briefly review how the introduction of supersymmetry (SUSY) changes the situation.

2.3   Quantum supergravity

Supergravity (SUGRA) is a supersymmetric theory of gravity encompassing GR. SUSY is a symmetry which mediates between bosons and fermions. It exhibits interesting features; for example, the running coupling constants in the Standard Model of particle physics can meet at an energy of around 10^{16} GeV if SUSY

---

[26]Incidentally, $m = |Q|/\sqrt{G}$ is the mass–charge relation of an extremal Reissner–Nordström black hole; see Chapter 7.

is added. SUGRA is a theory in its own right; see for example, van Nieuwen-huizen (1981) for a review. The main question of concern here is whether the perturbative UV behaviour of quantum gravity discussed in the last section can be improved by going over to SUGRA.

SUSY arose from the question whether the Poincaré group (and therefore space–time symmetries) can be unified with an internal (compact) group such as SO(3). A no-go theorem states that in a relativistic quantum field theory, given 'natural' assumptions of locality, causality, positive energy, and a finite number of elementary particles, such an invariance group can only be the direct product of the Poincaré group with a compact group, preventing a real unification. There is, however, a loophole. A true unification is possible if *anti*commutators are used instead of commutators in the formulation of a symmetry, leading to a 'graded Lie algebra'.[27] It was shown by Haag *et al.* (1975) that, with the above assumptions of locality etc., the algebraic structure is essentially unique.

The SUSY algebra is given by the anticommutator

$$[Q^i_\alpha, \bar{Q}^j_\beta]_+ = 2\delta^{ij}(\gamma^n)_{\alpha\beta}P_n \ , \ i,j = 1,\dots,N \ , \tag{2.126}$$

where Q^i_α denotes the corresponding generators, also called spinorial charges, $\bar{Q}^i_\alpha = Q^i_\alpha \gamma^0$ with γ^0 being one of Dirac's gamma matrices, N is the number of SUSY generators, and all anticommutators among the Qs and the $\bar{Q}$s themselves vanish. There are also the commutators

$$[P_n, Q_\alpha] = 0 \ , \ [P_m, P_n] = 0 \ . \tag{2.127}$$

(P_n denotes the energy–momentum four vector, the generator of space–time translations.) In addition, there are the remaining commutators of the Poincaré group, Eqns (2.34)–(2.36), as well as

$$[Q^i_\alpha, J_{mn}] = (\sigma_{mn})^\beta_\alpha Q^i_\beta \ , \tag{2.128}$$

where here $\sigma_{mn} = \mathrm{i}[\gamma_m, \gamma_n]$; cf. also Section 1.1. More details can be found, for example, in Weinberg (2000). The SUSY algebra is compatible with relativistic quantum field theory, that is, one can write the spinorial charges as an integral over a conserved current,

$$Q^i_\alpha = \int \mathrm{d}^3x \ J^i_{0\alpha}(x) \ , \quad \frac{\partial J^i_{m\alpha}(x)}{\partial x^m} = 0 \ . \tag{2.129}$$

Fermions and bosons are combined into 'super-multiplets' by irreducible representations of this algebra. There would be a fermionic super-partner to each boson and vice versa. One would thus expect that the partners should have the same mass. Since this is not observed in Nature, SUSY must be broken. The

---

[27]Anticommutators have of course been used before the advent of SUSY, in order to describe fermions, but not yet in the context of symmetries.

presence of SUSY would guarantee that there are an equal number of bosonic and fermionic degrees of freedom. For this reason, several divergences cancel due to the presence of opposite signs (e.g. the 'vacuum energy'). This gave rise to the hope that SUSY might generally improve the UV behaviour of quantum field theories.

Performing now an independent SUSY transformation at each space–time point one arrives at a corresponding *gauge symmetry*. Because the anticommutator (2.126) of two SUSY generators closes on the space–time momentum, this means that space–time translations are performed independently at each space–time point—these are nothing but general coordinate transformations. The gauge theory therefore contains GR and is called SUGRA.[28] To each generator one then finds a corresponding gauge field: P_n corresponds to the vierbein field e_μ^n (see Section 1.1), J_{mn} to the 'spin connection' ω_μ^{mn}, and Q_α^i to the 'Rarita–Schwinger fields' $\psi_\mu^{\alpha,i}$. The latter are fields with spin 3/2 and describe the fermionic superpartners to the graviton—the *gravitinos*. They are a priori massless, but can acquire a mass by a Higgs mechanism. For $N = 1$ (simple SUGRA), one has a single gravitino which sits together with the spin-2 graviton in one multiplet. The cases $N > 1$ are referred to as 'extended supergravities'. In the case $N = 2$, for example, the photon, the graviton, and two gravitinos together form one multiplet, yielding a 'unified' theory of gravity and electromagnetism. One demands that $0 \leq N \leq 8$ because otherwise there would be more than one graviton and also particles with spin higher than two (for which no satisfactory coupling exists).

For $N = 1$, the SUGRA action is the sum of the Einstein–Hilbert action and the Rarita–Schwinger action for the gravitino,

$$S = \frac{1}{16\pi G} \int d^4x \, (\det e_\mu^n)(R - 2\Lambda) + \frac{1}{2} \int d^4x \, \epsilon^{\mu\nu\rho\sigma} \bar\psi_\mu \gamma_5 \gamma_\nu D_\rho \psi_\sigma \qquad (2.130)$$

(recall $\det e_\mu^n = \sqrt{-g}$, and we have $\gamma_5 = i\gamma^0\gamma^1\gamma^2\gamma^3$), where we have introduced here the spinorial covariant derivative

$$D_\mu = \partial_\mu - 1/2 \, \omega_\mu^{nm} \sigma_{nm} \ ,$$

cf. (1.19). The action (2.130) is not only invariant under general coordinate transformations and local Poincaré transformations, but also under local SUSY transformations which for vierbein and gravitino field read

$$\delta e_\mu^m = 1/2 \, \sqrt{8\pi G} \bar\epsilon^\alpha \gamma_{\alpha\beta}^m \psi_\mu^\beta \ ,$$

$$\delta \psi_\mu^\alpha = \frac{1}{\sqrt{8\pi G}} D_\mu \epsilon^\alpha \ , \qquad (2.131)$$

where ϵ^α is an anticommuting parameter function and $\bar\epsilon^\alpha$ its complex conjugate. Note that the factors $\sqrt{G}$ are needed already for dimensional reasons.

---

[28]The gauging of the Poincaré group leads in fact to the Einstein–Cartan theory, which besides curvature contains also torsion.

What can now be said about the divergence properties of a quantum SUGRA perturbation theory? The situation is improved, but basically the same features as in Section 2.2 hold: the theory is non-renormalizable (the occurrence of the dimensionful coupling G due to the equivalence principle), and there is in general no cancellation of divergences (Deser 2000). To give a short summary of the situation, in $n = 4$ there are no one-loop or two-loop counterterms (due to SUSY Ward identities), but divergences can occur in principle from three loops on. The calculation of counterterms was, however, only possible after powerful methods from string theory have been used, establishing a relation between gravity and Yang–Mills theory, see Bern (2002) and references therein. It turns out that in $n = 4$, $N < 8$-theories are three-loop infinite, while $N = 8$-theories are five-loop infinite. The same seems to be true for dimensions $4 < n < 11$. Dimension $n = 11$ plays a special role. It is the maximal possible dimension for SUGRA and only $N = 1$ is possible there. The theory is of importance in connection with 'M-theory', see Chapter 9. It turns out that there are infinities at two loops and that there is thus no 'magic' avoidance of divergences from M-theory.

Therefore, as far as full quantum gravity is concerned (beyond its use in the framework of effective theories), one must either try to construct a full non-perturbative theory of the quantized gravitational field or a unified theory beyond field theory. The following chapters are devoted to these directions.

3

PARAMETRIZED AND RELATIONAL SYSTEMS

In this chapter, we shall consider some models that exhibit certain features of GR but which are much easier to discuss. In this sense they constitute an important conceptual preparation for the canonical quantization of GR, which is the topic of the next chapters. In addition, they are of interest in their own right.

The central aspect is *reparametrization invariance* and the ensuing existence of constraints, see for example, Sundermeyer (1982) and Henneaux and Teitelboim (1992) for a general introduction into constrained systems and Blagojević (2002) for an introduction with particular emphasis on gravitation. Kuchař (1973) gives a detailed discussion of reparametrization-invariant systems, which we shall partly follow in this chapter. Such invariance properties are often called 'general covariance' because they refer to an invariance with respect to a relabelling of the underlying space–time manifold. A more precise formulation has been suggested by Ehlers (1995) who states that an invariance group is a subgroup of the full covariance group which leaves the absolute, non-dynamical, elements of a theory invariant. Such an absolute element would, for example, be the conformal group in the scalar theory of gravity mentioned at the end of Section 2.1.1. In GR, the full metric is dynamical and the invariance group coincides with the covariance group, the group of all diffeomorphisms. Whereas dynamical elements are subject to quantization, absolute elements remain classical; see Section 1.3. Absolute elements can also appear in 'disguised form' if a theory has been reparametrized artificially. This is the case in the models of the non-relativistic particle and the parametrized field theory to be discussed below, but not in GR or the other dynamical systems considered in this chapter.

3.1 Particle systems

3.1.1 *Parametrized non-relativistic particle*

Consider the action for a point particle in classical mechanics

$$S[q(t)] = \int_{t_1}^{t_2} \mathrm{d}t \, L\left(q, \frac{\mathrm{d}q}{\mathrm{d}t}\right) . \tag{3.1}$$

It is only for simplicity that a restriction to one particle is being made. The following discussion can be easily generalized to n particles. For simplicity, the Lagrangian in (3.1) has been chosen t-independent.

We introduce now a formal time parameter τ ('label time') and elevate t (Newton's 'absolute time') formally to the rank of a dynamical variable (this is an example for an absolute structure in disguise, as mentioned above). We therefore write $q(\tau)$ and $t(\tau)$. Derivatives with respect to τ will be denoted by a dot, and restriction to $\dot{t} > 0$ is made. The action (3.1) can then be rewritten as

$$S[q(\tau), t(\tau)] = \int_{\tau_1}^{\tau_2} d\tau \, \dot{t} L\left(q, \frac{\dot{q}}{\dot{t}}\right) \equiv \int_{\tau_1}^{\tau_2} d\tau \, \tilde{L}(q, \dot{q}, \dot{t}) \,. \tag{3.2}$$

The Lagrangian $\tilde{L}$ possesses the important property that it is *homogeneous* (of degree one) in the velocities, that is

$$\tilde{L}(q, \lambda\dot{q}, \lambda\dot{t}) = \lambda\tilde{L}(q, \dot{q}, \dot{t}) \,, \tag{3.3}$$

where $\lambda \neq 0$ can be an arbitrary function of τ. Homogeneous Lagrangians lead to actions that are invariant under time reparametrizations $\tau \to \tilde{\tau} \equiv f(\tau)$ in the sense that they can be written as a $\tilde{\tau}$-integral over the same Lagrangian. Assuming $\dot{f} > 0$ gives

$$S = \int_{\tau_1}^{\tau_2} d\tau \, L(q, \dot{q}) = \int_{\tilde{\tau}_1}^{\tilde{\tau}_2} \frac{d\tilde{\tau}}{\dot{f}} \, L\left(q, \frac{dq}{d\tilde{\tau}}\dot{f}\right) = \int_{\tilde{\tau}_1}^{\tilde{\tau}_2} d\tilde{\tau} \, L\left(q, \frac{dq}{d\tilde{\tau}}\right) \,. \tag{3.4}$$

The canonical momentum for q is found from (3.2) to read

$$\tilde{p}_q = \frac{\partial\tilde{L}}{\partial\dot{q}} = \dot{t}\frac{\partial L}{\partial\left(\frac{\dot{q}}{\dot{t}}\right)}\frac{1}{\dot{t}} = p_q \,, \tag{3.5}$$

thus coinciding with the momentum corresponding to (3.1). But now there is also a momentum canonically conjugate to t,

$$p_t = \frac{\partial\tilde{L}}{\partial\dot{t}} = L\left(q, \frac{\dot{q}}{\dot{t}}\right) + \dot{t}\frac{\partial L\left(q, \frac{\dot{q}}{\dot{t}}\right)}{\partial\dot{t}}$$

$$= L\left(q, \frac{dq}{dt}\right) - \frac{dq}{dt}\frac{\partial L(q, dq/dt)}{\partial(dq/dt)} = -H \,. \tag{3.6}$$

Therefore, t and $-H$ (the negative of the Hamiltonian corresponding to the original action (3.1)) are canonically conjugate pairs. The Hamiltonian belonging to $\tilde{L}$ is found as

$$\tilde{H} = \tilde{p}_q\dot{q} + p_t\dot{t} - \tilde{L} = \dot{t}(H + p_t) \,. \tag{3.7}$$

But because of (3.6), this is constrained to vanish. It is appropriate at this stage to introduce a new quantity called 'super-Hamiltonian'. It is defined as

$$H_S \equiv H + p_t \,, \tag{3.8}$$

and one has the *constraint*

$$H_S \approx 0 \,. \tag{3.9}$$

The $\approx$ in this (and further) equation(s) means 'to vanish as a constraint' or 'weak equality' in the sense of Dirac (1964). It defines a subspace in phase space and can be set to zero only *after* all Poisson brackets etc. have been evaluated. One can now use instead of (3.1), the new action principle

$$S = \int_{\tau_1}^{\tau_2} d\tau \ (p_q \dot{q} + p_t \dot{t} - N H_S) \ , \tag{3.10}$$

where all quantities (including N) have to be varied; N is a Lagrange multiplier and variation with respect to it just yields the constraint (3.9). From Hamilton's equations, one has

$$\dot{t} = \frac{\partial(N H_S)}{\partial p_t} = N \ . \tag{3.11}$$

Therefore, N is called the *lapse function* because it gives the rate of change of Newton's time t with respect to label time τ.

The existence of the constraint (3.9) is a consequence of the reparametrization invariance with respect to τ. To see this explicitly, it will be proven that having a Lagrangian being homogeneous in the velocities is equivalent to the corresponding Hamiltonian being zero. Homogeneity has been shown above to be equivalent to reparametrization invariance.

Given a homogeneous Lagrangian, one finds for the canonical Hamiltonian,

$$H_c = \frac{\partial L}{\partial \dot{q}} \dot{q} - L = \lambda^{-1} \left(\frac{\partial L(q, \lambda \dot{q})}{\partial(\lambda \dot{q})} \lambda \dot{q} - L(q, \lambda \dot{q}) \right) = \lambda^{-1} H_c \ .$$

Since λ is arbitrary, H_c must vanish. On the other hand, if H_c vanishes, one gets after substituting $\dot{q}$ by $\lambda \dot{q}$,

$$\frac{\partial L(q, \lambda \dot{q})}{\partial \dot{q}} \dot{q} = L(q, \lambda \dot{q}) \ .$$

The left-hand side can be written as $\lambda p_q \dot{q} = \lambda L$, and one gets $\lambda L(q, \dot{q}) = L(q, \lambda \dot{q})$, that is, L is homogeneous.

One can have reparametrization invariance without Hamiltonian constraint if the q and p do not transform as scalars under reparametrizations (Henneaux and Teitelboim 1992). This is, however, not a natural situation. In this case, the theorem just proven remains true, but the connection to reparametrization invariance is lost.

Although Newton's time has been mixed amongst the other dynamical variables, it can easily be recovered, for its momentum p_t enters *linearly* into (3.8) (it is assumed that H has the usual form $H = p_q^2/2m + V(q)$). Therefore, one can easily solve (3.9) to find $p_t = -H$, choose the label $\tau = t$ ('fixing the gauge'), and find from (3.10)

$$S = \int dt \ \left(p_q \frac{dq}{dt} - H \right) \ , \tag{3.12}$$

that is, just the standard action (the Hamiltonian form of (3.1)). This process is called *deparametrization*. We shall see in Chapter 4 that there is an analogue to

Eqns (3.8) and (3.9) in GR, but that, there in contrast to here, all momenta occur quadratically. This leads to the interesting question whether a deparametrization for GR, that is, the identification of a distinguished time-like variable, is possible. It should be remarked that every system can be transformed artificially into 'generally covariant' form; cf. Kretschmann (1917). But this is possible only at the price of disguising absolute structures which formally appear then as dynamical variables, such as Newton's absolute time t. The general covariance of GR is natural in the sense that the metric is fully dynamical.

How does one quantize a system given by a constraint such as (3.9)? A successful, although heuristic, procedure is the proposal made by Dirac (1964). A classical constraint is implemented in the quantum theory as a restriction on physically allowed wave functions. Thus, (3.9) is translated into

$$\hat{H}_S \psi = 0 \ , \tag{3.13}$$

where $\hat{H}_S$ denotes the super-Hamilton operator associated with the classical super-Hamiltonian H_S. In the position representation, the $\hat{q}$ are represented by multiplication with q and the momenta $\hat{p}$ are represented by derivatives $(\hbar/\mathrm{i})\partial/\partial q$. For the parametrized particle, this includes also $\hat{p}_t = (\hbar/\mathrm{i})\partial/\partial t$. Therefore, the quantum version of the constraint (3.8,3.9) reads

$$\left(\hat{H} - \mathrm{i}\hbar \frac{\partial}{\partial t} \right) \psi(q, t) = 0 \ , \tag{3.14}$$

which is just the Schrödinger equation. Does this mean that t is a dynamical variable in quantum mechanics? The answer is no. We have already mentioned in Section 1.1.2 that time cannot be represented by an operator (e.g. it would be in contradiction with the boundedness of energy). This is the consequence of having an absolute structure in disguise—it remains an absolute structure in quantum theory, in spite of its formal appearance as a quantum variable.

3.1.2 *Some remarks on constrained systems*

Since constraints play a crucial role in this and the following chapters, a brief recapitulation of some of the basic properties of constrained systems are in order, see for example, Dirac (1964), Hanson *et al.* (1976), Sundermeyer (1982), and Henneaux and Teitelboim (1992). Starting from early work by Leon Rosenfeld, this formalism has been developed mainly by Peter Bergmann and Paul Dirac; cf. Bergmann (1989) and Rovelli (2000).

Assume that $\phi_a(q, p), a = 1, \ldots, n$ is a set of constraints,

$$\phi_a(q, p) \approx 0 \ , \tag{3.15}$$

where q and p represent positions and momenta for N particles. In Dirac's terminology, they are assumed to be of *first class*, which means that they obey the Poisson-bracket relations

$$\{\phi_a, \phi_b\} = f_{ab}^c \phi_c \tag{3.16}$$

and are compatible with the time evolution,

$$\{\phi_a, H\} = d_a^b \phi_b \ . \tag{3.17}$$

Constraints which do not obey these relations are called *second-class constraints*. They play a role, for example, in supergravity; cf. Section 5.3.6.

We now add the first-class constraints to the action with Lagrange multipliers λ_a,

$$S = \int d\tau \ (p\dot{q} - H - \lambda_a \phi_a) \ . \tag{3.18}$$

(In the example given by (3.10) one has instead of $-H - \lambda_a \phi_a$ only the term $-N H_S$ in the action, H_S being the only constraint.) Therefore, the time evolution of a function $A(q, p)$ reads

$$\dot{A}(q, p) = \{A, H\} + \lambda_a \{A, \phi_a\} \ . \tag{3.19}$$

The Lagrange parameter λ_a, therefore, introduces an arbitrariness into the time evolution. In fact, first-class constraints generate *gauge transformations*: expanding $A(q(\tau), p(\tau))$ around $\tau = 0$ up to order $\Delta\tau$ for two different values $\lambda_a^{(1)}$ and $\lambda_a^{(2)}$ and performing the difference δA, one obtains the 'gauge transformation'

$$\delta A = \epsilon_a \{A, \phi_a\} \ , \tag{3.20}$$

where $\epsilon_a = \Delta\tau(\lambda_a^{(1)}(0) - \lambda_a^{(2)}(0))$. The constraints (3.15) define a hypersurface Γ_c in phase space, the 'constraint hypersurface', and generate the gauge transformations (3.20) on this hypersurface.[1]

Functions $A(q, p)$ for which $\{A, \phi_a\} \approx 0$ holds are often called 'observables' because they do not change under a gauge transformation. It must be emphasized that there is no *a priori* relation of these observables to observables in an operational sense. This notion had been introduced by Bergmann in the hope that these quantities might play the role of the standard observables in quantum theory.

In order to select one physical representative amongst all equivalent configurations, one frequently employs 'gauge conditions'. This is important, for example, in path-integral quantization; see Section 2.2.3. A gauge should be chosen in such a way that there is no further gauge freedom left and that any configuration can be transformed in one satisfying the gauge. The first condition is sometimes violated ('Gribov ambiguities'), but this is irrelevant for infinitesimal gauge transformations. Instead of gauge fixing, one can keep the gauge freedom in the classical theory and perform a quantization by implementing the constraints in the way done in (3.13).

---

[1]For second-class constraints, which do not fulfil (3.16), the Lagrange parameters can be determined by choosing $A = \phi_b$ in (3.19).

3.1.3 *The relativistic particle*

We consider a relativistic particle with mass $m \neq 0$. In units where $c = 1$, its action can be taken to be proportional to the total proper time along its worldline,

$$S = -m \int_{s_1}^{s_2} \mathrm{d}s \ . \tag{3.21}$$

Using instead of proper time an arbitrary parameter τ for the worldline $x^\mu(\tau)$, the action reads

$$S = -m \int_{\tau_1}^{\tau_2} \mathrm{d}\tau \ \sqrt{-\dot{x}^2} \ , \tag{3.22}$$

where $\dot{x}^\mu \equiv \mathrm{d}x^\mu/\mathrm{d}\tau$ and $\eta_{\mu\nu}\dot{x}^\mu\dot{x}^\nu < 0$ (tangent vector is time-like). One immediately recognizes that the Lagrangian is homogeneous in the velocities and that, therefore, the action is invariant under $\tau \to f(\tau)$. The canonical momenta read

$$p_\mu = \frac{m\dot{x}_\mu}{\sqrt{-\dot{x}^2}} \ . \tag{3.23}$$

From this expression, it follows immediately that the momenta obey the 'mass-shell condition'

$$p^2 + m^2 = 0 \ . \tag{3.24}$$

In fact, this is a constraint in phase space and thus should be more properly written as $p^2 + m^2 \approx 0$. Because of reparametrization invariance, the canonical Hamiltonian vanishes,

$$H_\mathrm{c} = p_\mu\dot{x}^\mu - L = \frac{m\dot{x}^2}{\sqrt{-\dot{x}^2}} + m\sqrt{-\dot{x}^2} = 0 \ .$$

In fact,

$$H_\mathrm{c}(x,p) = -p^0\dot{x}^0 + \mathbf{p}\dot{\mathbf{x}} - L = -p^0\dot{x}^0 + \frac{\mathbf{p}^2\dot{x}^0}{\sqrt{\mathbf{p}^2 + m^2}} + m\sqrt{(\dot{x}^0)^2 - \dot{\mathbf{x}}^2}$$

$$= \dot{x}^0(-p^0 + \sqrt{\mathbf{p}^2 + m^2}) \approx 0 \ , \tag{3.25}$$

where the positive square root has been chosen, $p^0 = \sqrt{\mathbf{p}^2 + m^2}$, in order to render the energy positive. Analogously to (3.10), one can transform the action into Hamiltonian form,

$$S = \int_{\tau_1}^{\tau_2} \mathrm{d}\tau \ (p_\mu\dot{x}^\mu - NH_\mathrm{S}) \ , \tag{3.26}$$

where here

$$H_\mathrm{S} \equiv \eta^{\mu\nu}p_\mu p_\nu + m^2 \approx 0 \tag{3.27}$$

plays the role of the super-Hamiltonian which is constrained to vanish. The interpretation of the Lagrange multiplier N can be gained from Hamilton's equations,

$$\dot{x}^0 = \frac{\partial (N H_S)}{\partial p_0} = -2 N p_0 \ ,$$

to give

$$N = \frac{\dot{x}^0}{2\sqrt{\mathbf{p}^2 + m^2}} = \frac{\dot{x}^0}{2m\gamma} = \frac{1}{2m} \frac{\mathrm{d}s}{\mathrm{d}\tau} \ , \tag{3.28}$$

where γ is the standard relativistic factor. In contrast to (3.11), the lapse function N here is proportional to the rate of change of proper time (not x^0) with respect to parameter time.[2]

If we apply Dirac's quantization rule on the classical constraint (3.27), we get

$$\hat{H}_S \psi(x^\mu) \equiv \left(\Box - \frac{m^2}{\hbar^2} \right) \psi(x^\mu) = 0 \ . \tag{3.29}$$

This is the Klein–Gordon equation for relativistic one-particle quantum mechanics (spinless particles). We emphasize that the classical parameter τ has completely disappeared since particle trajectories do not exist in quantum theory.

With regard to the Hamiltonian action (3.26), the question arises how x, p, and N must transform under time reparametrizations in order to leave the action invariant. Since the first-class constraint H_S generates gauge transformations in the sense of (3.20), one has

$$\delta x(\tau) = \epsilon(\tau)\{x, H_S\} = \epsilon \frac{\partial H_S}{\partial p} \ , \tag{3.30}$$

$$\delta p(\tau) = \epsilon(\tau)\{p, H_S\} = -\epsilon \frac{\partial H_S}{\partial x} \ . \tag{3.31}$$

But how does the Lagrange multiplier N transform? We calculate for this purpose

$$\delta S = \int_{\tau_1}^{\tau_2} \mathrm{d}\tau \ (\dot{x}\delta p + p\delta\dot{x} - H_S\delta N - N\delta H_S) \ .$$

The last term is zero, and partial integration of the second term leads to

$$\delta S = \int_{\tau_1}^{\tau_2} \mathrm{d}\tau \ \left(-\epsilon \frac{\partial H_S}{\partial x} \dot{x} - \epsilon \frac{\partial H_S}{\partial p} \dot{p} - H_S \delta N \right) + \left[p\epsilon \frac{\partial H_S}{\partial p} \right]_{\tau_1}^{\tau_2} \ .$$

In order that only a surface term remains one has to choose

$$\delta N(\tau) = \dot{\epsilon}(\tau) \ . \tag{3.32}$$

This leads to

$$\delta S = \epsilon(\tau) \left[p\frac{\partial H_S}{\partial p} - H_S \right]_{\tau_1}^{\tau_2} \ . \tag{3.33}$$

Since the term in brackets gives $p^2 - m^2 \neq 0$, one must demand

$$\epsilon(\tau_1) = 0 = \epsilon(\tau_2) \ , \tag{3.34}$$

that is, the boundaries must not be transformed.

---

[2]If we had chosen (3.25) instead of (3.27), we would have found $N = \dot{x}^0$.

We note that for a constraint of the form $H_S = \alpha(x)p$, the term in brackets would vanish and there would be in this case no restriction at the boundaries. Constraints of this form arise in electrodynamics and Yang–Mills theories (Gauß constraint) provided the sources are treated dynamically too (otherwise, the constraint would no longer be homogeneous in the momenta, see e.g. $\nabla \mathbf{E} = \rho$ in electrodynamics).

We shall now show how the gauge can be fixed for the relativistic particle. If a gauge is independent of the lapse function N, it is called 'canonical gauge', otherwise it is called 'non-canonical'. Consider first a canonical gauge,

$$\chi(x, p, \tau) \approx 0 \ . \tag{3.35}$$

An example would be $x^0 - \tau \approx 0$ (such a gauge was used in the deparametrization of the non-relativistic particle, see the paragraph before (3.12)). A potential problem is that (3.35) holds at all times, including the endpoints, and may thus be in conflict with $\epsilon(\tau_1) = \epsilon(\tau_2) = 0$—since there is no gauge freedom at the endpoints, $\chi \approx 0$ could restrict physically relevant degrees of freedom.

For reparametrization-invariant systems, a canonical gauge must depend explicitly on τ. From the condition that (3.35) be invariant under time evolution,

$$0 \approx \frac{\mathrm{d}\chi}{\mathrm{d}\tau} = \frac{\partial\chi}{\partial\tau} + N\{\chi, H_S\} \ ,$$

a τ-independent gauge χ would lead to the unacceptable value $N = 0$ ('freezing' of the motion). (In order for the gauge to break the reparametrization invariance generated by H_S, $\{\chi, H_S\}$ must be non-vanishing.) For the relativistic particle, this yields

$$0 \approx \frac{\partial\chi}{\partial\tau} + N\frac{\partial\chi}{\partial x^\mu}\frac{\partial H_S}{\partial p_\mu} = \frac{\partial\chi}{\partial\tau} + 2Np^\mu\frac{\partial\chi}{\partial x^\mu} \ .$$

For the example $x^0 - \tau \approx 0$, one has $N = 1/2p^0$, in accordance with (3.28).

To avoid potential problems with the boundary, one can look for an equation of second order in ϵ (since there are two conditions $\epsilon(\tau_1) = 0 = \epsilon(\tau_2)$). As x and p transform proportional to ϵ, one would have to involve $\ddot{x}$ or $\ddot{p}$, which would render the action functional unnecessarily complicated. Therefore, since $\delta N = \dot{\epsilon}$, see (3.32), one can choose the 'non-canonical gauge'

$$\dot{N} = \chi(p, x, N) \ . \tag{3.36}$$

In electrodynamics, A^0 plays the role of N. Therefore, the Lorenz gauge $\partial_\mu A^\mu = 0$ is a non-canonical gauge, whereas the Coulomb gauge $\nabla \mathbf{A} = 0$ would be an example of a canonical gauge.

Some final remarks are in order; cf. Henneaux and Teitelboim (1992). First, the restriction $\epsilon(\tau_1) = 0 = \epsilon(\tau_2)$ only holds if the action is an integral over the Lagrangian without additional boundary terms. If appropriate boundary terms are present in the action principle, one can relax the condition on ϵ (but to determine these boundary terms, one has to solve first the equations of motion). Second, if such boundary terms are present, one can even choose τ-independent canonical gauges (an extreme choice would be $x^0(\tau) = 0$ for all τ).

3.2 The free bosonic string

Nowadays superstring theory (or 'M-theory') is considered to be a candidate for a unified theory of all interactions including quantum gravity. This aspect will be discussed in Chapter 9. In this section, we shall consider the free bosonic string as a model for (canonical) quantum gravity. However, the bosonic string (where no supersymmetry is included) is also used in a heuristic way in the development of superstring theory itself.

In the case of the relativistic particle, the action is proportional to the proper time, see (3.21). A straightforward generalization to the string would thus be to use an action proportional to the area of the worldsheet $\mathcal{M}$,

$$ S = -\frac{1}{2\pi\alpha'} \int_{\mathcal{M}} \mathrm{d}^2\sigma \sqrt{|\det G_{\alpha\beta}|} \; . \tag{3.37} $$

Here, $\mathrm{d}^2\sigma \equiv \mathrm{d}\sigma\mathrm{d}\tau$ denotes the integration over the parameters of the worldsheet (with both the space part σ and the time part τ chosen to be dimensionless), and $G_{\alpha\beta}$ is the metric on the worldsheet. The string tension is $(2\pi\alpha')^{-1}$, that is, there is a new fundamental parameter α' with dimension length/mass. In the quantum version, the fundamental string length

$$ l_{\mathrm{s}} = \sqrt{2\alpha'\hbar} \tag{3.38} $$

will occur. The string propagates in a higher dimensional space–time, and the worldsheet metric $G_{\alpha\beta}$ ($\alpha, \beta = 1, 2$) is induced by the metric of the embedding space–time. In the following, we shall assume that the string propagates in D-dimensional Minkowski space, with the worldsheet given by $X^\mu(\sigma, \tau)$, where $\mu = 0, \ldots, D - 1$. Denoting the derivative with respect to $\tau \equiv \sigma^0$ by a dot and the derivative with respect to $\sigma \equiv \sigma^1$ by a prime, one has

$$ G_{\alpha\beta} = \eta_{\mu\nu} \frac{\partial X^\mu}{\partial \sigma^\alpha} \frac{\partial X^\nu}{\partial \sigma^\beta} \; , \tag{3.39} $$

$$ |\det G_{\alpha\beta}| = -\det G_{\alpha\beta} = (\dot{X}X')^2 - \dot{X}^2(X')^2 \; . \tag{3.40} $$

The canonical momenta conjugate to the embeddings $X^\mu(\sigma, \tau)$ then read

$$ P_\mu = -\frac{1}{2\pi\alpha'\sqrt{-\det G_{\alpha\beta}}} \left[(\dot{X}X')X'_\mu - (X')^2\dot{X}_\mu \right] \; . \tag{3.41} $$

From this one gets the conditions

$$ P_\mu(X^\mu)' = -\frac{1}{2\pi\alpha'\sqrt{-\det G_{\alpha\beta}}} \left[(\dot{X}X')(X')^2 - (X')^2(\dot{X}X') \right] = 0 \tag{3.42} $$

as well as

$$P_\mu P^\mu = -\frac{(X')^2}{4\pi^2(\alpha')^2} \ . \tag{3.43}$$

In fact, the last two conditions are just constraints—a consequence of the reparametrization invariance

$$\tau \to \tau'(\tau, \sigma) \ , \quad \sigma \to \sigma'(\tau, \sigma) \ .$$

The constraint (3.43), in particular, is a direct analogue of (3.24).

As expected from the general considerations in Section 3.1.1, the Hamiltonian is constrained to vanish. For the Hamiltonian *density* $\mathcal{H}$, one finds that

$$\mathcal{H} = N\mathcal{H}_\perp + N^1\mathcal{H}_1 \ , \tag{3.44}$$

where N and N^1 are Lagrange multipliers, and

$$\mathcal{H}_\perp = \frac{1}{2}\left(P^2 + \frac{(X')^2}{4\pi^2(\alpha')^2}\right) \approx 0 \ , \tag{3.45}$$

$$\mathcal{H}_1 = P_\mu(X^\mu)' \approx 0 \ . \tag{3.46}$$

Quantization of these constraints is formally achieved by imposing the commutation relations

$$[X^\mu(\sigma), P_\nu(\sigma')]|_{\tau=\tau'} = -i\hbar\,\delta^\mu_\nu\delta(\sigma - \sigma') \tag{3.47}$$

and implementing the constraints à la Dirac as restrictions on physically allowed wave functionals,

$$\hat{\mathcal{H}}_\perp\Psi[X^\mu(\sigma)] \equiv \frac{1}{2}\left(-\hbar^2\frac{\delta^2\Psi}{\delta X^2} + \frac{(X')^2\Psi}{4\pi^2(\alpha')^2}\right) = 0 \ , \tag{3.48}$$

and

$$\hat{\mathcal{H}}_1\Psi[X^\mu(\sigma)] \equiv (X^\mu)'\frac{\delta\Psi}{\delta X^\mu} = 0 \ . \tag{3.49}$$

Note that in contrast to the examples in Section 3.1, one has now to deal with *functional* derivatives, defined by the Taylor expansion

$$\Psi[\phi(\sigma) + \eta(\sigma)] = \Psi[\phi(\sigma)] + \int d\sigma\,\frac{\delta\Psi}{\delta\phi(\sigma)}\eta(\sigma) + \cdots \ . \tag{3.50}$$

The above implementation of the constraints is only possible if there are no *anomalies*, see the end of this section and Section 5.3.5. An important property of the quantized string is that such anomalies in fact occur, *preventing* the validity of all quantum equations (3.48) and (3.49). Equations such as (3.48) and (3.49) will occur at several places later and will be further discussed there, for example in the context of parametrized field theories (Section 3.3).

Note that this level of quantization corresponds to a 'first-quantized string' in analogy to first quantization of point particles (Section 3.1). The usual 'second quantization' would mean to elevate the wave functions $\Psi[X^\mu(\sigma)]$ themselves into operators ('string field theory'). It must also be emphasized that 'first' and 'second' are at best heuristic notions since there is just one quantum theory (cf. in this context Zeh (2003)).

In the following, we shall briefly discuss the connection with the standard textbook treatment of the bosonic string, see for example, Polchinski (1998a). This will also be a useful preparation for the discussion of string theory in Chapter 9. Usually one starts with the Polyakov action for the bosonic string,

$$S_{\rm P} = -\frac{1}{4\pi\alpha'} \int_{\mathcal{M}} {\rm d}^2\sigma \, \sqrt{h} h^{\alpha\beta}(\sigma,\tau)\partial_\alpha X^\mu \partial_\beta X_\mu \,, \tag{3.51}$$

where $h_{\alpha\beta}$ denotes the *intrinsic* (not induced) metric on the worldsheet, and $h \equiv |{\rm det} h_{\alpha\beta}|$. In contrast to the induced metric, it consists of independent degrees of freedom with respect to which the action can be varied. The action (3.51) can be interpreted as describing 'two-dimensional gravity coupled to D scalar fields'. Since the Einstein–Hilbert action is a topological invariant in two dimensions, there is no pure gravity term present, and only the coupling of the metric to the X^μ remains in (3.51). One can also take into account a cosmological term, see Chapter 9. In contrast to (3.37), the Polyakov action is much easier to handle, especially when used in a path integral.

The Polyakov action has many invariances. First, it is invariant with respect to diffeomorphisms on the worldsheet. Second, and most importantly, it possesses *Weyl invariance*, that is, an invariance under the transformations

$$h_{\alpha\beta}(\sigma,\tau) \to {\rm e}^{2\omega(\sigma,\tau)} h_{\alpha\beta}(\sigma,\tau) \tag{3.52}$$

with an arbitrary function $\omega(\sigma,\tau)$. This is a special feature of two dimensions where $\sqrt{h} h^{\alpha\beta} \to \sqrt{h} h^{\alpha\beta}$. In addition, there is the Poincaré symmetry of the background Minkowski space–time, which is of minor interest here.

Defining the two-dimensional energy–momentum tensor according to[3]

$$T_{\alpha\beta} = -\frac{4\pi}{\sqrt{h}}\frac{\delta S_{\rm P}}{\delta h^{\alpha\beta}} = \frac{1}{\alpha'}\left(\partial_\alpha X^\mu \partial_\beta X_\mu - \frac{1}{2}h_{\alpha\beta}h^{\gamma\delta}\partial_\gamma X^\mu \partial_\delta X_\mu\right) \,, \tag{3.53}$$

one finds

$$h^{\alpha\beta}T_{\alpha\beta} = 0 \,. \tag{3.54}$$

This tracelessness of the energy–momentum tensor is a consequence of Weyl invariance. Using the field equations $\delta S_{\rm P}/\delta h_{\alpha\beta} = 0$, one finds

$$T_{\alpha\beta} = 0 \,. \tag{3.55}$$

In a sense, these are the Einstein equations with the left-hand side missing, since the Einstein–Hilbert action is a topological invariant. As (3.55) has no second

---

[3]Compared to GR there is an additional factor -2π here, which is introduced for convenience.

time derivatives, it is in fact a constraint—a consequence of diffeomorphism invariance. From (3.55), one can easily derive that

$$\det G_{\alpha\beta} = \frac{h}{4} (h^{\alpha\beta} G_{\alpha\beta})^2 \ , \tag{3.56}$$

where $G_{\alpha\beta}$ is the induced metric (3.39). Inserting this into (3.51) gives back the action (3.37). Therefore, 'on-shell' both actions are equivalent.

The constraints (3.42) and (3.43) can also be found directly from (3.51)—defining the momenta conjugate to X^μ in the usual manner—after use has been made of (3.55). One can thus formulate instead of (3.55) an alternative canonical action principle

$$S = \int_{\mathcal{M}} \mathrm{d}^2\sigma \ (P_\mu \dot{X}^\mu - N\mathcal{H}_\perp - N^1\mathcal{H}_1) \ , \tag{3.57}$$

where $\mathcal{H}_\perp$ and $\mathcal{H}_1$ are given by (3.45) and (3.46), respectively.

In the standard treatment of the bosonic string, the 'gauge freedom' (with respect to two diffeomorphisms and one Weyl transformation) is fixed by the choice $h_{\alpha\beta} = \eta_{\alpha\beta} = \mathrm{diag}(-1, 1)$. Instead of (3.51) one has then

$$S_{\mathrm{P}} = -\frac{1}{4\pi\alpha'} \int \mathrm{d}^2\sigma \ \eta^{\alpha\beta} \partial_\alpha X^\mu \partial_\beta X_\mu \ , \tag{3.58}$$

that is, the action for free scalar fields. In two dimensions there is a remaining symmetry which leaves the gauge-fixed action invariant—the *conformal transformations*. These are angle-preserving coordinate transformations[4] which change the metric by a factor $\mathrm{e}^{2\omega(\sigma,\tau)}$; they can therefore be compensated by a Weyl transformation, and the action (3.58) is invariant under this combined transformations. Field theories with this invariance are called 'conformal field theories'. A particular feature of two dimensions is that the conformal group is infinite-dimensional, giving rise to infinitely many conserved charges (see below).

Consider in the following the case of open strings where $\sigma \in (0, \pi)$. (Closed strings ensue a doubling of degrees of freedom corresponding to left- and right-movers.) The Hamiltonian of the gauge-fixed theory reads

$$H = \frac{1}{4\pi\alpha'} \int_0^\pi \mathrm{d}\sigma \left(\dot{X}^2 + (X')^2 \right) \ . \tag{3.59}$$

Introducing the components of the energy–momentum tensor with respect to the lightcone coordinates $\sigma^- = \tau - \sigma$ and $\sigma^+ = \tau + \sigma$, it is convenient to define the quantities

$$L_m = \frac{1}{2\pi\alpha'} \int_0^\pi \mathrm{d}\sigma \ \left(\mathrm{e}^{\mathrm{i}m\sigma} T_{++} + \mathrm{e}^{-\mathrm{i}m\sigma} T_{--} \right) \ . \tag{3.60}$$

---

[4]In GR, the term 'conformal transformation' is usually employed for what is here called a Weyl transformation.

One recognizes that $L_0 = H$. Because the energy–momentum tensor vanishes as a constraint, this holds also for the L_m, $L_m \approx 0$. The L_m obey the classical *Virasoro algebra*

$$\{L_m, L_n\} = -\mathrm{i}(m - n)L_{m+n} , \tag{3.61}$$

exhibiting that they generate the group of conformal transformations (the residual symmetry of the gauge-fixed action). The $\{L_n\}$ are the infinitely many conserved charges mentioned above. It turns out that quantization does not preserve this algebra but yields an additional term called 'anomaly', 'central term', or 'Schwinger term',[5]

$$[L_m, L_n] = (m - n)\hbar L_{m+n} + \frac{c\hbar^2}{12}(m^3 - m)\delta_{m+n,0} , \tag{3.62}$$

where c is the *central charge*. For the case of the free fields X^μ, it is equal to the number of space–time dimensions, $c = D$. Due to the presence of this extra term, one cannot implement the constraints $L_m \approx 0$ in the quantum theory as restrictions on wave functions. Instead, one can choose

$$L_n|\psi\rangle = 0, \; n > 0 , \quad L_0|\psi\rangle = a\hbar|\psi\rangle . \tag{3.63}$$

It turns out that to avoid states with negative norm, one must have $a = 1$ and $D = 26$. Negative-norm states are unwanted because they are in conflict with the probability interpretation of quantum theory. This choice of a and D corresponds to the preservation of Weyl invariance at the quantum level, see for example, Polchinski (1998a) and Chapter 9. This is achieved by the presence of Faddeev–Popov ghost degrees of freedom whose central charge cancels against the central charge c of the fields X^μ precisely for $D = 26$. It is most elegantly treated by 'BRST quantization' (see Section 9.2), leading to an equation of the form $Q_B|\Psi_{\text{tot}}\rangle = 0$, where Q_B is the BRST charge. This weaker condition replaces the direct quantum implementation of the constraints.

Defining now quantities

$$\tilde{L}_n = \frac{1}{2} \int \mathrm{d}\sigma \, \mathrm{e}^{\mathrm{i}n\sigma} \left(\sqrt{(X')^2}\mathcal{H}_\perp + \mathcal{H}_1 \right) \tag{3.64}$$

and using the Poisson-bracket relations between the constraints $\mathcal{H}_\perp$ and $\mathcal{H}_1$ (see in particular Section 3.3), one can show that

$$\{\tilde{L}_m, \tilde{L}_n\} = -\mathrm{i}(m - n)\tilde{L}_{m+n} , \tag{3.65}$$

which coincides with the Virasoro algebra (3.61). In fact, for the gauge fixing considered here—leading to (3.58)—one has $\tilde{L}_n = L_n$. The result (3.62) then shows that the naive implementation of the constraints (3.48) and (3.49) may be inconsistent. This is a general problem in the quantization of constrained

---

[5]Here and in the following, we shall omit 'hats' on operators.

systems and will be discussed further in Section 5.3. Kuchař and Torre (1991) have treated the bosonic string as a model for quantum gravity. They have shown that a covariant (covariant with respect to diffeomorphisms of the worldsheet) quantization is possible, that is, there exists a quantization procedure in which the algebra of constraints contains no anomalous terms. This is achieved by extracting internal time variables ('embeddings') which are not represented as operators.[6] A potential problem is the dependence of the theory on the choice of embedding. This is in fact a general problem; see Section 5.2. Kuchař and Torre make use of the fact that string theory is an 'already parametrized theory', which brings us to a detailed discussed of parametrized field theories in the next section.

3.3 Parametrized field theories

This example is a generalization of the parametrized non-relativistic particle discussed in Section 3.1.1. As it will be field theoretic by nature, it has similarities with the bosonic string discussed in the last section, but with notable differences. A general reference to parametrized field theories is Kuchař (1973, 1981) from which the following material is partially drawn.

Starting point is a real scalar field in Minkowski space, $\phi(X^\mu)$, where the standard inertial coordinates are here called $X^\mu \equiv (T, X^a)$. We introduce now arbitrary (in general curved) coordinates $x^\mu \equiv (t, x^a)$ and let the X^μ depend parametrically on x^μ. This is analogous to the dependence $t(\tau)$ in Section 3.1.1. The functions $X^\mu(x^\nu)$ describe a family of hypersurfaces in Minkowski space parametrized by $x^0 \equiv t$ (we shall restrict ourselves to the space-like case). Analogously to Section 3.1.1, the standard action for a scalar field is rewritten in terms of the arbitrary coordinates x^μ. This yields

$$S = \int \mathrm{d}^4 X \; \mathcal{L}\left(\phi, \frac{\partial \phi}{\partial X^\mu}\right) \equiv \int \mathrm{d}^4 x \; \tilde{\mathcal{L}} \,, \tag{3.66}$$

where

$$\tilde{\mathcal{L}}(\phi, \phi_{,a}, \dot{\phi}; X^\mu_{,a}, \dot{X}^\mu) = J\mathcal{L}\left(\phi, \phi_{,\nu}\frac{\partial x^\nu}{\partial X^\mu}\right) \,, \tag{3.67}$$

and J denotes the Jacobi determinant of the X with respect to the x (a dot is a differentiation with respect to t, and a comma is a differentiation with respect to the x). Instead of calculating directly the momentum canonically conjugate to X^μ, it is more appropriate to consider first the Hamiltonian density $\tilde{\mathcal{H}}$ corresponding to $\tilde{\mathcal{L}}$ with respect to ϕ,

$$\tilde{\mathcal{H}} = \tilde{p}_\phi \dot{\phi} - \tilde{\mathcal{L}} = J\frac{\partial \mathcal{L}}{\partial \dot{\phi}}\dot{\phi} - J\mathcal{L}$$

$$= J\frac{\partial x^0}{\partial X^\mu}\left(\frac{\partial \mathcal{L}}{\partial\left(\partial \phi / \partial X^\mu\right)}\frac{\partial \phi}{\partial X^\nu} - \delta^\mu_\nu \mathcal{L}\right)\dot{X}^\nu$$

---

[6]The anomaly is still present in a subgroup of the conformal group, but it does not disturb the Dirac quantization of the constraints.

$$\equiv J \frac{\partial x^0}{\partial X^\mu} T^\mu{}_\nu \dot{X}^\nu \; . \tag{3.68}$$

Both J and the canonical energy–momentum tensor $T^\mu{}_\nu$ do not, in fact, depend on the 'kinematical velocities' $\dot{X}^\mu$. This can be seen as follows. The Jacobi determinant J can be written as

$$J = \epsilon_{\rho\nu\lambda\sigma} \frac{\partial X^\rho}{\partial x^0} \frac{\partial X^\nu}{\partial x^1} \frac{\partial X^\lambda}{\partial x^2} \frac{\partial X^\sigma}{\partial x^3} \; ,$$

from which one gets

$$J \frac{\partial x^0}{\partial X^\mu} = \epsilon_{\mu\nu\lambda\sigma} \frac{\partial X^\nu}{\partial x^1} \frac{\partial X^\lambda}{\partial x^2} \frac{\partial X^\sigma}{\partial x^3} \; ,$$

which is just the vectorial surface element on $t = $ constant, which does not depend on the $\dot{X}^\mu$. For the same reason, the energy–momentum tensor does not depend on these velocities.

As a generalization of (3.9), one may envisage to introduce the kinematical momenta Π_ν via the constraint

$$\mathcal{H}_\nu \equiv \Pi_\nu + J \frac{\partial x^0}{\partial X^\mu} T^\mu{}_\nu \approx 0 \; . \tag{3.69}$$

Taking then the action

$$S = \int \mathrm{d}^4 x \; (\tilde{p}_\phi \dot{\phi} - \tilde{\mathcal{H}}) \; ,$$

inserting (3.68) for $\tilde{\mathcal{H}}$ and adding the constraints (3.69) with Lagrange multipliers N^ν, one gets the action principle

$$S = \int \mathrm{d}^4 x \; (\tilde{p}_\phi \dot{\phi} + \Pi_\nu \dot{X}^\nu - N^\nu \mathcal{H}_\nu) \; . \tag{3.70}$$

This is the result that one would also get from defining the kinematical momenta directly via (3.67). It is analogous to (3.10).

It is convenient to decompose (3.69) into components orthogonal and parallel to the hypersurfaces $x^0 = $ constant. Introducing the normal vector n^μ (with $\eta_{\mu\nu} n^\mu n^\nu = -1$) and the tangential vectors $X^\nu_{,a}$ (obeying $n_\nu X^\nu_{,a} = 0$), one gets the constraints

$$\mathcal{H}_\perp \equiv \mathcal{H}_\nu n^\nu \approx 0 \; , \tag{3.71}$$
$$\mathcal{H}_a \equiv \mathcal{H}_\nu X^\nu_{,a} \approx 0 \; . \tag{3.72}$$

Equations (3.71) and (3.72) are called the Hamiltonian constraint and the momentum (or diffeomorphism) constraint, respectively. They are similar to the corresponding constraints (3.45) and (3.46) in string theory.

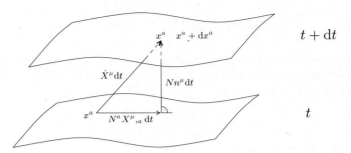

FIG. 3.1. Geometric interpretation of lapse and shift.

The action (3.70) then reads

$$S = \int d^4x \; (\tilde{p}_\phi \dot{\phi} + \Pi_\nu \dot{X}^\nu - N\mathcal{H}_\perp - N^a \mathcal{H}_a) \; . \tag{3.73}$$

To interpret the Lagrange multipliers N and N^a, we vary this action with respect to Π_ν and obtain

$$\dot{X}^\nu \equiv t^\nu = Nn^\nu + N^a X^\nu_{,a} \; . \tag{3.74}$$

The geometric interpretation is depicted in Fig. 3.1: $\dot{X}^\nu$ is a vector that points from a point with (spatial) coordinates x^a on $t = $ constant to a point with the *same* coordinates on a neighbouring hypersurface $t + dt = $ constant. The purely temporal distance between the hypersurfaces is given by N, and therefore N is called the *lapse function*. Similarly, N^a is a vector that points from the point with coordinates x^a on $t = $ constant to the point on the same hypersurface from which the normal is erected to reach the point with the same coordinates x^a on $t + dt = $ constant. It is called the *shift vector*.

Instead of Minkowski space one can also choose an arbitrary curved background for the embedding. Denoting the spatial metric by h_{ab}, that is

$$h_{ab} = g_{\mu\nu} \frac{\partial X^\mu}{\partial x^a} \frac{\partial X^\nu}{\partial x^b} \; , \tag{3.75}$$

the four-dimensional line element can be decomposed as follows,

$$\begin{aligned} ds^2 = g_{\mu\nu} dx^\mu dx^\nu &= -N^2 dt^2 + h_{ab}(dx^a + N^a dt)(dx^b + N^b dt) \\ &= (h_{ab} N^a N^b - N^2)dt^2 + 2h_{ab} N^a dx^b dt + h_{ab} dx^a dx^b \; . \end{aligned} \tag{3.76}$$

The action (3.73) is invariant under the reparametrizations

$$\begin{aligned} x^0 \to x^{0'} &= x^0 + f(x^a) \; , \\ x^a \to x^{a'} &= g(x^b) \end{aligned} \tag{3.77}$$

with arbitrary functions (obeying standard differentiability conditions) f and g. This is not equivalent to the full set of space–time diffeomorphisms; see the discussion at the end of this section.

A simple example of the above procedure is the case of a massless scalar field on (1+1)-dimensional Minkowski space–time (Kuchař 1973, 1981). Its Lagrangian reads

$$\mathcal{L}\left(\phi, \frac{\partial \phi}{\partial T}, \frac{\partial \phi}{\partial X}\right) = -\frac{1}{2}\eta^{\mu\nu}\frac{\partial \phi}{\partial X^\mu}\frac{\partial \phi}{\partial X^\nu} = \frac{1}{2}\left[\left(\frac{\partial \phi}{\partial T}\right)^2 - \left(\frac{\partial \phi}{\partial X}\right)^2\right] . \quad (3.78)$$

If the above procedure of parametrization is followed, one finds for the constraints (3.71) and (3.72) the expressions (dots denote derivatives with respect to $x^0 \equiv t$, primes denote derivatives with respect to $x^1 \equiv x$),

$$\mathcal{H}_\perp = \frac{1}{\sqrt{(X')^2 - (T')^2}}\left(X'\Pi_T + T'\Pi_X + \frac{1}{2}(\tilde{p}_\phi^2 + \phi'^2)\right) , \quad (3.79)$$

$$\mathcal{H}_1 = T'\Pi_T + X'\Pi_X + \phi'\tilde{p}_\phi . \quad (3.80)$$

(The space-like nature of the hypersurfaces guarantees that the discriminant of the square root is positive.) One recognizes from (3.79) that the kinematical momenta enter the Hamiltonian constraint *linearly*. This is different from the string case (3.45) and distinguishes, in fact, a parametrized theory from a theory which is *intrinsically* reparametrization invariant.

Quantization is performed by imposing the formal commutators

$$[X^\mu(\mathbf{x}), \Pi_\nu(\mathbf{y})] = i\hbar\, \delta_\nu^\mu \delta(\mathbf{x} - \mathbf{y}) \quad (3.81)$$

and

$$[\phi(\mathbf{x}), \tilde{p}_\phi(\mathbf{y})] = i\hbar\, \delta(\mathbf{x} - \mathbf{y}) . \quad (3.82)$$

From (3.71) and (3.72), one then finds according to Dirac's prescription, the quantum constraints

$$\mathcal{H}_\perp\Psi[\phi(x), X^\mu(x)] = 0 , \quad (3.83)$$

$$\mathcal{H}_a\Psi[\phi(x), X^\mu(x)] = 0 . \quad (3.84)$$

In the above example of the free scalar field, the constraints read

$$\mathcal{H}_\perp\Psi = \frac{1}{\sqrt{(X')^2 - (T')^2}}\left(-i\hbar X'(x)\frac{\delta}{\delta T(x)} - i\hbar T'(x)\frac{\delta}{\delta X(x)}\right.$$
$$\left. +\frac{1}{2}\left[-\hbar^2\frac{\delta^2}{\delta\phi(x)\delta\phi(x)} + \phi'^2(x)\right]\right)\Psi[\phi(x), T(x), X(x)] = 0 , \quad (3.85)$$

$$\mathcal{H}_1\Psi = \frac{\hbar}{i}\left(T'(x)\frac{\delta}{\delta T(x)} + X'(x)\frac{\delta}{\delta X(x)} + \phi'(x)\frac{\delta}{\delta\phi(x)}\right)\Psi = 0 . \quad (3.86)$$

The functional derivatives occurring in these equations are treated formally, that is, as if they were ordinary derivatives, and all derivatives are put to the right. One should be aware that there is always the problem of factor ordering (as in quantum mechanics) and that singularities arise if a functional derivative is taken of an ordinary function with respect to the same argument. This is a general problem of functional differential equations and will be further discussed in Section 5.3.

The above equations are very different from the equations that one would get from the standard Lagrangian (3.78) or the corresponding action. The reason is that the wave functional is usually evolved along *flat* hypersurfaces $T = \text{constant}$ only, whereas in the parametrized version it can be evolved along *any* family of space-like hypersurfaces. The latter description needs two functions $X(x)$ and $T(x)$, also called 'many-fingered time' or 'bubble-time' description. Like for the non-relativistic particle, the parametrized theory can easily be deparametrized. Choosing $X = x$ as a coordinate on the hypersurfaces and evolving the wave functionals along flat hypersurfaces described by $T(x) = T_0 \in (-\infty, \infty)$, one finds (Kuchař 1973)

$$i\hbar \frac{\partial \Psi}{\partial T_0} = \frac{1}{2} \int dX \left(-\hbar^2 \frac{\delta^2}{\delta \phi^2(X)} + \left[\frac{\partial \phi}{\partial X} \right]^2 \right) \Psi , \qquad (3.87)$$

which is just the ordinary (functional) Schrödinger equation for the massless scalar field. Note that this is only one equation instead of the infinitely many equations (3.85) and (3.86).

The general interpretation of the momentum constraint can be easily recognized from the example (3.86). Performing an infinitesimal coordinate transformation on $T = \text{constant}$, $x \to \bar{x} = x + \delta N^1(x)$, one gets

$$T(x) \to T(x + \delta N^1(x)) = T(x) + T'(x)\delta N^1(x)$$

and similar equations for $X(x)$ and $\phi(x)$. For the wave functional, the transformation yields

$$\Psi \to \Psi[T(x) + T'(x)\delta N^1(x), \ldots] = \Psi[T(x), \ldots]$$
$$+ \int dx \left(T'(x) \frac{\delta \Psi}{\delta T(x)} + \cdots \right) \delta N^1(x) . \qquad (3.88)$$

Therefore, the momentum constraint (3.86) enforces the independence of Ψ under infinitesimal coordinate transformations on the hypersurfaces.

Going back to the general action (3.73), one finds that the Hamiltonian is, as expected, a linear combination of constraints,

$$H = \int d^3x \ (N\mathcal{H}_\perp + N^a \mathcal{H}_a) . \qquad (3.89)$$

Dynamical consistency of a constrained system is only gained if the constraints are preserved in time (here, with respect to the time parameter x^0). This is the case only if the Poisson brackets between all constraints are combinations of the constraints themselves. One finds in fact the Poisson-bracket algebra (Dirac 1964)

$$\{\mathcal{H}_\perp(\mathbf{x}), \mathcal{H}_\perp(\mathbf{y})\} = -\sigma\delta_{,a}(\mathbf{x}, \mathbf{y}) \left(h^{ab}(\mathbf{x})\mathcal{H}_b(\mathbf{x}) + h^{ab}(\mathbf{y})\mathcal{H}_b(\mathbf{y})\right) , \quad (3.90)$$
$$\{\mathcal{H}_a(\mathbf{x}), \mathcal{H}_\perp(\mathbf{y})\} = \mathcal{H}_\perp \delta_{,a}(\mathbf{x}, \mathbf{y}) , \quad (3.91)$$
$$\{\mathcal{H}_a(\mathbf{x}), \mathcal{H}_b(\mathbf{y})\} = \mathcal{H}_b(\mathbf{x})\delta_{,a}(\mathbf{x}, \mathbf{y}) + \mathcal{H}_a(\mathbf{y})\delta_{,b}(\mathbf{x}, \mathbf{y}) , \quad (3.92)$$

with the derivatives all acting on $\mathbf{x}$. We have here introduced a space–time metric with signature diag $(\sigma, 1, 1, 1)$ in order to exhibit the difference between the Lorentzian case ($\sigma = -1$, the relevant case here) and the Euclidean case ($\sigma = 1$). This algebra will play a crucial role in canonical gravity, see Chapters 4–6. It is often convenient to work with a 'smeared-out' version of the constraints, that is,

$$\mathcal{H}[N] = \int d^3x \, N(\mathbf{x})\mathcal{H}_\perp(\mathbf{x}) , \quad \mathcal{H}[N^a] = \int d^3x \, N^a(\mathbf{x})\mathcal{H}_a(\mathbf{x}) . \quad (3.93)$$

The constraint algebra then reads

$$\{\mathcal{H}[N], \mathcal{H}[M]\} = \mathcal{H}[K^a] , \quad K^a = -\sigma h^{ab}(NM_{,b} - N_{,b}M) , \quad (3.94)$$
$$\{\mathcal{H}[N^a], \mathcal{H}[N]\} = \mathcal{H}[M] , \quad M = N^a N_{,a} \equiv \mathcal{L}_\mathbf{N} N , \quad (3.95)$$
$$\{\mathcal{H}[N^a], \mathcal{H}[M^b]\} = \mathcal{H}[K] , \quad \mathbf{K} = [\mathbf{N}, \mathbf{M}] \equiv \mathcal{L}_\mathbf{N} \mathbf{M} , \quad (3.96)$$

where $\mathcal{L}$ denotes here the Lie derivative. Some remarks are in order:

1. This algebra is *not* a Lie algebra, since (3.94) contains the (inverse) three-metric $h^{ab}(\mathbf{x})$ of the hypersurfaces $x^0 = $ constant (i.e. one has structure functions depending on the canonical variables instead of structure constants). An exception is two-dimensional space–time (Teitelboim 1984).
2. The signature σ of the embedding space–time can be read off directly from (3.94).
3. The algebra is the same as for the corresponding constraints in the case of the bosonic string, that is, it is in two dimensions equivalent to the Virasoro algebra (3.61). The reason is its general geometric interpretation to be discussed in the following.

It turns out that the above algebra has a purely kinematical interpretation. It is just the algebra of surface deformations for hypersurfaces which are embedded in a Riemannian (or pseudo-Riemannian) space. If a hypersurface is again described by $X^\mu(x^a)$, the generators of coordinate transformations *on* the hypersurface are given by

$$X_{ax} \equiv X^\mu_{,a}(\mathbf{x})\frac{\delta}{\delta X^\mu(\mathbf{x})} ,$$

while the generators of the normal deformations are given by

$$X_x \equiv n^\mu(\mathbf{x})\frac{\delta}{\delta X^\mu(\mathbf{x})}$$

with the normal vector obeying the normalization condition $n^\mu n_\mu = \sigma$. Kuchař (1973) calculated the algebra,

$$[X_x, X_y] = \sigma\delta_{,a}(\mathbf{x}, \mathbf{y})\left(h^{ab}(\mathbf{x})X_{bx} + h^{ab}(\mathbf{y})X_{by}\right) , \qquad (3.97)$$

$$[X_{ax}, X_y] = -X_x\delta_{,a}(\mathbf{x}, \mathbf{y}) , \qquad (3.98)$$

$$[X_{ax}, X_{by}] = -X_{bx}\delta_{,a}(\mathbf{x}, \mathbf{y}) - X_{ay}\delta_{,b}(\mathbf{x}, \mathbf{y}) . \qquad (3.99)$$

Up to a sign, this algebra has the same structure as the constraint algebra (3.90)–(3.92). The reason for the different sign is the relation

$$[X_f, X_g] = -X_{\{f,g\}} .$$

The constraints $\mathcal{H}_\perp$ and $\mathcal{H}_a$ generate, in fact, the algebra of hypersurface deformations, which are given by (3.77) and which are *not* identical to the algebra of space–time diffeomorphisms; cf. also Chapter 4. That surface deformations form a larger class of transformations than space–time diffeomorphism can be seen if one considers surfaces which intersect each other at a point P. Under surface deformations this point is shifted to two *different* points, depending on which of the two surfaces one starts. On the other hand, a spacetime diffeomorphism shifts each point in a unique way independent of the surface on which it lies; space–time diffeomorphisms thus induce only special surface deformations.

In Section 4.1, we will show that one can construct the theory of GR from the above constraint algebra provided the three-metric h_{ab} and its canonical momentum are the only canonical variables. Before this will be done, we shall discuss in the next section a 'relational' mechanical model which exhibits some interesting features being of relevance for the quantization of gravity.

3.4   Relational dynamical systems

Newtonian mechanics needs for its formulation the concepts of absolute space and absolute time. This was criticized already by some of Newton's contemporaries, notably Leibniz, who insisted that only observable quantities should appear in the fundamental equations. In the nineteenth century Ernst Mach emphasized that the concepts of absolute space and time should be abandoned altogether and that physics should only use relational concepts.

Consider in a gedanken experiment two successive 'snapshots' of the universe within a short time interval (Barbour 1986). The universe is considered for simplicity as a collection of n particles with masses m_i, $i = 1, \ldots, n$, evolving in Euclidean space under the influence of Newtonian gravity. Due to the short time interval, the relative distances will be only slightly different. Can one predict the future evolution from these two observations? The definite answer is *no* because the two sets of relative separations give no information about the angular momentum or kinetic energy of the system, both of which affect the future evolution.

This 'Poincaré defect' (because Poincaré pronounced this lack of predictability) motivated Barbour and Bertotti (1982) to look for a slight generalization of

Newtonian mechanics in which the future can be predicted solely on the basis of *relative* separations (and their rates of change). The key idea is to introduce a 'gauge freedom' with respect to translations and rotations (because these transformations leave the relative distances invariant) and the choice of the time parameter τ. The theory should thus be invariant under the following gauge transformations,

$$\mathbf{x}_k \rightarrow \mathbf{x}'_k = \mathbf{x}_k + \mathbf{a}(\tau) + \alpha(\tau) \times \mathbf{x}_k \ , \tag{3.100}$$

where $\mathbf{a}$ parametrizes translations, α rotations, and $\mathbf{x}_k$ is the position vector of particle k. They depend on the 'label time' τ which can be arbitrarily reparametrized,

$$\tau \rightarrow f(\tau) \ , \quad \dot{f} > 0 \ . \tag{3.101}$$

Due to (3.100) one has instead of the original $3n$ only $3n - 6$ parameters to describe the relative distances. Equations (3.100) and (3.101) define the 'Leibniz group' (Barbour and Bertotti 1982; Barbour 1986). One can now define a total velocity for each particle according to

$$\frac{D\mathbf{x}_k}{D\tau} \equiv \frac{\partial \mathbf{x}_k}{\partial \tau} + \dot{\mathbf{a}}(\tau) + \dot{\alpha}(\tau) \times \mathbf{x}_k \ , \tag{3.102}$$

in which the first term on the right-hand side denotes the rate of change in some chosen frame, and the second and third terms the rate of change due to a τ-dependent change of frame. This velocity is not yet gauge invariant. A gauge-invariant quantity can be constructed by minimizing the 'kinetic energy'

$$\sum_{k=1}^{n} \frac{D\mathbf{x}_k}{D\tau} \frac{D\mathbf{x}_k}{D\tau}$$

with respect to $\mathbf{a}$ and α. This procedure is also called 'horizontal stacking' (Barbour 1986). Intuitively it can be understood as putting two slides with the particle positions marked on them on top of each other and moving them relative to each other until the centres of mass coincide and there is no overall rotation. The result of the horizontal stacking is a gauge-invariant 'intrinsic velocity', $d\mathbf{x}/d\tau$. Having these velocities for each particle at one's disposal one can construct the kinetic term

$$T = \frac{1}{2} \sum_{k=1}^{n} m_k \left(\frac{d\mathbf{x}_k}{d\tau} \right)^2 \ . \tag{3.103}$$

The potential is the standard Newtonian potential

$$V = -G \sum_{k<l} \frac{m_k m_l}{r_{kl}} \ , \tag{3.104}$$

where $r_{kl} = |\mathbf{x}_k - \mathbf{x}_l|$ is the relative distance (more generally, one can take any potential $V(r_{kl})$). With this information at hand, one can construct the following action,

$$S = 2 \int d\tau \sqrt{-VT} \ , \tag{3.105}$$

which is homogeneous in the velocities $d\mathbf{x}/d\tau$ and therefore reparametrization-invariant with respect to τ; cf. Section 3.1.1. After the horizontal stacking is performed, one is in a preferred frame in which the intrinsic velocities coincide with the ordinary velocities.

The equations of motion constructed from the action (3.105) read

$$\frac{d}{d\tau} \left(\sqrt{\frac{-V}{T}} m_k \frac{d\mathbf{x}_k}{d\tau} \right) = -\sqrt{\frac{T}{-V}} \frac{\partial V}{\partial \mathbf{x}_k} \ . \tag{3.106}$$

Note that this is a *non-local* equation because the frame is determined by the global stacking procedure and also the total kinetic and potential energy of the universe occur explicitly. The gauge invariance with respect to translations and rotations leads to the constraints

$$\mathbf{P} = \sum_k \mathbf{p}_k = 0 \ , \quad \mathbf{L} = \sum_k \mathbf{x}_k \times \mathbf{p}_k = 0 \ , \tag{3.107}$$

that is, the total momentum and angular momentum of the universe is constrained to vanish. Since the momentum of the kth particle is given by

$$\mathbf{p}_k = \frac{\partial L}{\partial \dot{\mathbf{x}}} = m_k \sqrt{\frac{-V}{T}} \dot{\mathbf{x}}_k \ ,$$

one finds the Hamiltonian constraint

$$H \equiv \sum_{k=1}^{n} \frac{\mathbf{p}_k^2}{2m_k} + V = 0 \ , \tag{3.108}$$

which is a consequence of reparametrization invariance with respect to τ; see Section 3.1.1. Equation (3.106) can be drastically simplified if a convenient gauge choice is being made for τ: it is chosen such as to make the total energy vanish,

$$T + V = 0 \ . \tag{3.109}$$

Note that (3.109) is not the usual energy equation, since there is no external time present here. On the contrary, this equation is used to *define* time.

With (3.109) one then gets from (3.106) just Newton's equations. Therefore, only after this choice the connection with Newtonian mechanics has been established. The in-principle observational difference to Newtonian mechanics is that here the total energy, momentum, and angular momentum must vanish. We note in this connection that in 1905 Henri Poincaré argued for a definition of time that makes the equations of motion assume their simplest form. He writes[7]

[7]'Le temps doit être défini de telle façon que les équations de la mécanique soient aussi simples que possible. En d'autres termes, il n'y a pas une manière de mesurer le temps qui soit plus vraie qu' une autre; celle qui est généralement adoptée est seulement plus *commode*.' (Poincaré 1970)

Time must be defined in such a way that the equations of mechanics are as simple as possible. In other words, there is no way to measure time that is more true than any other; the one that is usually adopted is only more *convenient*.

It is, however, a fact that the choice (3.109) is not only distinguished because then the equations of motion (3.106) take their simplest form but also because only such a choice will ensure that the various clocks of (approximately isolated) subsystems march in step, since $\sum_k (T_k + V_k) = \sum_k E_k = 0$ (Barbour 1994). The only truly isolated system is the universe as a whole and to determine time it is (in principle) necessary to monitor the whole universe. In practice this is done even when atomic clocks are employed, for example, in the determination of the pulse arrival times from binary pulsars (Damour and Taylor 1991).

In this approach, the inertial frame and absolute time of Newtonian mechanics are *constructed* from observations through the minimization of the kinetic energy and the above choice of τ. One could call this a Leibnizian or Machian point of view. The operational time defined by (3.109) corresponds to the notion of 'ephemeris time' used in astronomy. That time must be defined such that the equations of motion be simple was already known by Ptolemy (Barbour 1989). His theory of eclipses only took a simple form if sidereal time (defined by the rotation of the heavens, i.e. the rotation of the Earth) is used. This choice corresponds to a 'uniform flow of time'.

Time-reparametrization invariant systems have already been discussed in Section 3.1.1 in connection with the parametrized non-relativistic particle. In contrast to there, however, *no* absolute time is present here and the theory relies exclusively on observational elements.

A formal analogy of the action (3.105) is given by Jacobi's action in classical mechanics (Barbour 1986; Lanczos 1986; Brown and York 1989),

$$S_J = \int \mathrm{d}s \, \sqrt{E - V} \, , \tag{3.110}$$

where E is the total energy. Writing $\mathrm{d}s = \dot{s}\mathrm{d}\tau$, one has

$$\mathrm{d}s^2 = \sum_{k=1}^{n} m_k \mathrm{d}\mathbf{x}_k \mathrm{d}\mathbf{x}_k \, ,$$

and one gets $\dot{s} = \sqrt{2T}$; if $E = 0$, then, Jacobi's action coincides with the action (3.105).

This 'timeless' description of mechanics employs only paths in configuration space. A 'speed' is determined later by solving the energy equation $T + V = E$ (T contains the velocities $\dot{\mathbf{x}}_k$). Barbour (1986) argues that for an isolated system, such as the universe is assumed to be, this demonstrates the redundancy of the notion of an independent time. All the essential dynamical content is already contained in the timeless paths. We shall see in Chapter 4 that GR can be described by an action similar to (3.105) and thus can be interpreted as 'already parametrized'. In this sense, the constraints of relational mechanics correspond to those of GR (see Chapter 4) more closely than the other examples considered in this chapter and suggest that GR is time-less in a significant sense.

Quantization of the Barbour–Bertotti model then follows Dirac's prescription, leading to the quantized constraints

$$\mathcal{H}\psi(\mathbf{x}_k) = 0 , \quad \mathbf{P}\psi(\mathbf{x}_k) = 0 , \quad \mathbf{L}\psi(\mathbf{x}_k) = 0 . \tag{3.111}$$

The wave function is actually defined on the relative configuration space, the space of relative distances.

4

HAMILTONIAN FORMULATION OF GENERAL RELATIVITY

4.1 The seventh route to geometrodynamics

It is the purpose of this chapter to develop the Hamiltonian formulation of general relativity, which will serve as the starting point for quantization in Chapters 5 and 6. In the present section, we shall derive it directly from the algebra of surface deformations (Section 3.3), while in the next section, it will be recovered from the Einstein–Hilbert action through a 3+1 decomposition. In these two sections, the formalism will be applied to the traditional metric formulation, while in the last section, a more recent formalism using connections will be used. The quantization of the latter version will be treated in Chapter 6.

That the theory of GR in its Hamiltonian version can be constructed directly from the algebra of surface deformations (3.90)–(3.92) was shown by Hojman *et al.* (1976), and we shall follow their exposition in this section. They call their approach the 'seventh route to geometrodynamics', complementing the six routes presented in box 17.6 of Misner *et al.* (1973).

4.1.1 *Principle of path independence*

Starting point is the assumption that the *only* gravitational canonical pair of variables on the spatial hypersurfaces consists of the three-dimensional metric $h_{ab}(\mathbf{x})$ and its conjugate momentum $p^{cd}(\mathbf{y})$. In addition, matter fields may be present. The constraints will not be imposed—in fact, they will be *derived* from the algebra. Different classical theories of gravity typically contain additional degrees of freedom in the configurations space of the gravitational sector (such as a scalar field in the Brans–Dicke case or the extrinsic curvature for R^2-theories). The central idea in the derivation is the use of a 'principle of path independence'. Let us assume the presence of two different three-dimensional geometries (three-geometries for short) $\mathcal{G}_{\mathrm{in}}$ and $\mathcal{G}_{\mathrm{fin}}$ and a set of observers distributed on $\mathcal{G}_{\mathrm{in}}$; the observers bifurcate and follow different evolutions of intermediate hypersurface, making records of them as well as the fields on them, until they all end up on $\mathcal{G}_{\mathrm{fin}}$. The principle of path independence states that the change in all field variables must be independent of the route that is chosen between $\mathcal{G}_{\mathrm{in}}$ and $\mathcal{G}_{\mathrm{fin}}$. Only in this case can the evolution of three-geometries be interpreted as arising from different foliations through the same four-dimensional space–time.[1]

The evolution of a function F of the gravitational canonical variables is given by, cf. (3.73),

---

[1] This holds only for cases where the sandwich conjecture is satisfied, that is, for which two three-geometries uniquely determine the space–time 'sandwiched' between them.

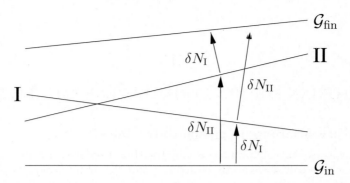

FIG. 4.1. Normal   deformations   along   two   different   intermediate three-geometries.

$$\dot{F}(h_{ab}(x), p^{cd}(x)) = \int \mathrm{d}x' \; (\{F, \mathcal{H}_\perp(x')\}N(x') + \{F, \mathcal{H}_a(x')\}N^a(x'))$$

$$\equiv \int \mathrm{d}x' \; \{F, \mathcal{H}_\mu(x')\}N^\mu(x') \;, \tag{4.1}$$

where $\mathrm{d}x'$ is a shorthand for d^3x' and x for $\mathbf{x}$, etc. Consider now an infinitesimal evolution along two different intermediate hypersurfaces (Fig. 4.1): first, a normal deformation from $\mathcal{G}_{\mathrm{in}}$ to I and then from I to $\mathcal{G}_{\mathrm{fin}}$; second, a normal deformation from $\mathcal{G}_{\mathrm{in}}$ to II and then from II to $\mathcal{G}_{\mathrm{fin}}$. Evolving F along the first possibility yields

$$F\,[\mathcal{G}_{\mathrm{fin}}] = F\,[\mathcal{G}_{\mathrm{in}}] + \int \mathrm{d}x' \; \{F, \mathcal{H}_\perp(x')\}\delta N_{\mathrm{I}}(x')$$

$$+ \int \mathrm{d}x' \; \{F, \mathcal{H}_\perp(x')\}\delta N_{\mathrm{II}}(x')$$

$$+ \int \mathrm{d}x'\mathrm{d}x'' \; \{\{F, \mathcal{H}_\perp(x')\}\delta N_{\mathrm{I}}(x'), \mathcal{H}_\perp(x'')\}\, \delta N_{\mathrm{II}}(x'') \;.$$

The first two terms on the right-hand side describe the evolution of F from $\mathcal{G}_{\mathrm{in}}$ to I, and the last two terms the evolution from I to $\mathcal{G}_{\mathrm{fin}}$. The option to reach $\mathcal{G}_{\mathrm{fin}}$ via hypersurface II gives an analogous result. Performing the difference between both expressions and using the Jacobi identity leads one to

$$\delta F = - \int \mathrm{d}x'\mathrm{d}x'' \; \{\{\mathcal{H}_\perp(x'), \mathcal{H}_\perp(x'')\}, F\}\, \delta N_{\mathrm{I}}(x'')\delta N_{\mathrm{II}}(x') \;. \tag{4.2}$$

From (3.97) or (3.94) one knows that the difference between these two normal deformations must be given by a tangential deformation,[2]

---

[2]Recall that $\mathcal{H}_a$ generates *minus* the surface deformations.

$$\delta F = - \int \mathrm{d}x' \, \{F, \mathcal{H}_a(x')\} \delta N^a(x') \,, \tag{4.3}$$

where

$$\delta N^a = -\sigma h^{ab} \left(\delta N_{\mathrm{I}} \delta N_{\mathrm{II},b} - \delta N_{\mathrm{II}} \delta N_{\mathrm{I},b} \right) \,. \tag{4.4}$$

Inserting for calculational purposes a delta function into (4.3) and performing a partial integration, one finds

$$- \int \mathrm{d}x' \, \{F, \mathcal{H}_a(x')\} \delta N^a(x') =$$
$$-\sigma \int \mathrm{d}x' \mathrm{d}x'' \, \frac{\partial}{\partial x''^b} \delta(x' - x'') \{F, \mathcal{H}_a(x'')\} h^{ab}(x'') \delta N_{\mathrm{I}}(x'') \delta N_{\mathrm{II}}(x')$$
$$-\sigma \int \mathrm{d}x' \mathrm{d}x'' \, \frac{\partial}{\partial x''^b} \delta(x' - x'') \{F, \mathcal{H}_a(x')\} h^{ab}(x') \delta N_{\mathrm{I}}(x'') \delta N_{\mathrm{II}}(x') \,. \tag{4.5}$$

Setting this equal to (4.2) and using the arbitrariness of $\delta N_{\mathrm{I}}(x'') \delta N_{\mathrm{II}}(x')$, one finds

$$\{F, \{\mathcal{H}_\perp(x'), \mathcal{H}_\perp(x'')\}\} = -\sigma \delta_{,b}(x' - x'') h^{ab}(x'') \{F, \mathcal{H}_a(x'')\}$$
$$-\sigma \delta_{,b}(x' - x'') h^{ab}(x') \{F, \mathcal{H}_a(x')\} \,. \tag{4.6}$$

Inserting the Poisson bracket (3.90) on the left-hand side, one finds the condition

$$\frac{\partial}{\partial x'^a} \delta(x' - x'') \left(\{F, h^{ab}(x')\} \mathcal{H}_b(x') + \{F, h^{ab}(x'')\} \mathcal{H}_b(x'') \right) = 0 \,. \tag{4.7}$$

As this should hold for all F, the generators $\mathcal{H}_a$ themselves must vanish as constraints, $\mathcal{H}_a \approx 0$. This result from the principle of path independence only follows because the Poisson bracket (3.90) depends on the metric h^{ab}. Since $\mathcal{H}_a \approx 0$ must hold on every hypersurface, it must be conserved under a normal deformation. From (3.91) one then finds that $\mathcal{H}_\perp$ must also vanish, $\mathcal{H}_\perp = 0$. We have thus shown that the algebra of surface deformations, together with the principle of path independence (equivalent to the principle of embeddability), enforces the constraints

$$\mathcal{H}_\perp \approx 0 \,, \quad \mathcal{H}_a \approx 0 \,. \tag{4.8}$$

This result follows for all number of space–time dimensions different from two. For two dimensions (after a suitable definition of the $\mathcal{H}_\mu$), the metric does not appear on the right-hand side of the algebra (Teitelboim 1984). This leads to the possibility of having path independence without constraints, giving one the option to have Schwinger terms in the quantum theory (cf. Sections 3.2 and 5.3.5).

4.1.2 *Explicit form of generators*

How does one find the explicit form of $\mathcal{H}_\perp$ and $\mathcal{H}_a$? The constraints $\mathcal{H}_a$ generate coordinate transformations on a hypersurface. If the transformation law of certain fields is given, $\mathcal{H}_a$ follows (or, alternatively, the transformation properties are determined by a given $\mathcal{H}_a$). Knowing $\mathcal{H}_a$, one can consider the Poisson bracket (3.90) as a system of infinitely many equations to determine $\mathcal{H}_\perp$, which must then depend on h_{ab} because the right-hand side of (3.90) does. If, moreover, the assumption is made that $\mathcal{H}_\perp$ depends *only* on h_{ab} (and its momentum) as gravitational degrees of freedom, GR will follow.

Before this is done, some general properties will be briefly discussed, see Teitelboim (1980) for proofs and details. Writing $\mathcal{H}_\perp$ as the sum of a gravitational and a matter part (with 'matter' referring here to all non-gravitational bosonic fields[3] symbolically denoted by ϕ),

$$\mathcal{H}_\perp = \mathcal{H}_\perp^{\mathrm{g}}[h_{ab}, p^{cd}] + \mathcal{H}_\perp^{\mathrm{m}}[h_{ab}, p^{cd}; \phi, p_\phi] \ , \tag{4.9}$$

it follows from (3.90) that $\mathcal{H}_\perp^{\mathrm{m}}$ *must* depend on the gravitational degrees of freedom, that is, gravity couples to all forms of matter. This is the Hamiltonian version of Weinberg's result discussed in Section 2.1. More restrictions can be obtained if one assumes that $\mathcal{H}_\perp^{\mathrm{m}}$ does not depend on the gravitational momenta p^{cd}. This corresponds to the presence of 'non-derivative couplings' only, that is, there are no gravitational velocities on the Lagrangian level and therefore there is no modification of the relationship between momentum and velocity. Then one can show that $\mathcal{H}_\perp^{\mathrm{m}}[h_{ab}; \phi, p_\phi]$ depends on the h_{ab} only ultralocally, that is, no derivatives or integrals of h_{ab} appear. From this one can infer that the terms $\mathcal{H}_\perp^{\mathrm{m}}$ obey the relation (3.90) separately, that is,

$$\{\mathcal{H}_\perp^{\mathrm{m}}(x), \mathcal{H}_\perp^{\mathrm{m}}(y)\} = -\sigma \delta_{,a}(x, y) \left(h^{ab}(x)\mathcal{H}_b^{\mathrm{m}}(x) + h^{ab}(y)\mathcal{H}_b^{\mathrm{m}}(y) \right) \ . \tag{4.10}$$

The general form of $\mathcal{H}_a$ can be restricted by the following two requirements:

1. It must be linear in the momenta in order to generate transformation of the fields under coordinate transformations (and not mix fields and momenta).
2. It should contain the momenta only up to the first spatial derivatives because it should generate first-order derivatives in the fields (Taylor expansion to first order).

Therefore,

$$\mathcal{H}_a = b_a{}^b{}_B(\phi_C)p^B{}_{,b} + a_{aB}(\phi_C)p^B \ , \tag{4.11}$$

where ϕ_A now is a symbolic notation for all fields, including gravity, and p^A denotes the corresponding momenta. If one demands the existence of ultralocal solutions to (3.90) (i.e. a deformation localized at a point $\mathbf{x}_0$ can change the field only at $\mathbf{x}_0$), one finds that $\mathcal{H}_\perp$ must depend ultralocally on the momenta (Teitelboim 1980), and that

$$b^{ab}{}_B = b^{ba}{}_B \ . \tag{4.12}$$

---

[3]For the fermionic case, see the remarks below.

We now investigate the explicit form of $\mathcal{H}_a$ for various fields. The first case is a scalar field ϕ. Under an infinitesimal coordinate transformation $x'^a = x^a - \delta N^a(x)$, it transforms as

$$\delta\phi(x) \equiv \phi'(x) - \phi(x) \approx \frac{\partial\phi}{\partial x^a}\delta N^a \equiv \mathcal{L}_{\delta\mathbf{N}}\phi \ , \tag{4.13}$$

where $\mathcal{L}$ denotes the Lie derivative. This is generated by

$$\mathcal{H}_a = p_\phi\phi_{,a} \ . \tag{4.14}$$

Comparison with (4.11) shows that $b^{ab} = 0 = b^{ba}$, so that (4.12) is fulfilled and ultralocality holds.

For a vector field, $A_a(x)$, the transformation is

$$\delta A_a = A_{a,b}\delta N^b + A_b\delta N^b_{,a} \equiv (\mathcal{L}_{\delta\mathbf{N}}A)_a \ , \tag{4.15}$$

which is generated by

$$\mathcal{H}_a = -p^b{}_{,b}A_a + (A_{b,a} - A_{a,b})p^b \ . \tag{4.16}$$

Comparison with (4.11) shows that

$$b_a{}^b{}_C = -A_a\delta^b{}_C \ . \tag{4.17}$$

Therefore, the condition for ultralocality (4.12) is not fulfilled for vector fields. Its restoration will lead to the concept of gauge theories (Section 4.1.3).

For a covariant tensor field of second rank (not necessarily symmetric), $t_{ab}(x)$, one has

$$\delta t_{ab} = t_{ab,c}\delta N^c + t_{ac}\delta N^c{}_{,b} + t_{cb}\delta N^c{}_{,a} \equiv (\mathcal{L}_{\delta\mathbf{N}}t)_{ab} \ , \tag{4.18}$$

which is generated by

$$\mathcal{H}_a = t_{bc,a}p^{bc} - (t_{ab}p^{cd})_{,c} - (t_{ca}p^{cb})_{,b} \ . \tag{4.19}$$

It turns out that in order for (4.12) to be fulfilled, one must have

$$t_{ab} = f(x)h_{ab} \tag{4.20}$$

with an arbitrary function $f(x)$, that is, the tensor field must be proportional to the metric itself. Choosing in particular $t_{ab} = h_{ab}$, one finds for the generator (4.19) the expression

$$\mathcal{H}^{\mathrm{g}}_a = -2p_a{}^c{}_{,c} + 2\Gamma^d_{ac}p_d{}^c \equiv -2D_bp_a{}^b \equiv -2p_a{}^b{}_{|b} \ . \tag{4.21}$$

The last two terms denote the covariant derivative in three dimensions (recall that p^{ab} is a tensor density of weight one).

Using the result (4.21) for $\mathcal{H}_a^g$, one can construct from (3.90) the explicit expression for $\mathcal{H}_\perp^g$. A rather lengthy but straightforward calculation leads to (Hojman *et al.* 1976; Teitelboim 1980)

$$\mathcal{H}_\perp^g = 16\pi G\, G_{abcd} p^{ab} p^{cd} + V[h_{ab}]\,, \tag{4.22}$$

with

$$G_{abcd} = \frac{1}{2\sqrt{h}}(h_{ac}h_{bd} + h_{ad}h_{bc} - h_{ab}h_{cd}) \tag{4.23}$$

as the (inverse) 'DeWitt metric',[4] h denoting the determinant of h_{ab}, and

$$V = \frac{\sigma\sqrt{h}}{16\pi G}(\,^{(3)}R - 2\Lambda)\,, \tag{4.24}$$

where $^{(3)}R$ is the three-dimensional Ricci scalar.[5] The inverse of (4.23) is called 'DeWitt' metric because it plays the role of a 'supermetric' in the space of all metrics (DeWitt 1967a). The explict expression reads

$$G^{abcd} = \frac{\sqrt{h}}{2}(h^{ac}h^{bd} + h^{ad}h^{bc} - 2h^{ab}h^{cd}) \tag{4.25}$$

(the last term here is the same in all space dimensions), obeying

$$G^{abcd}G_{cdef} = 1/2\left(\delta_e^a\delta_f^b + \delta_f^a\delta_e^b\right)\,. \tag{4.26}$$

The interpretation of the Poisson-bracket relation (3.91) is that $\mathcal{H}_\perp$ transforms as a scalar density under coordinate transformations; this is explicitly fulfilled by (4.22) (G_{abcd} has weight -1, p^{ab} and V have weight 1, so $\mathcal{H}_\perp^g$ has weight 1). Explicitly,

$$\delta\mathcal{H}_\perp^g(x) = \int \mathrm{d}y\, \{\mathcal{H}_\perp^g(x), \mathcal{H}_a^g(y)\}\delta N^a(y) = \frac{\partial}{\partial x^a}\left(\mathcal{H}_\perp^g(x)\delta N^a(x)\right)\,. \tag{4.27}$$

It will be shown in Section 4.2 that $\mathcal{H}_\perp^g$ and $\mathcal{H}_a^g$ uniquely characterize GR, that is, they follow from the Einstein–Hilbert action (1.1).

4.1.3 *Geometrodynamics and gauge theories*

We have seen that for vector fields $\mathcal{H}_a$ is of such a form that the condition of ultralocality for $\mathcal{H}_\perp$ would be violated, see (4.16). Since vector fields are an important ingredient in the description of nature, the question arises whether a different formulation can be found that is in accordance with ultralocality. For

---

[4] In d space dimensions, the last term reads $-2/(d-1)h_{ab}h_{cd}$.

[5] G and Λ are at this stage just free parameters. They will later be identified with the gravitational constant and the cosmological constant, respectively.

this purpose it is suggestive to omit the term $-p^b{}_{,b}A_a$ in (4.16) because then the $b_a{}^b{}_B$ will become zero. This leads to the replacement (Teitelboim 1980)

$$\mathcal{H}_a \to \bar{\mathcal{H}}_a \equiv \mathcal{H}_a + p^b{}_{,b}A_a = (A_{b,a} - A_{a,b})p^b \ . \tag{4.28}$$

What happens with the Poisson-bracket relation (3.92) after this modification? A brief calculation shows

$$\{\bar{\mathcal{H}}_a(x), \bar{\mathcal{H}}_b(y)\} = \bar{\mathcal{H}}_b(x)\delta_{,a}(x,y) + \bar{\mathcal{H}}_a(y)\delta_{,b}(x,y) - F_{ab}(x)p^c{}_{,c}(x)\delta(x,y) \ , \tag{4.29}$$

where $F_{ab} = \partial_a A_b - \partial_b A_a$. The new term in (4.29) will only be harmless if it generates physically irrelevant transformations ('gauge transformations'). This is the case if the new term actually vanishes as a constraint. Since setting F_{ab} to zero would appear too strong (leaving only the restricted option $A_a = \partial_a \varphi$), it is suggestive to demand that $p^a{}_{,a} = 0$. One therefore introduces the constraint

$$\mathcal{G}(x) \equiv -\frac{1}{e}p^a{}_{,a}(x) \equiv -\frac{1}{e}E^a{}_{,a}(x) \equiv -\frac{1}{e}\nabla\mathbf{E}(x) \ . \tag{4.30}$$

The constraint $\mathcal{G} \approx 0$ is just *Gauß' law* of electrodynamics (in the sourceless case) with the momentum being equal to the electric field $\mathbf{E}$ (the electric charge, e, has been introduced for convenience). As usually, Gauß' law generates gauge transformations,

$$\delta A_a(x) = \int dy \ \{A_a(x), \mathcal{G}(y)\}\xi(y) = \frac{1}{e}\partial_a \xi(x) \ , \tag{4.31}$$

$$\delta p^a(x) = \int dy \ \{p^a(x), \mathcal{G}(y)\}\xi(y) = 0 \ . \tag{4.32}$$

The electric field is of course gauge invariant, and so is the field strength F_{ab}. In the modified constraint (4.28), the first term generates the usual transformations for a vector field, see (4.15), while the second term generates gauge transformations for the 'vector potential' $A_a(x)$. Therefore, $A_a(x)$ transforms under $\bar{\mathcal{H}}_a$ not like a covariant vector but only like a vector modulo a gauge transformation. This fact was already encountered in the space–time picture; see Section 2.1. The electric field, however, transforms as a contravariant tensor density, since the additional term in (4.28) has no effect.

The above introduction of the gauge principle can be extended in a straightforward manner to the non-Abelian case. Consider instead of a single $A_a(x)$ now a set of several fields, $A_a^i(x)$, $i = 1, \dots, N$. The simplest generalization of the Abelian case consists in the assumptions that $A_a^i(x)$ should not mix with its momentum $p_i^a(x)$ under a gauge transformation, that the momenta should transform homogeneously, and that the gauge constraint (the non-Abelian version of Gauß' law) is local. This then leads to (Teitelboim 1980)

$$\mathcal{G}_i = -\frac{1}{f}p_i{}^a{}_{,a} + C_{ij}{}^k A_a^j p_k{}^a \approx 0 \ , \tag{4.33}$$

where f and $C_{ij}{}^k$ are constants. Demanding that the commutation of two gauge transformations be again a transformation it follows that the $C_{ij}{}^k$ are the structure constants of a Lie algebra. One then has

$$\{\mathcal{G}_i(x), \mathcal{G}_j(y)\} = C_{ij}{}^k \mathcal{G}_k(x)\delta(x,y) \ , \tag{4.34}$$

which characterizes a *Yang–Mills theory*.

As in the case of the gravitational field, one can construct the corresponding part of the Hamiltonian constraint, $\mathcal{H}_\perp^{\text{YM}}$, from the Poisson-bracket relation (3.90). Writing

$$\mathcal{H}_\perp = \mathcal{H}_\perp^{\text{g}} + \mathcal{H}_\perp^{\text{YM}} \ , \tag{4.35}$$

and demanding that the Yang–Mills part be independent of the gravitational momenta (so that $\mathcal{H}_\perp^{\text{YM}}$ depends only ultralocally on the metric and must therefore obey (3.90) separately), one is led to the form

$$\mathcal{H}_\perp^{\text{YM}} = \frac{1}{2\sqrt{h}} \left(h_{ab}\gamma^{ij} p_i^a p_j^b - \sigma h^{ab}\gamma_{ij} B_a^i B_b^j \right) \ , \tag{4.36}$$

where $\gamma_{ij} = C_{ik}{}^l C_{jl}{}^k$ is the 'group metric' (γ^{ij} being its inverse), and $B_a^i = \frac{1}{2}\epsilon_{abc}F^{ibc}$ are the non-Abelian 'magnetic fields'. The non-Abelian field strength is given by

$$F_{ab}^i = \partial_a A_b^i - \partial_b A_a^i + fC^i{}_{jk}A_a^j A_b^k \ .$$

The Hamiltonian (4.36) can be found from the action

$$S_{\text{YM}} = -\frac{\sigma}{4} \int \mathrm{d}^4x \ \sqrt{-g}\gamma_{ij}F_{\mu\nu}^i F^{j\mu\nu} \ , \tag{4.37}$$

which is the usual Yang–Mills action. To summarize, the principle of path independence together with the demand that $\mathcal{H}_\perp$ be ultralocal[6] in the momenta leads to the concept of gauge theories in a natural way.

What about fermionic fields? Recalling that the Dirac equation is the 'square root' of the Klein–Gordon equation, one may try to construct a similar 'square root' for the generators of surface deformations. This has been done by Tabensky and Teitelboim (1977); it leads to spin-3/2 fields and the concept of supergravity (cf. Section 2.3) but *not* to spin-1/2. This could be a hint that the usual spin-1/2 fields only emerge through the use of superstrings (Chapter 9). The Hamiltonian formalism for supergravity will be discussed in Section 5.2.

4.2   The 3+1 decomposition of general relativity

It will be shown in this section that GR is characterized by having (4.21) and (4.22) as the constraints. This is achieved by choosing appropriate canonical variables and casting the Einstein–Hilbert action (1.1) into Hamiltonian form.

---

[6]As one knows from the discussion of the Aharonov–Bohm effect, a formulation without the vector potential can only be obtained in a non-local way.

4.2.1 *The canonical variables*

The Hamiltonian formalism starts from the choice of a configuration variable and the definition of its momentum. Since the latter requires a time coordinate ('$p = \partial L/\partial \dot{q}$'), one must cast GR in a form where it exhibits a 'distinguished' time. This is achieved by *foliating* the space–time described by $(\mathcal{M}, g)$ into a set of three-dimensional space-like hypersurfaces Σ_t; cf. also Section 3.3. The covariance of GR is preserved by allowing for the possibility to consider *all* possible foliations of this type.

This is not only of relevance for quantization (which is our motivation here), but also for important applications in the classical theory. For example, numerical relativity needs a description in terms of foliations in order to describe the dynamical evolution of events, for example, the coalescence of black holes and their emission of gravitational waves (Seidel 1998).

As a necessary condition we want to demand that $(\mathcal{M}, g)$ be globally hyperbolic, that is, that it possesses a Cauchy surface Σ (an 'instant of time') on which initial data can be described to determine uniquely the whole space–time, see for example, Wald (1984) or Hawking and Ellis (1973) for details. In such cases, the classical initial value formulation makes sense, and the Hamiltonian form of GR can be constructed. An important theorem states that for a global hyperbolic space–time $(\mathcal{M}, g)$ there exists a global 'time function' f such that each surface $f = $ constant is a Cauchy surface; therefore, $\mathcal{M}$ can be foliated into Cauchy hypersurfaces, and its topology is a direct product,

$$\mathcal{M} \cong \mathbb{R} \times \Sigma . \tag{4.38}$$

The topology of space–time is thus fixed. This may be a reasonable assumption in the classical theory, since topology change is usually connected with singularities or closed time-like curves. In the quantum theory, topology change may be a viable option and its absence in the formalism could be a possible weakness of the canonical approach.[7] Nevertheless, the resulting quantum theory is general enough to cope with most of the interesting situations.

One therefore starts with performing a foliation of space–time into Cauchy surfaces Σ_t, with t denoting the global time function ('3+1 decomposition'). The corresponding vector field ('flow of time') is denoted by t^μ, obeying $t^\mu \nabla_\mu t = 1$. The relation between infinitesimally neighboured hypersurfaces is the same as shown in Fig. 3.1.[8] The space–time metric $g_{\mu\nu}$ induces a three-dimensional metric on each Σ_t according to

$$h_{\mu\nu} = g_{\mu\nu} + n_\mu n_\nu , \tag{4.39}$$

where n_μ denotes again the unit normal to Σ_t, with $n^\mu n_\mu = -1$.

---

[7]A more general formulation allowing topology change to occur in principle is the path-integral approach of Section 2.2.

[8]The vector field t^μ was called $\dot{X}^\mu$ in Fig. 3.1 and the relation (3.74).

This is in accordance with the earlier definition (3.75): multiplication of (4.39) with $X^{\mu}_{,a}X^{\nu}_{,b}$ and using $X^{\mu}_{,a}n_{\mu} = 0$ leads to (3.75). In fact, $h_{\mu\nu}$ is a three-dimensional object only, since it acts as a projector on Σ_t, $h_{\mu\nu}n^{\nu} = 0$, $h_{\mu\nu}h^{\nu\rho} = h_{\mu}^{\rho}$. It is therefore really the three-dimensional metric, and we shall write below for it h_{ab}, since there is an isomorphism between tensor fields on $\mathcal{M}$ that are orthogonal to n^{μ} in each index and tensor fields on Σ_t.

As in (3.74), one can decompose t^{μ} into its components normal and tangential to Σ_t,

$$t^{\mu} = Nn^{\mu} + N^{\mu} , \tag{4.40}$$

where N is the lapse function and N^{μ} (called $N^a X^{\nu}_{,a}$ in (3.74)) the shift vector. In fact, N^{μ} is a three-dimensional object and can be identified with N^a. The lapse function can be written as $N = -t^{\mu}n_{\mu}$ from which one can infer

$$N = \frac{1}{n^{\mu}\nabla_{\mu}t} . \tag{4.41}$$

Similar to (3.28), one can interpret this expression as the ratio between proper time (given by $t^{\mu}\nabla_{\mu}t = 1$) and coordinate time $n^{\mu}\nabla_{\mu}t$. As in Section 3.3, the four-metric can be decomposed into spatial and temporal components,

$$g_{\mu\nu} = \begin{pmatrix} N_aN^a - N^2 & N_b \\ N_c & h_{ab} \end{pmatrix} . \tag{4.42}$$

Its inverse reads

$$g^{\mu\nu} = \begin{pmatrix} -\frac{1}{N^2} & \frac{N^b}{N^2} \\ \frac{N^c}{N^2} & h^{ab} - \frac{N^aN^b}{N^2} \end{pmatrix} . \tag{4.43}$$

Here, h^{ab} is the inverse of the three-metric (i.e. obeying $h^{ab}h_{bc} = \delta^a_c$), and one recognizes that the spatial part of $g^{\mu\nu}$ is not identical with h^{ab} but contains an additional term involving the shift vector. The components of the normal vector can be found from the one-form $n_{\mu}dx^{\mu} = -Ndt$ to read

$$n^{\mu} = g^{\mu\nu}n_{\nu} = \left(\frac{1}{N}, -\frac{\mathbf{N}}{N} \right) . \tag{4.44}$$

The various hypersurfaces Σ_t can be identified by a diffeomorphism that is generated by the integral curves of t^{μ}. The globally hyperbolic space–time $(\mathcal{M}, g)$ can thus be interpreted as the time evolution of a Riemannian metric on a *fixed* manifold Σ, that is, as an evolution from $h_{ab}(t_0)$ to $h_{ab}(t)$. This suggests the use of the three-metric h_{ab} as the appropriate dynamical variable for the canonical formalism.

In order to introduce the corresponding 'velocity' for h_{ab}, one can start by considering the following tensor field,[9]

---

[9]Sometimes a different sign is used in this definition.

FIG. 4.2. Geometric interpretation of extrinsic curvature.

$$K_{\mu\nu} = h_\mu{}^\rho \nabla_\rho n_\nu . \tag{4.45}$$

Since $K_{\mu\nu} n^\mu = 0 = K_{\mu\nu} n^\nu$, this tensor field is a purely spatial quantity and can be mapped to its spatial version K_{ab} (with indices being moved by the three-metric). One can prove, using Frobenius' theorem for the hypersurface-orthogonal vector field n^μ, that this tensor field is symmetric, $K_{\mu\nu} = K_{\nu\mu}$.

Its geometric interpretation can be inferred from Fig. 4.2. Consider the normal vectors at two different points P and Q of a hypersurface. Be $\tilde{n}^\mu$ the vector at P resulting from parallel transporting n^μ along a geodesic from Q to P. The difference between n^μ and $\tilde{n}^\mu$ is a measure for the embedding curvature of Σ into $\mathcal{M}$ at P. One therefore recognizes that the tensor field (4.45) can be used to describe this embedding curvature, since it vanishes for $n^\mu = \tilde{n}^\mu$. One can also rewrite $K_{\mu\nu}$ in terms of a Lie derivative,

$$K_{\mu\nu} = \frac{1}{2}\mathcal{L}_\mathbf{n} h_{\mu\nu} , \tag{4.46}$$

where $\mathbf{n}$ denotes the normal vector field. Therefore, K_{ab} can be interpreted as the 'velocity' associated with h_{ab}. It is called 'extrinsic curvature' or 'second fundamental form'. Its trace,

$$K \equiv K_a{}^a = h^{ab} K_{ab} \equiv \theta \tag{4.47}$$

can be interpreted as the 'expansion' of a geodesic congruence orthogonal to Σ.[10] In terms of lapse and shift, the extrinsic curvature can be written as

$$K_{ab} = \frac{1}{2N}\left(\dot{h}_{ab} - D_a N_b - D_b N_a\right) , \tag{4.48}$$

and the two terms involving the spatial covariant derivative are equivalent to $-\mathcal{L}_\mathbf{N} h_{ab}$.

4.2.2 Hamiltonian form of the Einstein–Hilbert action

One can now reformulate the Einstein–Hilbert action (1.1) in terms of the three-dimensional variables h_{ab} and K_{ab}. For this purpose one needs the relationship

[10]For a Friedmann universe, cf. Section 8.1.2, K is -3 times the Hubble parameter.

between the four-dimensional and the three-dimensional curvatures. This is given by the generalized Gauß equation,

$$^{(3)}R_{\mu\nu\lambda}{}^{\rho} = h_\mu{}^{\mu'} h_\nu{}^{\nu'} h_\lambda{}^{\lambda'} h^\rho{}_{\rho'} R_{\mu'\nu'\lambda'}{}^{\rho'} - K_{\mu\lambda}K_\nu{}^{\rho} + K_{\nu\lambda}K_\mu{}^{\rho} , \tag{4.49}$$

and the generalized Codazzi equation,

$$D_\mu K_{\nu\lambda} - D_\nu K_{\mu\lambda} = h_\mu{}^{\mu'} h_\nu{}^{\nu'} h_\lambda{}^{\lambda'} R_{\mu'\nu'\lambda'\rho} n^\rho . \tag{4.50}$$

Contraction of (4.50) with $g^{\mu\nu}$ gives

$$D_\mu K^\mu{}_\nu - D_\nu K = R_{\rho\lambda} n^\lambda h^\rho{}_\nu . \tag{4.51}$$

In the much simpler case of a two-dimensional hypersurface embedded in three-dimensional flat euclidean space, (4.49) is the famous *theorema egregium* of Gauß (cf. the discussion by Kuchař 1993). In this case, the first term on the right-hand side is zero, and the *theorema* connects the only independent component of the two-dimensional Riemann tensor, $^{(2)}R_{2112}$, with the extrinsic curvature of the hypersurface,

$$^{(2)}R_{2112} = K_{11}K_{22} - K_{21}K_{12} \equiv \det K_{ab} . \tag{4.52}$$

With the aid of the two principal curvatures κ_1 and κ_2, this can be written as

$$^{(2)}R = 2\kappa_1\kappa_2 . \tag{4.53}$$

This gives the connection between intrinsic and extrinsic geometry, and its exact form holds because the embedding three-dimensional space is flat. If this embedding space has Lorentzian signature, one gets instead

$$^{(2)}R = -2\kappa_1\kappa_2 . \tag{4.54}$$

Kuchař (1993) expresses this in the form that the 'law of the instant' (because the hypersurface refers to $t = $ constant) implies the 'dynamical law' (expressing the flatness of the whole embedding space–time).

 In 3+1 dimensions the situation is more complicated. Addressing the vacuum Einstein equations,[11] $G_{\mu\nu} = 0$, one finds for its 'space–time component'

$$0 = h^\mu{}_\rho G_{\mu\nu} n^\nu = h^\mu{}_\rho R_{\mu\nu} n^\nu ,$$

which can be rewritten with the help of (4.51) as

$$D_b K^b{}_a - D_a K = 0 . \tag{4.55}$$

For the 'time–time component' one has

$$0 = G_{\mu\nu} n^\mu n^\nu = R_{\mu\nu} n^\mu n^\nu + \frac{R}{2} . \tag{4.56}$$

---

[11] The cosmological constant here is neglected for simplicity.

From (4.49) one finds upon contraction of indices,

$$^{(3)}R + K_\mu{}^\mu K_\nu{}^\nu - K_{\mu\nu}K^{\mu\nu} = h^{\mu\mu'} h_\nu{}^{\nu'} h_\mu{}^{\lambda'} h^\nu{}_{\rho'} R_{\mu'\nu'\lambda'}{}^{\rho'} \; .$$

The right-hand side is equal to

$$R + 2R_{\mu\nu}n^\mu n^\nu = 2G_{\mu\nu}n^\mu n^\nu \; ,$$

and so the 'time–time component' of Einstein's equations reads

$$K^2 - K_{ab}K^{ab} + {}^{(3)}R = 0 \; . \tag{4.57}$$

This is the (3+1)-dimensional version of the *theorema egregium*. Both (4.55) and (4.57) are *constraints*—they only contain first-order time derivatives. These constraints play a crucial role in the initial value formulation of classical GR, see for example, Choquet-Bruhat and York (1980) for a review. There, one can specify initial data (h_{ab}, K_{cd}) on Σ, where h_{ab} and K_{cd} satisfy the constraints (4.55) and (4.57). One can then prove that there exists one globally hyperbolic space–time obeying Einstein's equations (i.e. a unique solution for the four-metric up to diffeomorphisms), which has a Cauchy surface on which the induced metric and the extrinsic curvature are just h_{ab} and K_{cd}, respectively.

In electrodynamics, for comparison, one has to specify $\mathbf{A}$ and $\mathbf{E}$ on Σ satisfying the constraint (Gauß' law (4.30)) $\nabla\mathbf{E} = 0$. One then gets in space–time a solution of Maxwell's equations that is unique up to gauge transformation. The important point is that the space–time is fixed in this case, whereas in the gravitational case it is part of the solution.

That the dynamical laws follow from the laws of the instant can be inferred from the validity of the following 'interconnection theorems', cf. Kuchař (1981),

1. If the constraints are valid on an initial hypersurface and if $G_{ab} = 0$ (pure spatial components of the vacuum Einstein equations) on space–time, the constraints hold on *every* hypersurface.

2. If the constraints hold on every hypersurface, the equations $G_{ab} = 0$ hold on space–time.

In order to reformulate the Einstein–Hilbert action (1.1), one has to express the volume element and the Ricci scalar in terms of h_{ab} and K_{cd}. For the volume element one finds

$$\sqrt{-g} = N\sqrt{h} \; . \tag{4.58}$$

This can be seen as follows. Defining the three-dimensional volume element as (see e.g. Wald 1984)

$$^{(3)}e_{\mu\nu\lambda} = e_{\rho\mu\nu\lambda}t^\rho \; ,$$

with t^ρ according to (4.40) and $e_{\rho\mu\nu\lambda}$ denoting the time-independent four-dimensional volume element, one has by using $\epsilon_{\rho\mu\nu\lambda} = \sqrt{-g}\,e_{\rho\mu\nu\lambda}$,

$$\epsilon_{\rho\mu\nu\lambda}t^{\rho} = \sqrt{-g}\, e_{\mu\nu\lambda} = \sqrt{-\frac{g}{h}}\epsilon_{\mu\nu\lambda} \ ,$$

from which (4.58) follows after using (4.40) and taking purely spatial components. Eqn (4.58) can also be found from (4.42).

We shall now assume in the following that Σ is compact without boundary; the boundary terms for the non-compact case will be discussed separately in Section 4.2.4. In order to rewrite the curvature scalar, we use first (4.56) in the following form,

$$R = {}^{(3)}R + K^2 - K_{ab}K^{ab} - 2R_{\mu\nu}n^{\mu}n^{\nu} \ . \tag{4.59}$$

Using the definition of the Riemann tensor in terms of second covariant derivatives,

$$R^{\rho}{}_{\mu\rho\nu}n^{\mu} = \nabla_{\rho}\nabla_{\nu}n^{\rho} - \nabla_{\nu}\nabla_{\rho}n^{\rho} \ ,$$

the second term on the right-hand side can be written as

$$\begin{aligned} -2R_{\mu\nu}n^{\mu}n^{\nu} &= 2(\nabla_{\rho}n^{\nu})(\nabla_{\nu}n^{\rho}) - 2\nabla_{\rho}(n^{\nu}\nabla_{\nu}n^{\rho}) \\ &\quad -2(\nabla_{\nu}n^{\nu})(\nabla_{\rho}n^{\rho}) + 2\nabla_{\nu}(n^{\nu}\nabla_{\rho}n^{\rho}) \ . \end{aligned} \tag{4.60}$$

The second and fourth term are total divergences. They can thus be cast into surface terms at the temporal boundaries. The first surface term yields $-2(n^{\nu}\nabla_{\nu}n^{\rho})n_{\rho} = 0$, while the second one gives $2\nabla_{\mu}n^{\mu} = -2K$ (recall (4.45)). The two remaining terms in (4.60) can be written as $2K_{ab}K^{ab}$ and $-2K^2$, respectively. Inspecting the Einstein–Hilbert action (1.1), one recognizes that the temporal surface term is cancelled, and that the action now reads

$$\begin{aligned} 16\pi G\, S_{\mathrm{EH}} &= \int_{\mathcal{M}} \mathrm{d}t\mathrm{d}^3x\ N\sqrt{h}(K_{ab}K^{ab} - K^2 + {}^{(3)}R - 2\Lambda) \\ &\equiv \int_{\mathcal{M}} \mathrm{d}t\mathrm{d}^3x\ N\left(G^{abcd}K_{ab}K_{cd} + \sqrt{h}[\,{}^{(3)}R - 2\Lambda]\right) \ , \end{aligned} \tag{4.61}$$

where in the second line DeWitt's metric (4.25) was introduced. The action (4.61) is also called the 'ADM action' in recognition of the work by Arnowitt, Deser, and Misner, see Arnowitt *et al.* (1962). It has the classic form of kinetic energy minus potential energy, since the extrinsic curvature contains the 'velocities' $\dot{h}_{ab}$, see (4.48). Writing

$$S_{\mathrm{EH}} \equiv \int_{\mathcal{M}} \mathrm{d}t\mathrm{d}^3x\ \mathcal{L}^{\mathrm{g}} \ ,$$

one gets for the canonical momenta the following expressions. First,

$$p_N \equiv \frac{\partial \mathcal{L}^{\mathrm{g}}}{\partial \dot{N}} = 0 \ , \quad p_a \equiv \frac{\partial \mathcal{L}^{\mathrm{g}}}{\partial \dot{N}^a} = 0 \ . \tag{4.62}$$

Because lapse function and shift vector are only Lagrange multipliers (similar to A^0 in electrodynamics), these are constraints (called 'primary constraints'

according to Dirac (1964), since they do not involve the dynamical equations).
Second,

$$p^{ab} \equiv \frac{\partial \mathcal{L}^g}{\partial \dot{h}_{ab}} = \frac{1}{16\pi G} G^{abcd} K_{cd} = \frac{\sqrt{h}}{16\pi G} \left(K^{ab} - K h^{ab} \right) . \tag{4.63}$$

Note that the gravitational constant G appears here explicitly, although no coupling to matter is involved. This is the reason why it will appear in vacuum quantum gravity; see Section 5.2. One therefore has the Poisson-bracket relation[12]

$$\{h_{ab}(x), p^{cd}(y)\} = \delta^c_{(a} \delta^d_{b)} \delta(x, y) . \tag{4.64}$$

Recalling (4.48) and taking the trace of (4.63), one can express the velocities in terms of the momenta,

$$\dot{h}_{ab} = \frac{32\pi G N}{\sqrt{h}} \left(p_{ab} - \frac{1}{2} p h_{ab} \right) + D_a N_b + D_b N_a , \tag{4.65}$$

where $p \equiv p^{ab} h_{ab}$. One can now calculate the canonical Hamiltonian density

$$\mathcal{H}^g = p^{ab} \dot{h}_{ab} - \mathcal{L}^g ,$$

for which one gets the expression[13]

$$\mathcal{H}^g = 16\pi G N G_{abcd} p^{ab} p^{cd} - N \frac{\sqrt{h} (\,^{(3)}R - 2\Lambda)}{16\pi G} - 2N_b (D_a p^{ab}) . \tag{4.66}$$

The full Hamiltonian is found by integration,

$$H^g \equiv \int d^3x \, \mathcal{H}^g \equiv \int d^3x \, (N \mathcal{H}^g_\perp + N^a \mathcal{H}^g_a) . \tag{4.67}$$

The action (4.61) can be written in the form

$$16\pi G \, S_{EH} = \int d^3x \, \left(p^{ab} \dot{h}_{ab} - N \mathcal{H}^g_\perp - N^a \mathcal{H}^g_a \right) . \tag{4.68}$$

Variation with respect to the Lagrange multipliers N and N^a yields the constraints[14]

$$\mathcal{H}^g_\perp = 16\pi G \, G_{abcd} p^{ab} p^{cd} - \frac{\sqrt{h}}{16\pi G} (\,^{(3)}R - 2\Lambda) \approx 0 , \tag{4.69}$$

$$\mathcal{H}^g_a = -2 D_b p_a{}^b \approx 0 . \tag{4.70}$$

In fact, the constraint (4.69) is equivalent to (4.57), and (4.70) is equivalent to (4.55)—they are called *Hamiltonian constraint* and *diffeomorphism (or momentum) constraint*, respectively. From its structure, (4.69) has some similarity to

---

[12]This is formal at this stage since it does not take into account that $\sqrt{h} > 0$.

[13]This holds modulo a total divergence which does not contribute in the integral because Σ is compact.

[14]These also follow from the preservation of the primary constraints, $\{p_N, H^g\} = 0 = \{p_a, H^g\}$.

the constraint for the relativistic particle, Eqn (3.24), while (4.70) is similar to (4.30). It can now be seen explicitly that these constraints are equivalent to the results from the 'seventh route to geometrodynamics', see (4.21) and (4.22). The total Hamiltonian is thus constrained to vanish, a result that is in accordance with our general discussion of reparametrization invariance of Section 3.1. In the case of non-compact space, boundary terms are present in the Hamiltonian; see Section 4.2.4.

In addition to the constraints, one has the six dynamical equations, the Hamiltonian equations of motion. The first half, $\dot{h}_{ab} = \{h_{ab}, H^g\}$, just gives (4.65). The second half, $\dot{p}^{ab} = \{p_{ab}, H^g\}$, yields a lengthy expression (see e.g. Wald 1984) that is not needed in the following.

If non-gravitational fields are coupled, the constraints acquire extra terms. In (4.56) one has to use that

$$2G_{\mu\nu}n^\mu n^\nu = 16\pi G T_{\mu\nu}n^\mu n^\nu \equiv 16\pi G\rho .$$

Instead of (4.69) one now has the following expression for the Hamiltonian constraint,

$$\mathcal{H}_\perp = 16\pi G\, G_{abcd}p^{ab}p^{cd} - \frac{\sqrt{h}}{16\pi G}(\,^{(3)}R - 2\Lambda) + \sqrt{h}\rho \approx 0 . \tag{4.71}$$

Similarly, one has instead of (4.70) for the diffeomorphism constraints,

$$\mathcal{H}_a = -2D_b p_a{}^b + \sqrt{h}J_a \approx 0 , \tag{4.72}$$

where $J_a \equiv h_a{}^\mu T_{\mu\nu}n^\nu$ is the 'Poynting vector'. Consider as special examples the cases of a scalar field and the electromagnetic field. With the Lagrange density

$$\mathcal{L} = \sqrt{-g}\left(-1/2\ g^{\mu\nu}\phi_{,\mu}\phi_{,\nu} - 1/2\ m^2\phi^2\right) \tag{4.73}$$

for the scalar field one finds for its Hamiltonian

$$H_\phi = \int d^3x\ N\left(\frac{p_\phi^2}{2\sqrt{h}} + \frac{\sqrt{h}}{2}h^{ab}\phi_{,a}\phi_{,b} + \frac{1}{2}\sqrt{h}m^2\phi^2\right)$$
$$+ \int d^3x\ N^a p_\phi \phi_{,a} . \tag{4.74}$$

The term in parentheses in the first integral has to be added to $\mathcal{H}_\perp^g$, while the term $p_\phi\phi_{,a}$ (which we have already encountered in (4.14)) must be added to $\mathcal{H}_a^g$. For the electromagnetic field, starting from

$$\mathcal{L} = -1/4\ \sqrt{-g}g^{\mu\rho}g^{\nu\sigma}F_{\mu\nu}F_{\rho\sigma} , \tag{4.75}$$

one gets for the Hamiltonian

$$H_{\mathrm{EM}} = \int d^3x\ N\left(\frac{h_{ab}}{2\sqrt{h}}p^a p^b + \frac{\sqrt{h}}{2}h^{ab}B_a B_b\right)$$

$$+ \int d^3x \; N^a \left(\partial_a A_b - \partial_b A_a \right) p^b - \int d^3x \; A_0 \partial_a p^a \; , \qquad (4.76)$$

where $p^a = \partial \mathcal{L}/\partial \dot{A}_a$ is the electric field and $B_a = (1/2)\epsilon_{abc}F^{bc}$, the magnetic field. Note that the term in parentheses in the second integral is just the $\mathcal{H}_a$ from (4.28), and variation with respect to the Lagrange multiplier A_0 yields Gauß' law (4.30). If fermionic degrees of freedom are present, one must perform the 3+1 decomposition with respect to the vierbein instead of the four-metric; see for example, Ashtekar (1991). The classical canonical formalism for the gravitational field as discussed up to now was pioneered by Peter Bergmann, Paul Dirac, 'ADM', and others in the 1950s. For a historical account and references, see for example, Bergmann (1989) and Rovelli (2000).

4.2.3 Discussion of the constraints

The presence of the constraints derived in the last subsection means that only part of the variables constitute physical degrees of freedom (cf. Section 3.1.2). How may one count them? The three-metric $h_{ab}(x)$ is characterized by six numbers per space point (often symbolically denoted as $6 \times \infty^3$). The diffeomorphism constraints (4.70) generate coordinate transformations on three-space. These are characterized by three numbers, so $6 - 3 = 3$ numbers per point remain. The constraint (4.69) corresponds to one variable per space point describing the location of Σ in space–time (since Σ changes under normal deformations). In a sense, this one variable therefore corresponds to 'time', and $2 \times \infty^3$ degrees of freedom remain. Baierlein et al. (1962) have interpreted this as the 'three-geometry carrying information about time'.

The gravitational field thus seems to be characterized by $2 \times \infty^3$ intrinsic degrees of freedom. This is consistent with the corresponding number found in linear quantum gravity—the two spin-2 states of the graviton (Section 2.1). One can alternatively perform the following counting in phase space: the canonical variables $(h_{ab}(x), p^{cd}(y))$ are $12 \times \infty^3$ variables. Due to the presence of the four constraints in phase space, $4 \times \infty^3$ variables have to be subtracted. The remaining $8 \times \infty^3$ variables define the constraint hypersurface Γ_c. Since the constraints generate a four-parameter set of gauge transformations on Γ_c (see Section 3.1.2), $4 \times \infty^3$ degrees of freedom must be subtracted in order to 'fix the gauge'. The remaining $4 \times \infty^3$ variables define the reduced phase space Γ_r and correspond to $2 \times \infty^3$ degrees of freedom in configuration space—in accordance with the counting above. It must be emphasized that this counting always holds modulo a *finite* number of degrees of freedom—an example is gravity in 2+1 dimensions (Section 8.1.3).

Does a three-dimensional geometry indeed contain information about time? Consider a situation in non-gravitational physics, for example, electrodynamics. There the specification of the, say, magnetic field on two hypersurfaces does not suffice to determine the field everywhere. In addition, the time parameters of the two surfaces must be specified for an appropriate boundary-value problem. In contrast to the gravitational case, the background space–time is fixed here (i.e., it is non-dynamical). As we have seen in Section 3.4, two configurations (e.g. of a clock) in classical mechanics do not suffice to determine the motion—one needs in addition the two times of the clock configurations or its speed.

The situation in the gravitational case is related to the 'sandwich conjecture'. This states that two three-geometries do (in the generic case) determine the temporal separation (the proper times) along each time-like worldline connecting them in the resulting space–time. Whereas not much is known about the finite version of this conjecture, results are available for the infinitesimal case. In this 'thin-sandwich conjecture', one specifies on one hypersurface the three-metric h_{ab} and its 'time derivative' $\partial h_{ab}/\partial t$—the latter is only required up to a numerical factor, since the 'speed' itself is meaningless; only the 'direction' in configuration space is of significance. The thin-sandwich conjecture holds if one can determine from these initial conditions lapse and shift from the constraints.[15] It has been shown that this can be done locally for 'generic' situations; see Bartnik and Fodor (1993) for pure gravity and Giulini (1999) for gravity plus matter.[16]

The 'temporal' degree of freedom of the three-geometry cannot in general be separated from other variables, that is, all three degrees of freedom contained in h_{ab} (after the diffeomorphism constraints have been considered) should be interpreted as physical variables, and be treated on equal footing. In the special case of linear gravity (Section 2), a background structure is present. This enables one to separate a distinguished time and to regard the remaining variables, the two degrees of freedom of the graviton, as the only physical variables. Torre (1993) has shown that GR cannot globally be equivalent to a deparametrized theory in the sense of Section 3.3, that is, no distinguished time variable is available.

Starting from the ADM action (4.61), one may find an alternative formulation by first varying the action with respect to N and then inserting the ensuing solution back into it. This corresponds to the solution of the Hamiltonian constraint. Writing for the DeWitt metric

$$G^{abcd} \equiv \sqrt{h}\mathcal{G}^{abcd} ,$$

a variation with respect to N yields (for $\Lambda = 0$)

$$N = \frac{1}{2}\sqrt{\frac{\mathcal{G}^{abcd}(\dot{h}_{ab} - 2D_{(a}N_{b)})(\dot{h}_{cd} - 2D_{(c}N_{d)})}{{}^{(3)}R}} . \qquad (4.77)$$

This corresponds to the solution $N = \sqrt{T/ - V}$ in the Barbour–Bertotti model (3.105). Re-inserting (4.77) into (4.61), one finds the 'Baierlein–Sharp–Wheeler (BSW)' form of the action,

---

[15] This boundary-value problem must be distinguished from the one above where h_{ab} and p^{cd} are specified on Σ and a space–time can be chosen after lapse and shift are *freely* chosen.

[16] The condition is that the initial speed must have at each space point a negative square with respect to the DeWitt metric.

$$16\pi G S_{\text{BSW}} = \int \mathrm{d}t\mathrm{d}^3x \sqrt{h}\sqrt{^{(3)}R\mathcal{G}^{abcd}(\dot{h}_{ab} - 2D_{(a}N_{b)})(\dot{h}_{cd} - 2D_{(c}N_{d)})} \; . \quad (4.78)$$

A justification of S_{BSW} from a 'Machian' viewpoint can be found in Barbour *et al.* (2002). This action resembles the Jacobi-type action (3.105), but is much more sophisticated. Now only the shift functions N^a have to be varied. If a unique solution existed, one could employ the thin-sandwich conjecture, that is, one could construct the space–time out of initial data h_{ab} and $\dot{h}_{ab}$. This procedure corresponds to the 'horizon stacking' mentioned in Section 3.4. In such an approach, GR can be derived without any prior assumptions of a space–time nature such as the path-independence embeddability derivation described in Section 4.1. It is not necessary to require the constraints to close in the specific manner of the algebra of surface deformations (3.90)–(3.92). Mere closure in any fashion is sufficient to derive GR, universality of the lightcone and gauge theory (Barbour *et al.* 2002).

In this connection, Barbour (1994) argues that the Hamiltonian constraint does not further restrict the number of physical variables (i.e. does not restrict them from $3 \times \infty^3$ to $2 \times \infty^3$), but is an identity that reflects the fact that only the *direction* of the initial velocity matters, not its absolute value. In the model of Section 3.4, one has for the canonical momentum the expression

$$\mathbf{p}_k = \sqrt{-V} \frac{m_k \dot{\mathbf{x}}_k}{\sqrt{\frac{1}{2}\sum_k m_k \dot{\mathbf{x}}_k^2}} \; , \quad (4.79)$$

leading to the constraint (3.108). The second factor in (4.79) describes 'direction cosines', their usual relation—the sum of their squares is equal to one—being equivalent to (3.108). Here one finds from (4.78) for the momentum[17]

$$p_{\text{BSW}}^{ab} = \frac{\sqrt{h}}{16\pi G}\sqrt{\frac{^{(3)}R}{T}}\mathcal{G}^{abcd}(\dot{h}_{cd} - 2D_{(c}N_{d)}) \; , \quad (4.80)$$

where the 'kinetic term' is given by

$$T = \mathcal{G}^{abcd}(\dot{h}_{ab} - 2D_{(a}N_{b)})(\dot{h}_{cd} - 2D_{(c}N_{d)}) \; . \quad (4.81)$$

Similar to (4.79), these momenta define direction cosines which due to the infinite dimensions of configuration space are 'local' direction cosines.

The somewhat ambiguous nature of the Hamiltonian constraint (4.69) leads to the question whether it really generates gauge transformations. The answer should be 'yes' in view of the general fact (Section 3.1.2) that first-class constraints have this property. On the other hand, $\mathcal{H}_\perp^{\text{g}}$ is also responsible for the time evolution, mediating between different hypersurfaces. Can this time evolution be interpreted as the 'unfolding' of a gauge transformation? This is indeed

---

[17]This is of course equal to p^{ab} after the identification (4.77) for the lapse has been made.

possible because the presence of the constraint $\mathcal{H}_\perp^g \approx 0$ expresses the fact that evolutions along different foliations are equivalent.

A related issue concerns the notion of an 'observable'. This was defined in Section 3.1.2 as a variable that weakly commutes with the constraints. In the present situation, an observable $\mathcal{O}$ should then satisfy

$$\{\mathcal{O}, \mathcal{H}_a^g\} \approx 0 \ , \tag{4.82}$$

$$\{\mathcal{O}, \mathcal{H}_\perp^g\} \approx 0 \ . \tag{4.83}$$

While the first condition is certainly reasonable (observables should not depend on the chosen coordinates of Σ), the situation is not so clear for the second condition. To emphasize the difference between both equations, Kuchař (1993) refers to quantities obeying (4.82) already as observables and to variables that obey *in addition* (4.83) as 'perennials'. Since perennials weakly commute with the full Hamiltonian, they are constants of motion. The insistence on perennials as the only allowed quantities in the theory would correspond to an (unphysical) 'outside' view of the world and would not be in conflict with our experience of the world evolving in time. The latter arises from tracking one part of the variables with respect to the remaining part ('inside view'). There can thus be a sensible notion of time evolution with respect to intrinsic observers. We shall say more about these interpretational issues in the context of the quantum theory; see Chapter 5.

We have already seen in Section 4.1 that the transformations generated by the constraints (4.69) and (4.70) are different from the original space–time diffeomorphisms of GR. The formal reason is that $\mathcal{H}_\perp^g$ is non-linear in the momenta, so the transformations in the phase space Γ spanned by (h_{ab}, p^{cd}) cannot be reduced to space–time transformations. What, then, is the relation between both types of transformations? Let $(\mathcal{M}, g)$ be a globally hyperbolic space–time. We shall denote by Riem $\mathcal{M}$ the space of all (pseudo-) Riemannian metrics on $\mathcal{M}$. Since the group of space–time diffeomorphisms, Diff $\mathcal{M}$, does not act transitively, there exist non-trivial orbits in Riem $\mathcal{M}$. One can make a projection down to the space of all four-geometries, Riem $\mathcal{M}/$Diff $\mathcal{M}$. By considering a particular section,

$$\sigma : \text{Riem } \mathcal{M}/\text{Diff } \mathcal{M} \mapsto \text{Riem } \mathcal{M} \ , \tag{4.84}$$

one can choose a particular representative metric on $\mathcal{M}$ for each geometry. In this way one can define formal points of the 'background manifold' $\mathcal{M}$, which a priori have no meaning (in GR, points cannot be disentangled from the metric fields). The map between different sections is *not* a single diffeomorphism, but a more complicated transformation (an element of the 'Bergmann–Komar group', see Bergmann and Komar 1972). Hájíček and Kijowski (2000) have shown (see also Hájíček and Kiefer 2001 *a* and Section 7.2) that there exists a map from

$$\text{Riem } \mathcal{M}/\text{Diff } \mathcal{M} \times \text{Emb}(\Sigma, \mathcal{M}) \ ,$$

where $\text{Emb}(\Sigma, \mathcal{M})$ denotes the space of all embeddings of Σ into $\mathcal{M}$, into the phase space Γ but *excluding* points where the constructed space–times admit an isometry. Therefore, the identification between space–time diffeomorphisms and the transformations in phase space proceeds via whole 'histories'. The necessary exclusion of points representing Cauchy data for space–times with Killing vectors from Σ is one of the reasons why GR cannot be equivalent globally to a deparametrized theory (Torre 1993).

One interesting limit for the Hamiltonian constraint (4.69) is the 'strong-coupling limit' defined by setting formally $G \to \infty$. This is the limit opposite to the weak-coupling expansion of Chapter 2. It also corresponds formally to the limit $c \to 0$, that is, the limit opposite to the Galileian case of infinite speed of light. This can be seen by noting that the constant in front of the potential term in (4.69) in fact reads $c^4/16\pi G$. Therefore, in this limit, the lightcones effectively collapse to the axes $x =$ constant; different spatial point decouple because all spatial derivative terms being present in $^{(3)}R$ have disappeared. One can show that this situation corresponds to having a 'Kasner universe' at each space point; see Pilati (1982, 1983) for details. Since the potential term also carries the signature σ, this limit also corresponds formally to $\sigma = 0$, that is, the Poisson bracket between the Hamiltonian constraint (3.90) becomes zero.

4.2.4 The case of open spaces

Up to now we have neglected the presence of possible spatial boundary terms in the Hamiltonian. In this subsection, we shall briefly discuss the necessary modifications for the case of open spaces (where 'open' means 'asymptotically flat'). The necessary details can be found in Regge and Teitelboim (1974) and Beig and Ó Murchadha (1987).

Variation of the full Hamiltonian H^g with respect to the canonical variables h_{ab} and p^{cd} yields

$$\delta H^g = \int d^3x \left(A^{ab}\delta h_{ab} + B_{ab}\delta p^{ab} \right) - \delta C , \qquad (4.85)$$

where δC denotes surface terms. Because H^g must be a differentiable function with respect to h_{ab} and p^{cd} (otherwise Hamilton's equations of motion would not make sense), δC must be *cancelled* by introducing explicit surface terms to H^g. For the derivation of such surface terms, one must impose fall-off conditions for the canonical variables. For the three-metric they read

$$h_{ab} \overset{r\to\infty}{\sim} \delta_{ab} + \mathcal{O}\left(\frac{1}{r}\right) , \quad h_{ab,c} \overset{r\to\infty}{\sim} \mathcal{O}\left(\frac{1}{r^2}\right) , \qquad (4.86)$$

and analogously for the momenta. The lapse and shift, if again combined to the four-vector N^μ, are supposed to obey

$$N^\mu \overset{r\to\infty}{\sim} \alpha^\mu + \beta^\mu_a x^a , \qquad (4.87)$$

where α^μ describe space–time translations, $\beta_{ab} = -\beta_{ba}$ spatial rotations, and $\beta^\perp_a$ boosts. Together, they form the Poincaré group of Minkowski space–time, which is a symmetry in the asymptotic sense. The procedure mentioned above then leads to the following expression for the total Hamiltonian,

$$H^g = \int d^3x \ (N\mathcal{H}_\perp^g + N^a\mathcal{H}_a^g) + \alpha E_{ADM} - \alpha^a P_a + 1/2 \ \beta_{\mu\nu} M^{\mu\nu} \ , \qquad (4.88)$$

where E_{ADM} (also called 'ADM' energy, see Arnowitt *et al.* 1962), P_a, and $M^{\mu\nu}$ are the total energy, the total momentum, and the total angular momentum plus the generators of boosts, respectively. Together they form the generators of the Poincaré group at infinity. For the ADM energy, in particular, one finds the expression

$$E_{ADM} = \frac{1}{16\pi G} \oint_{r\to\infty} d^2\sigma_a(h_{ab,b} - h_{bb,a}) \ . \qquad (4.89)$$

Note that the total energy is defined by a surface integral over a sphere for $r \to \infty$ and *not* by a volume integral. One can prove that $E_{ADM} \geq 0$.

The integral in (4.88) is the same integral as in (4.67). Because of the constraints (4.69) and (4.70), H^g is numerically equal to the surface terms. For vanishing asymptotic shift and lapse equal to one, it is just given by the ADM energy.

4.2.5 *Structure of configuration space*

An important preparation for the quantum theory is an investigation into the structure of the configuration space because this will be the space on which the wave functional will be defined.

We have seen that the canonical formalism deals with the set of all three-metrics on a given manifold Σ. We call this space Riem Σ (not to be confused with Riem $\mathcal{M}$ considered above). As configuration space of the theory we want to address the quotient space in which all metrics corresponding to the same three-geometry are identified. Following Wheeler (1968) we call this space 'superspace'. It is defined by

$$\mathcal{S}(\Sigma) \equiv \text{Riem } \Sigma/\text{Diff } \Sigma \ . \qquad (4.90)$$

By going to superspace the momentum constraints (4.70) are automatically fulfilled.

In general, Diff Σ can be divided into a 'symmetry part' and a 'redundancy part' (Giulini 1995a). Symmetries arise typically in the case of asymptotically flat spaces (Section 4.2.4). They describe, for example, rotations with respect to the rest of the universe ('fixed stars'). Since they have physical significance, they should not be factored out, and Diff Σ is then understood to contain only the 'true' diffeomorphisms (redundancies). In the closed case (relevant in particular for cosmology) only the redundancy part is present.

For closed Σ, $\mathcal{S}(\Sigma)$ has a non-trivial singularity structure due to the occurrence of metrics with isometries (Fischer 1970); at such singular points, superspace is not a manifold (a situation like e.g. at the tip of a cone). A proposal to avoid such singularities employing a 'resolution space' $\mathcal{S}_R(\Sigma)$ was made by DeWitt (1970).

In the open case, one can perform a 'one-point compactification', $\bar{\Sigma} \equiv \Sigma \cup \{\infty\}$. The corresponding superspace is then defined as

$$\mathcal{S}(\Sigma) \equiv \text{Riem } \bar{\Sigma}/\mathcal{D}_\text{F}(\bar{\Sigma}) , \tag{4.91}$$

where $\mathcal{D}_\text{F}(\bar{\Sigma})$ are all diffeomorphisms that fix the frames at infinity. The open and the closed case are closer to each other than expected. One can show (Fischer 1986) that $\mathcal{S}_\text{R}(\bar{\Sigma})$ and $\mathcal{S}(\Sigma)$ are diffeomorphic. For this reason, one can restrict oneself for topological investigations to the former space (Giulini 1995a).

The DeWitt metric G^{abcd}, see (4.25), plays the role of a metric on Riem Σ,

$$G(h, k) \equiv \int_\Sigma \mathrm{d}^3x \, G^{abcd} h_{ab} k_{cd} . \tag{4.92}$$

Due to its symmetry properties, it can formally be considered as a symmetric 6×6-matrix at each space point (DeWitt 1967a). At each point this matrix can therefore be diagonalized, and the signature turns out to read

$$\text{diag}(-, +, +, +, +, +) .$$

A quantity with similar properties is known in elasticity theory—the fourth-rank elasticity tensor c^{abcd} possesses the same symmetry properties as DeWitt's metric and, therefore, has (in three spatial dimensions) 21 independent components; see e.g. Marsden and Hughes (1983).

It must be emphasized that the negative sign in DeWitt's metric has nothing to do with the Lorentzian signature of space–time; in the Euclidean case, the minus sign stays and only the relative sign between potential and kinetic term will change ($\sigma = 1$ instead of -1 in (4.24)). Due to the presence of this minus sign, the kinetic term for the gravitational field is *indefinite*. To gain a deeper understanding of its meaning, consider the following class of (generalized) DeWitt metrics which exhaust (up to trivial transformations) the set of all ultralocal metrics (DeWitt 1967a),

$$G_\beta^{abcd} = \frac{\sqrt{h}}{2} \left(h^{ac} h^{bd} + h^{ad} h^{bc} - 2\beta h^{ab} h^{cd} \right) , \tag{4.93}$$

where β is any real number (the GR-value is $\beta = 1$). Its inverse is then given by

$$G_{abcd}^\alpha = \frac{1}{2\sqrt{h}} \left(h_{ac} h_{bd} + h_{ad} h_{bc} - 2\alpha h_{ab} h_{cd} \right) , \tag{4.94}$$

where

$$\alpha + \beta = 3\alpha\beta \tag{4.95}$$

(in GR, $\alpha = 0.5$).[18] What would be the meaning of the constraints $\mathcal{H}_\perp^\text{g}$ and $\mathcal{H}_a^\text{g}$, if the generalized metric (4.93) were used? Section 4.1 shows that for $\beta \neq 1$ the

[18]In d space dimensions, one has $\alpha + \beta = d\alpha\beta$.

principle of path independence must be violated, since the GR-value $\beta = 1$ follows uniquely from this principle. The theories defined by these more generalized constraints can thus not correspond to reparametrization-invariant ('covariant') theories at the Lagrangian level. In a sense, these would be genuinely Hamiltonian theories. One can perform the following coordinate transformation in Riem Σ,

$$\tau = 4\sqrt{|\beta - 1/3|}\,h^{1/4} \ , \quad \tilde{h}_{ab} = h^{-1/3}h_{ab} \ , \tag{4.96}$$

thus decomposing the three-metric into 'scale part' τ and 'conformal part' $\tilde{h}_{ab}$. The 'line element' in Riem Σ can then be written as

$$G_\beta^{abcd}\mathrm{d}h_{ab} \otimes \mathrm{d}h_{cd} = -\mathrm{sgn}\,(\beta - 1/3)\,\mathrm{d}\tau \otimes \mathrm{d}\tau + \frac{\tau^2}{16|\beta - 1/3|}\mathrm{tr}\left(\tilde{h}^{-1}\mathrm{d}\tilde{h} \otimes \tilde{h}^{-1}\mathrm{d}\tilde{h}\right) \tag{4.97}$$

(for the inverse metric entering (4.69) one must substitute β by α). It is evident that the line element becomes degenerate for $\alpha = 1/3$ (corresponding to $\alpha \to \infty$); for $\beta < 1/3$ it becomes positive definite, whereas for $\beta > 1/3$ it is indefinite (this includes the GR-case).

At each space point, G_β^{abcd} can be considered as a metric in the space of symmetric positive definite 3×3 matrices, which is isomorphic to $\mathbb{R}^6$. Thus,

$$\mathbb{R}^6 \cong \mathrm{GL}(3,\mathbb{R})/\mathrm{SO}\,(3) \cong \mathrm{SL}(3,\mathbb{R})/\mathrm{SO}\,(3) \times \mathbb{R}^+ \ . \tag{4.98}$$

The $\tilde{h}_{ab}$ are coordinates on $\mathrm{SL}(3,\mathbb{R})/\mathrm{SO}\,(3)$, and τ is the coordinate on $\mathbb{R}^+$. The relation (4.98) corresponds to the form (4.97) of the line element. All structures on $\mathrm{SL}(3,\mathbb{R})/\mathrm{SO}\,(3) \times \mathbb{R}^+$ can be transferred to Riem Σ, since G_β^{abcd} is ultralocal.

One can give an interpretation of one consequence of the signature change that occurs for $\beta = 1/3$ in (4.97). For this one calculates the acceleration of the three-volume $V = \int \mathrm{d}^3x\,\sqrt{h}$ (assuming it is finite) for $N = 1$. After some calculation, one finds the expression (Giulini and Kiefer 1994)

$$\frac{\mathrm{d}^2}{\mathrm{d}t^2}\int \mathrm{d}^3x\,\sqrt{h} = -3(3\alpha - 1)\int \mathrm{d}^3x\,\sqrt{h}\left(\frac{2}{3}\,{}^{(3)}R - 2\Lambda\right.$$
$$\left. -16\pi G\left[\mathcal{H}_\mathrm{m} - \frac{1}{3}h^{ab}\frac{\partial\mathcal{H}_\mathrm{m}}{\partial h^{ab}}\right]\right) \ . \tag{4.99}$$

We call gravity 'attractive' if the sign in front of the integral on the right-hand side is negative. This is because then

1. a positive ${}^{(3)}R$ contributes with a negative sign and leads to a *deceleration* of the three-volume;

2. a positive cosmological constant acts repulsively;

3. in the coupling to matter, an overall sign change corresponds to a sign change in G.[19]

---

[19] This does not of course necessarily mean that G changes sign in all relations.

In the Hamiltonian constraint (4.69), the inverse metric (4.94) enters. The critical value separating the positive definite from the indefinite case is thus $\alpha = 1/3$. One, therefore, recognizes that there is an intimate relation between the signature of the DeWitt metric and the attractivity of gravity: only for an indefinite signature is gravity attractive. From observations (primordial Helium abundance) one can estimate (Giulini and Kiefer 1994) that

$$0.4 \lesssim \alpha \lesssim 0.55 . \tag{4.100}$$

This is, of course, in accordance with the GR-value $\alpha = 0.5$.

We return now to the case of GR. The discussion so far concerns the metric on Riem Σ, given by the DeWitt metric (4.25). Does this metric also induce a metric on superspace (Giulini 1995b)? In Riem Σ, one can distinguish between 'vertical' and 'horizontal' directions. The vertical directions are the directions along the orbits generated by the three-dimensional diffeomorphisms. Metrics along a given orbit describe the same geometry. Horizontal directions are defined as being orthogonal to the orbits, where orthogonality holds with respect to the DeWitt metric. Since the latter is indefinite, the horizontal directions may also contain vertical directions (this happens in the 'light-like' case for zero norm). Calling V_h (H_h) the vertical (horizontal) subspace with respect to a given metric h_{ab}, one can show that

1. if $V_h \cap H_h = \{0\}$, then G^{abcd} can be projected to the horizontal subspace where it defines a metric;
2. if $V_h \cap H_h \neq \{0\}$, then there exist critical points in $\mathcal{S}(\Sigma)$ where the projected metric changes signature.

The task is then to classify the regions in Riem Σ according to these two cases. There exist some partial results (Giulini 1995b). For metrics obeying $^{(3)}R_{ab} = \lambda h_{ab}$ ('Einstein metrics'), one has $V_h \cap H_h = \{0\}$ and, consequently, a metric on $\mathcal{S}(\Sigma)$ exists. Moreover, one can show that for $^{(3)}R > 0$ (three-sphere) there remains only *one negative direction* in H_h out of the infinitely many negative directions in the DeWitt metric (in V_h there are infinitely many negative directions). The kinetic term in the Hamiltonian constraint is then of a truly hyperbolic nature (in contrast to the general hyperbolicity in the pointwise sense). This holds in particular in the vicinity of closed Friedmann universes which are therefore distinguished in this respect. A perturbation around homogeneity and isotropy has exhibited this property explicitly (Halliwell and Hawking 1985). It is an open question 'how far' one has to go from the hyperbolic case near $^{(3)}R > 0$ in order to reach the first points where the signature changes.

For Ricci-negative metrics[20] (i.e. all eigenvalues of $^{(3)}R_{ab}$ are negative), one finds that $V_h \cap H_h = \{0\}$, and that the projected metric on superspace contains infinitely many plus and minus signs. For flat metrics, one has $V_h \cap H_h \neq \{0\}$. Some of these results can explicitly be confirmed in Regge calculus (Williams 1997).

---

[20] Any Σ admits such metrics.

4.3   Canonical gravity with connections and loops

4.3.1   *The canonical variables*

One of the key ingredients in the canonical formalism is the choice of the symplectic structure, that is, the choice of the canonical variables. In the previous sections, we have chosen the three-metric h_{ab} and its momentum p^{cd}. In this section, we shall introduce different variables introduced by Ashtekar (1986) following earlier work by Sen (1982). These 'new variables' will exhibit their main power in the quantum theory; see Chapter 6. Since they are analogous to Yang–Mills variables (using connections), the name 'connection dynamics' is also used. A more detailed introduction into these variables can be found in Ashtekar (1988, 1991) and—taking into account more recent developments—Thiemann (2001).

The first step consists in the introduction of *triads* (or dreibeine). They will play the role of the canonical momentum. Similar to the tetrads (vierbeine) used in Section 1.1.4, they are given by the variables $e_i^a(x)$ which define an orthonormal basis at each space point. Here, $a = 1, 2, 3$ is the usual space index (referring to the tangent space $T_x(\Sigma)$ at x) and $i = 1, 2, 3$ are internal indices enumbering the vectors. The position of the internal indices is arbitrary. One has the orthonormality condition

$$h_{ab} e_i^a e_j^b = \delta_{ij} \ , \tag{4.101}$$

from which one gets

$$h^{ab} = \delta^{ij} e_i^a e_j^b \equiv e_i^a e^{bi} \ . \tag{4.102}$$

This introduces an SO(3) (or SU(2)) symmetry into the formalism, since the metric is invariant under local rotations of the triad. Associated with e_i^a is an orthonormal frame in the cotangent space $T_x^*(\Sigma)$, denoted by e_a^i (basis of one-forms). It obeys

$$e_a^i e_j^a = \delta_j^i \ , \quad e_a^i e_i^b = \delta_a^b \ . \tag{4.103}$$

The three-dimensional formalism using triads can be obtained from the corresponding space–time formalism by using the 'time gauge' $e_a^0 = -n_a = Nt_{,a}$ (Schwinger 1963) for the tetrads.

The variable of interest is not the triad itself, but its densitized version (because it will play the role of the momentum),

$$E_i^a(x) = \sqrt{h} e_i^a(x) \ , \tag{4.104}$$

where from (4.101) one has $\sqrt{h} = \det(e_a^i)$.

To find the canonically conjugate momentum, consider first the extrinsic curvature in the form

$$K_a^i(x) \equiv K_{ab}(x) e^{bi}(x) \ , \tag{4.105}$$

where $K_{ab}(x)$ denotes the previous expression for the extrinsic curvature; cf. (4.45) and (4.48). One can show that K_a^i is canonically conjugate to E_i^a,

$$K_a^i \delta E^{ia} = \frac{K_{ab}}{2\sqrt{h}} \delta \left(E^{ia} E^{ib} \right) = -\frac{\sqrt{h}}{2} \left(K^{ab} - K h^{ab} \right) \delta h_{ab} = -8\pi G p^{ab} \delta h_{ab} \ ,$$

where (4.102) has been used. The SO(3)-rotation connected with the introduction of the triads is generated by the constraint

$$\mathcal{G}_i(x) \equiv \epsilon_{ijk} K_a^j(x) E^{ka}(x) \approx 0 \ , \tag{4.106}$$

which has the structure of '$\mathbf{x} \times \mathbf{p}$' (generator of translations). It is called 'Gauß constraint'. Its presence also guarantees the symmetry of K_{ab}. (This can be seen by inserting (4.105) into (4.106), multiplying with ϵ^{ilm} and contracting.)

An arbitrary vector field can be decomposed with respect to the triad as

$$v^a = v^i e_i^a \ . \tag{4.107}$$

The covariant derivative with respect to internal indices is defined by

$$D_a v^i = \partial_a v^i + \omega_a{}^i{}_j v^j \ , \tag{4.108}$$

where $\omega_a{}^i{}_j$ are the components of the 'spin connection'; cf. also Section 1.1.4. For vanishing torsion one has the following relation between the spin connection and the Levi–Cività connection,

$$\omega_a{}^i{}_j = \Gamma_{kj}^i e_a^k \ , \tag{4.109}$$

where Γ_{kj}^i are the components of the Levi–Cività connection with respect to the triads. The usual coordinate components are found from

$$\Gamma_{kj}^i = e_k^d e_j^f e_c^i \Gamma_{df}^c - e_k^d e_j^f \partial_d e_f^i \ . \tag{4.110}$$

Inserting (4.109) and (4.110) into

$$D_a e_b^i = \partial_a e_b^i - \Gamma_{ab}^c e_c^i + \omega_a{}^i{}_j e_b^j \ ,$$

one finds the covariant constancy of the triads,

$$D_a e_b^i = 0 \ , \tag{4.111}$$

in analogy to $D_a h_{bc} = 0$. Parallel transport is defined by

$$dv^i = -\omega_a{}^i{}_j v^j dx^a \ .$$

Defining

$$\Gamma_a^i = -1/2 \, \omega_{ajk} \epsilon^{ijk} \ , \tag{4.112}$$

this parallel transport corresponds to the infinitesimal rotation of the vector v^i by an angle

$$\delta \omega^i = \Gamma_a^i dx^a \ . \tag{4.113}$$

From (4.111) one finds

$$\partial_{[a}e^i_{b]} = -\omega^{\ i}_{[a\ j}e^j_{b]} = -\epsilon^i_{jk}\Gamma^j_{[a}e^k_{b]} . \tag{4.114}$$

Parallel transport around a closed loop yields

$$dv^i = -R^i_{jab}v^j dx^a dx^b \equiv \epsilon^i_{jk}v^j \delta\omega^k ,$$

where R^i_{jab} are the components of the curvature two-form. The angle $\delta\omega^k$ can be written as

$$\delta\omega^k = -R^k_{ab}dx^a dx^b , \tag{4.115}$$

with $R^k_{ab}\epsilon^i_{jk} \equiv R^i_{jab}$. The curvature components R^i_{ab} obey (from Cartan's second equation)

$$R^i_{ab} = 2\partial_{[a}\Gamma^i_{b]} + \epsilon^i_{jk}\Gamma^j_a\Gamma^k_b \tag{4.116}$$

and the 'cyclic identity'

$$R^i_{ab}e^b_i = 0 . \tag{4.117}$$

The curvature scalar is given by

$$R[e] = -R^i_{ab}\epsilon_{abc}\epsilon_i^{\ jk}e^a_j e^b_k = -R^j_{kab}e^a_j e^{bk} . \tag{4.118}$$

The triad e^a_i (and similarly K^i_a) contains nine variables instead of the six variables of h_{ab}. The Gauß constraint (4.106) reduces the number again from nine to six.
     The formalism presented up to now was known long ago. The progress achieved by Ashtekar (1986) consists in the second step—the mixing of E^a_i and K^i_a into a connection A^i_a. This is defined by

$$GA^i_a(x) = \Gamma^i_a(x) + \beta K^i_a(x) , \tag{4.119}$$

where the 'Barbero–Immirzi' parameter β can be any (non-vanishing) complex number. It must be emphasized that the product GA^i_a has the dimension of an inverse length (like a Yang–Mills connection), but A^i_a itself has dimension mass over length squared. Therefore, GA is the relevant quantity ($\int GAdx$ is dimensionless). The important fact is that A^i_a and $E^b_j/8\pi\beta$ are canonically conjugate variables,

$$\{A^i_a(x), E^b_j(y)\} = 8\pi\beta\delta^i_j\delta^b_a\delta(x, y) . \tag{4.120}$$

In addition, one has

$$\{A^i_a(x), A^j_b(y)\} = 0 . \tag{4.121}$$

In the following, A^i_a will be considered as the new *configuration* variable and E^b_j will be the corresponding canonical momentum.

4.3.2   Discussion of the constraints

The task now is to rewrite all constraints in terms of the new variables. We start with the Gauß constraint which itself is only present due to the use of triads instead of metrics and the associated SO(3) redundancy. Using (4.119) one finds after some straightforward calculations[21]

$$\mathcal{G}_i = \partial_a E_i^a + G\epsilon_{ijk}A_a^j E^{ka} \equiv \mathcal{D}_a E_i^a \approx 0 \ . \tag{4.122}$$

The terminus 'Gauß constraint' now becomes evident since it has a form similar to the Gauß constraint of Yang–Mills theories; cf. (4.33) with $c_{ij}{}^k = -G\epsilon_{ij}{}^k/\beta f$. In (4.122) we have also defined the covariant derivative $\mathcal{D}_a$ associated with A_a^i. Its associated curvature is

$$F_{ab}^i = 2G\partial_{[a}A_{b]}^i + G^2\epsilon_{ijk}A_a^j A_b^k \ . \tag{4.123}$$

The Gauß constraint generates transformations similar to the Yang–Mills case,

$$\delta E_j^a(x) = \int dy \ \{E_j^a(x), \mathcal{G}_i(y)\}\xi^i(y) = 8\pi\beta G\epsilon_{ijk}E^{ka}\xi^i \ ,$$

and

$$\delta A_a^i(x) = \int dy \ \{A_a^i(x), \mathcal{G}_i(y)\}\xi^i(y) = -8\pi\beta \mathcal{D}_a\xi^i \ .$$

Sometimes it is also convenient to introduce su(2)-valued matrices

$$E^a = \tau_i E_i^a \ , \quad A_a = \tau_i A_a^i \ , \tag{4.124}$$

where $\tau_i = i\sigma_i/2$ with σ_i as the Pauli matrices. Under an SU(2)-transformation g, one then has

$$E^a \to gE^a g^{-1} \ , \quad A_a \to g(A_a + \partial_a)g^{-1} \ . \tag{4.125}$$

The next step is to rewrite the genuine gravitational constraints (4.69) and (4.70) in terms of the new variables. Introducing $\tilde{\mathcal{H}}_\perp = -8\pi G\beta^2\mathcal{H}_\perp^g$ (plus terms proportional to the Gauß constraint) and $\tilde{\mathcal{H}}_a = -8\pi G\beta\mathcal{H}_a^g$ (plus terms proportional to the Gauß constraint), the new form of the constraints is

$$\tilde{\mathcal{H}}_\perp = -\frac{\sigma}{2}\frac{\epsilon^{ijk}F_{abk}}{\sqrt{\det E_i^a}}E_i^a E_j^b$$

$$+\frac{\beta^2\sigma - 1}{\beta^2\sqrt{\det E_i^a}}E_{[i}^a E_{j]}^b(GA_a^i - \Gamma_a^i)(GA_b^j - \Gamma_b^j) \approx 0 \ , \tag{4.126}$$

and

$$\tilde{\mathcal{H}}_a = F_{ab}^i E_i^b \approx 0 \ . \tag{4.127}$$

Equation (4.127) has the form of the cyclic identity (4.117) with R_{ab}^i replaced by F_{ab}^i. If applied on A_a^i, the constraint (4.127) yields a transformation that can be written as a sum of a gauge transformation and a pure diffeomorphism.

---

[21] The constraint is also redefined through multiplication with β.

As in Section 4.1, one has $\sigma = -1$ for the Lorentzian and $\sigma = 1$ for the Euclidean case. One recognizes that (4.126) can be considerably simplified by choosing $\beta = i$ for the Lorentzian and $\beta = 1$ (or $\beta = -1$) for the Euclidean case, because then the second term vanishes. The potential term has disappeared, leading to a situation resembling the strong-coupling limit discussed at the end of Section 4.2.3 (see also section III.4 of Ashtekar (1988)). In fact, the original choice was $\beta = i$ for the relevant Lorentzian case. Then,

$$2\tilde{\mathcal{H}}_\perp = \epsilon^{ijk} F_{abk} E_i^a E_j^b \approx 0 \ .$$

This leads to a *complex* connection A_a^i, see (4.119), and makes it necessary to implement reality conditions in order to recover GR—a task that seems impossible to achieve in the quantum theory. However, this choice is geometrically preferred (Rovelli 1991a); A_a^i is then the three-dimensional projection of a four-dimensional self-dual spin connection A_μ^{IJ},

$$A_\mu^{IJ} = \omega_\mu^{IJ} - 1/2 \ i\epsilon^{IJ}{}_{MN}\omega_\mu^{MN} \ . \tag{4.128}$$

(It turns out that the curvature $F_{\mu\nu}^{IJ}$ of the self-dual connection is the self-dual part of the Riemann curvature.) To avoid the problems with the reality conditions, however, it is better to work with real variables. Barbero (1995) has chosen $\beta = -1$ for the Lorentzian case (which does not seem to have a special geometrical significance), so the Hamiltonian constraint reads

$$\tilde{\mathcal{H}}_\perp = \frac{e^{ijk} E_i^a E_j^b}{2\sqrt{h}} (F_{abk} - 2R_{abk}) \approx 0 \ . \tag{4.129}$$

An alternative form using (4.124) is

$$\tilde{\mathcal{H}}_\perp = \frac{1}{\sqrt{h}} \mathrm{tr}\left((F_{ab} - 2R_{ab})[E^a, E^b] \right) \ . \tag{4.130}$$

The constraint algebra (3.90)–(3.92) remains practically unchanged, but one should keep in mind that the constraints have been modified by a term proportional to the Gauß constraint $\mathcal{G}_i$; cf. also (4.28). In addition, one has of course the relation for the generators of SO(3),

$$\{\mathcal{G}_i(x), \mathcal{G}_j(y)\} = \epsilon_{ij}{}^k \mathcal{G}_k(x)\delta(x, y) \ . \tag{4.131}$$

Following Thiemann (1996), we shall now rewrite the Hamiltonian constraint in a way that will turn out to be very useful in the quantum theory (Section 6.3). Consider the 'Euclidean part' of $\tilde{\mathcal{H}}_\perp$,

$$\mathcal{H}_{\mathrm{E}} = \frac{\mathrm{tr}(F_{ab}[E^a, E^b])}{\sqrt{h}} \ . \tag{4.132}$$

Recalling (4.124), one finds

$$[E^a, E^b]_i = -\sqrt{h}\epsilon^{abc}e^i_c \ . \tag{4.133}$$

From the expression for the volume,

$$V = \int_\Sigma d^3x \ \sqrt{h} = \int_\Sigma d^3x \ \sqrt{\det E^a_i} \ , \tag{4.134}$$

one gets $2\delta V/\delta E^c_i(x) = e^i_c(x)$ and therefore,

$$\frac{[E^a, E^b]_i}{\sqrt{h}} = -2\epsilon^{abc}\frac{\delta V}{\delta E^c_i} = -2\frac{\epsilon^{abc}}{8\pi\beta}\{A^i_c, V\} \ .$$

This yields for $\mathcal{H}_E$ the expression

$$\mathcal{H}_E = -\frac{1}{4\pi\beta}\epsilon^{abc}\mathrm{tr}(F_{ab}\{A_c, V\}) \ . \tag{4.135}$$

Thiemann (1996) also considered the integrated trace of the extrinsic curvature,

$$T \equiv \int_\Sigma d^3x \ \sqrt{h}K = \int_\Sigma d^3x \ K^i_a E^a_i \ ,$$

for which one gets

$$\{A^i_a(x), T\} = 8\pi\beta K^i_a(x) \ . \tag{4.136}$$

For $H_E \equiv \int d^3x \ \mathcal{H}_E$, one finds, using (4.114),

$$\{H_E, V\} = 8\pi\beta^2 GT \ . \tag{4.137}$$

One now considers the following sum (written here for general β),

$$\tilde{\mathcal{H}}_\perp + \frac{1-\beta^2(\sigma+1)}{\beta^2}\mathcal{H}_E = \frac{\beta^2\sigma - 1}{2\beta^2\det E^a_i}\left(F^i_{ab} - R^i_{ab}\right)[E^a, E^b]_i \ . \tag{4.138}$$

The reason for performing this combination is to get rid of the curvature term. From (4.116) and using (4.119), one can write

$$R^i_{ab} = F^i_{ab} + \beta^2\epsilon^i_{jk}K^j_a K^k_b + 2\beta\mathcal{D}_{[b}K^i_{a]} \ .$$

With the help of (4.136) and (4.133), one then finds after some straightforward manipulations,

$$\tilde{\mathcal{H}}_\perp + \frac{1-\beta^2(\sigma+1)}{\beta^2}\mathcal{H}_E = \frac{\beta^2\sigma - 1}{2(4\pi\beta)^3}\epsilon^{abc}\mathrm{tr}\left(\{A_a, T\}\{A_b, T\}\{A_c, V\}\right) \ . \tag{4.139}$$

This will serve as the starting point for the discussion of the quantum Hamiltonian constraint in Section 6.3.

4.3.3　*Loop variables*

An alternative formulation that is closely related to the variables discussed in the last subsections employs so-called 'loop variables' introduced by Rovelli and Smolin (1990). Consider for this purpose a closed loop on Σ, that is, a continuous piecewise analytic map from the interval $[0, 1]$ to Σ,

$$\alpha : [0, 1] \to \Sigma , \quad s \mapsto \{\alpha^a(s)\} . \tag{4.140}$$

The *holonomy* $U[A, \alpha]$ corresponding to $A_a = A_a^i \tau_i$ along the curve α is given by

$$U[A, \alpha](s) \in SU(2) , \quad U[A, \alpha](0) = \mathbb{I} ,$$

$$\frac{\mathrm{d}}{\mathrm{d}s} U[A, \alpha](s) - G A_a(\alpha(s))\dot{\alpha}^a(s) U[A, \alpha](s) = 0 , \tag{4.141}$$

where $\dot{\alpha}^a(s) \equiv \mathrm{d}\alpha^a / \mathrm{d}s$ and $U[A, \alpha](s)$ is a shorthand for $U[A, \alpha](0, s)$. The formal solution for the holonomy reads

$$U[A, \alpha](0, s) = \mathcal{P} \exp \left(G \int_\alpha A \right) \equiv \mathcal{P} \exp \left(G \int_0^s \mathrm{d}\tilde{s}\, \dot{\alpha}^a A_a^i(\alpha(\tilde{s}))\tau_i \right) . \tag{4.142}$$

Here, $\mathcal{P}$ denotes path ordering which is necessary because the A are matrices (like in Yang–Mills theories). The holonomy is not yet gauge invariant with respect to SU(2)-transformations. Gauge invariance is achieved after performing the trace, thus arriving at the 'Wilson loop' known, for example, from lattice gauge theories,

$$T[\alpha] = \mathrm{tr}\, U[A, \alpha](0, 0) . \tag{4.143}$$

One can also define

$$T^a[\alpha](s) = \mathrm{tr}\, [U[A, \alpha](s, s) E^a(s)] , \tag{4.144}$$

where E^a is inserted at the point s of the loop. Analogously one can define higher 'loop observables',

$$T^{a_1 \cdots a_N}[\alpha](s_1, \ldots, s_n) ,$$

by inserting E^a at the corresponding points described by the s-values. These loop observables obey a closed Poisson algebra called the *loop algebra*. One has, for example,

$$\{T[\alpha], T^a[\beta](s)\} = \Delta^a[\alpha, \beta(s)] \left(T[\alpha \# \beta] - T[\alpha \# \beta^{-1}] \right) , \tag{4.145}$$

where

$$\Delta^a[\alpha, x] = \int \mathrm{d}s\, \dot{\alpha}^a(s)\delta(\alpha(s), x) , \tag{4.146}$$

and β^{-1} denotes the loop β with the reversed direction. The right-hand side of (4.146) is only non-vanishing if α and β have an intersection at a point P; $\alpha \# \beta$

is then defined as starting from P, going through the loop α, then through β, and ending at P. Of particular interest is the quantity

$$E[S, f] \equiv \int_S d\sigma_a \, E_i^a f^i \, , \qquad (4.147)$$

where S denotes a two-dimensional surface in Σ and $f = f^i \tau_i$. This will be used in the quantum theory; see Chapter 6.

5

QUANTUM GEOMETRODYNAMICS

5.1 The programme of canonical quantization

Given a classical theory, one cannot derive a unique 'quantum theory' from it. The only possibility is to 'guess' such a theory and to test it by experiment. For this purpose, one has devised sets of 'quantization rules' which turned out to be successful in the construction of quantum theories, for example, of quantum electrodynamics. Strictly speaking, the task is to construct a quantum theory from its classical limit.

In Chapter 4, we have developed a Hamiltonian formulation of GR. This is the appropriate starting point for a canonical quantization, which requires the definition of a configuration variable and its conjugate momentum. A special feature of GR is the fact, as is the case in all reparametrization-invariant systems, that the dynamics is entirely generated by constraints: the total Hamiltonian either vanishes as a constraint (for the spatially compact case) or solely consists of surfaces terms (in the asymptotically flat case). The central difficulty is thus, both conceptually and technically, the correct treatment of the quantum constraints, that is, the quantum version of the constraints (4.69) and (4.70).

In Chapter 3, we have presented a general procedure for the quantization of constrained systems. Following Dirac (1964), a classical constraint is turned into a restriction on physically allowed wave functionals, see Section 3.1,

$$\mathcal{G}_a \approx 0 \longrightarrow \hat{\mathcal{G}}_a \Psi = 0 \ . \tag{5.1}$$

At this stage, such a transition is only a heuristic recipe which has to be made more precise. Following Ashtekar (1991) and Kuchař (1993) I shall divide the 'programme of canonical quantization' into six steps which will be shortly presented here and then implemented in the following sections.

The **first step** consists in the identification of configuration variables and their momenta. In the language of geometric quantization (Woodhouse 1992), it is the choice of polarization. Together with the unit operator, these variables are called the 'fundamental variables' V_i. The implementation of Dirac's procedure is the translation of Poisson brackets into commutators for the fundamental variables, that is,

$$V_3 = \{V_1, V_2\} \longrightarrow \hat{V}_3 = -\frac{\mathrm{i}}{\hbar}[\hat{V}_1, \hat{V}_2] \ . \tag{5.2}$$

In the geometrodynamical formulation of GR, see Sections 4.1 and 4.2, the fundamental variables are, apart from the unit operator, the three-metric $h_{ab}(\mathbf{x})$ and its momentum $p^{cd}(\mathbf{x})$ (or a subset of them, see below). In the connection formulation of Section 4.3 one has the connection $A_a^i(\mathbf{x})$ and the densitized triad $E_j^b(\mathbf{y})$, while in the loop-space formulation one has $\mathcal{T}[\alpha]$ and $\mathcal{T}^a[\alpha](s)$. In this chapter, restriction will be made to the quantization of the geometrodynamical formulation, while quantum connection dynamics and quantum loop dynamics will be discussed in Chapter 6.

Application of (5.2) to (4.64) would yield

$$[\hat{h}_{ab}(\mathbf{x}), \hat{p}^{cd}(\mathbf{y})] = i\hbar \delta_{(a}^c \delta_{b)}^d \delta(\mathbf{x}, \mathbf{y}) \; , \tag{5.3}$$

plus vanishing commutators between, respectively, the metric components and the momentum components. Since p^{cd} is linearly related with the extrinsic curvature, describing the embedding of the three-geometry into the fourth dimension, the presence of the commutator (5.3) and the ensuing 'uncertainty relation' between intrinsic and extrinsic geometry means that the classical space–time picture has completely dissolved in quantum gravity. This is fully analogous to the disappearance of particle trajectories as fundamental concepts in quantum mechanics and constitutes one of the central interpretational ingredients of quantum gravity. The fundamental variables form a vector space that is closed under Poisson brackets and complete in the sense that every dynamical variable can be expressed as a sum of products of fundamental variables.

Equation (5.3) does not implement the positivity requirement $\det h > 0$ of the classical theory. But this could only be a problem if (the smeared version of) $\hat{p}^{ab}$ were self-adjoint and its exponentiation therefore a unitary operator, which could 'shift' the metric to negative values.

The **second step** addresses the quantization of a general variable, F, of the fundamental variables. From general theorems of quantum theory (going back to Groenewald and van Hove), one knows that it is impossible to achieve this purpose without violation of the transformation rule (5.2), while assuming an irreducible representation of the commutation rules. This is, of course, related to the problem of 'factor ordering'. Therefore, additional criteria must be invoked to find the 'correct' quantization, such as the demand for 'Dirac consistency' to be discussed in Section 5.3.

The **third step** concerns the construction of an appropriate representation space, $\mathcal{F}$, for the dynamical variables, on which they should act as operators. We shall usually employ the functional Schrödinger picture, in which the operators act on wave functionals defined in an appropriate functional space. For example, the implementation of (5.3) would be achieved by implementing

$$\hat{h}_{ab}(\mathbf{x})\Psi\left[h_{ab}(\mathbf{x})\right] = h_{ab}(\mathbf{x}) \cdot \Psi\left[h_{ab}(\mathbf{x})\right] \; , \tag{5.4}$$

$$\hat{p}^{cd}(\mathbf{x})\Psi\left[h_{ab}(\mathbf{x})\right] = \frac{\hbar}{i} \frac{\delta}{\delta h_{cd}(\mathbf{x})} \Psi\left[h_{ab}(\mathbf{x})\right] \; . \tag{5.5}$$

These relations do not define self-adjoint operators since there is no Lebesgue measure on Riem Σ (which would have to be invariant under translations in function space). Thus, one would not expect the fundamental relations (5.3) to be necessarily in conflict with $\det h > 0$. Other examples for the use of the functional Schrödinger picture have already been presented in Chapter 3.

The representation space $\mathcal{F}$ is only an auxiliary space: before the constraints are implemented, it does not necessarily contain only physical states. Therefore, neither does it have to be a Hilbert space nor do operators acting on $\mathcal{F}$ have to be self-adjoint. It might even be inconsistent to demand that the constraints be self-adjoint operators on an auxiliary Hilbert space $\mathcal{F}$.

The **fourth step** consists in the implementation of the constraints. According to (5.1), one would implement the classical constraints $\mathcal{H}_\perp \approx 0$ and $\mathcal{H}_a \approx 0$ as

$$\mathcal{H}_\perp \Psi = 0 \ , \tag{5.6}$$

$$\mathcal{H}_a \Psi = 0 \ . \tag{5.7}$$

These are infinitely many equations (one equation at each space point), collectively called $\mathcal{H}_\mu \Psi = 0$. Only solutions to these 'quantum constraints' can be regarded as candidates for physical states. The solution space shall be called $\mathcal{F}_0$. How the constraints (5.6) and (5.7) are written in detail depends on one's approach to the 'problem of time'; see Section 5.2. It has to be expected that the solution space is still too large; as in quantum mechanics, one may have to impose further conditions on the wave functions, such as normalizability. This requirement is needed in quantum mechanics because of the probability interpretation, but it is far from clear whether this interpretation can be maintained in quantum gravity; cf. Chapter 10. The physical space on which wave functionals act, $\mathcal{F}_{\text{phys}}$, is thus expected to form a genuine subspace, $\mathcal{F}_{\text{phys}} \subset \mathcal{F}_0 \subset \mathcal{F}$.

The **fifth step** concerns the role of observables. We have already mentioned in Section 3.1 that 'observables' are characterized by having weakly vanishing Poisson brackets with the constraints, $\{\mathcal{O}, \mathcal{G}_a\} \approx 0$. They should not be confused with observables in an operationalistic sense. In quantum mechanics, observables are associated in a somewhat vague manner with self-adjoint operators (only this concept is mathematically precise). In practice, however, only few operators correspond in fact with quantities that are 'measured'. Only the latter represent 'beables' in the sense of John Bell, supposingly describing 'reality', see Bell (1987).[1] For an operator corresponding to a classical observable satisfying $\{\mathcal{O}, \mathcal{H}_\mu\} \approx 0$, one would expect that in the quantum theory the relation

$$[\hat{\mathcal{O}}, \hat{\mathcal{H}}_\mu]\Psi = 0 \tag{5.8}$$

holds. For operators $\hat{F}$ with $[\hat{F}, \hat{\mathcal{H}}_\mu]\Psi \neq 0$ one would have $\hat{\mathcal{H}}_\mu(\hat{F}\Psi) \neq 0$. This is sometimes interpreted as meaning that the 'measurement' of the quantity being related to this operator leads to a state that is no longer annihilated by the constraints, 'throwing one out' of the solution space. This would, however, only be a problem for a 'collapse' interpretation of quantum gravity, an interpretation that seems to be highly unlikely to hold in quantum gravity; see Chapter 10.

---

[1] These are quantities that are subject to decoherence; see Chapter 10.

Since the classical Hamiltonian and diffeomorphism constraints differ from each other in their interpretation (Chapter 4), the same should hold for their quantum versions (5.6) and (5.7). This is, in fact, the case and will be discussed in detail in this chapter. The distinction between 'observables' and 'perennials', see (4.82) and (4.83), thus applies also to the quantum case.

The **sixth** (and last) **step** concerns the role of the physical Hilbert space (cf. also step 3). Do the observables have to be represented in some Hilbert space? If yes, which one? It can certainly not be the auxiliary space $\mathcal{F}$, but it is unclear whether it is $\mathcal{F}_0$ or only $\mathcal{F}_{\text{phys}} \subset \mathcal{F}_0$. To represent all perennials by self-adjoint operators in Hilbert space would be contradictory: be $\hat{F}$ and $\hat{G}$ self-adjoint perennials, then the product $\hat{F}\hat{G}$ is again a perennial, but no longer self-adjoint, since

$$\left(\hat{F}\hat{G}\right)^{\dagger} = \hat{G}^{\dagger}\hat{F}^{\dagger} = \hat{G}\hat{F} \overset{\text{in general}}{\neq} \hat{F}\hat{G} \,.$$

Since, moreover, the fundamental variables h_{ab} and p^{cd} are not perennials, one might, at this stage, forget about this notion. The 'problem of Hilbert space' is intimately connected with the 'problem of time' in quantum gravity, to which we shall now turn.

5.2 The problem of time

The concepts of time in GR and quantum theory differ drastically from each other. As already remarked in Section 1.1, time in quantum theory is an external parameter (an absolute element of the theory), whereas time in GR is dynamical. A consistent theory of quantum gravity should, therefore, exhibit a novel concept of time. The history of physics has shown that new theories often entail a new concept of space and time (Ehlers 1973). The same should happen again with quantum gravity.

The absolute nature of time in quantum mechanics is crucial for its interpretation. Matrix elements are usually evaluated at fixed t, and the scalar product is conserved in time ('unitarity'). Unitarity expresses the conservation of the total probability. 'Time' is part of the classical background which, according to the Copenhagen interpretation, is needed for the interpretation of measurements. As we have remarked at the end of Section 3.1., the introduction of a time operator in quantum mechanics is problematic. The time parameter t appears explicitly in the Schrödinger equation (3.14). Note that it comes together with the imaginary unit i, a fact that finds an explanation in the semiclassical approximation to quantum geometrodynamics (Section 5.4). The occurrence of the imaginary unit in the Schrödinger equation was already discussed in an interesting correspondence between Ehrenfest and Pauli; see Pauli (1985).

In GR, space–time is dynamical and therefore there is no absolute time. Space–time influences material clocks in order to allow them to show proper time. The clocks, in turn, react on the metric and change the geometry. In this sense, the metric itself *is* a clock (Zeh 2001). A quantization of the metric can thus be interpreted as a quantization of the concept of time. Since the nature of time in quantum gravity is not yet clear—the classical constraints do not contain any time parameter—one speaks of the 'problem of time'. One can distinguish basically three possible solutions of this problem, as reviewed, in particular, by Isham (1993) and Kuchař (1992):

1. choice of a concept of time *before* quantization;
2. identification of a concept of time *after* quantization;
3. *'timeless'* options.

The first two possibilities will be discussed in the following, while the third option will be addressed in Sections 5.3 and 5.4.

5.2.1   *Time before quantization*

In Section 4.2, it was argued that the three-dimensional geometry in GR contains information about time. Motivated by the parametrized theories discussed in Chapter 3, one can attempt to perform a canonical transformation aiming at an isolation of time from the 'true degrees of freedom'. Starting from the 'ADM variables' $h_{ab}(\mathbf{x})$ and $p^{cd}(\mathbf{x})$, one would like to perform the step

$$\left(h_{ab}(\mathbf{x}), p^{cd}(\mathbf{x})\right) \longrightarrow \left(X^A(\mathbf{x}), P_B(\mathbf{x}); \phi^r(\mathbf{x}), p_s(\mathbf{x})\right) \ , \tag{5.9}$$

where the $8 \times \infty^3$ variables X^A and P_B $(A, B = 0, 1, 2, 3)$ are the 'embedding variables' and their canonical momenta, while the $4 \times \infty^3$ variables ϕ^r and p_s $(r, s = 1, 2)$ denote the 'true' degrees of freedom of the gravitational field, cf. also Section 7.4. As already remarked in Section 4.2, GR is not equivalent to a deparametrized theory. Therefore, (5.9) is certainly non-unique and not valid globally (see in this context Hájíček and Kijowski 2000). The next step is the elimination of $4 \times \infty^3$ of the $8 \times \infty^3$ variables by casting the classical constraints $\mathcal{H}_\mu \approx 0$ into the form[2]

$$P_A(\mathbf{x}) + h_A(\mathbf{x}; X^B, \phi^r, p_s] \approx 0 \ . \tag{5.10}$$

This is referred to as 'solving the constraints on the classical level' or 'reduced quantization' and corresponds in the case of particle systems to (3.9). As in Section 3.1, the remaining $4 \times \infty^3$ variables are eliminated by inserting (5.10) into the action

$$S = \int \mathrm{d}t \int_\Sigma \mathrm{d}^3x \ \left(P_A \dot{X}^A + p_r \dot{\phi}^r - N\mathcal{H}_\perp - N^a\mathcal{H}_a\right) \ , \tag{5.11}$$

where all fields are functions of $\mathbf{x}$ and t, and going to the constraint hypersurface ('deparametrization'), yielding

---

[2](\mathbf{x} etc. means: dependence on $\mathbf{x}$ as a function, while $p_s]$ etc. means: dependence on $p_s(\mathbf{x})$ as a functional.

$$S = \int \mathrm{d}t \int_\Sigma \mathrm{d}^3 x \; \left(p_r \dot{\phi}^r - h_A(\mathbf{x}; X_t^B, \phi^r, p_s] \dot{X}_t^A(\mathbf{x}) \right) \; , \qquad (5.12)$$

where $\dot{X}_t^A(\mathbf{x})$ is now a prescribed function of $\mathbf{x}$, which must not be varied. This action corresponds to the action (3.12) in which the prescribed function is $t(\tau)$. The action (5.12) describes an ordinary canonical system with a 'true', that is, un-constrained Hamiltonian given by

$$H_{\text{true}}(t) = \int_\Sigma \mathrm{d}^3 x \; h_A(\mathbf{x}; X_t^B, \phi^r, p_s] \dot{X}_t^A(\mathbf{x}) \; . \qquad (5.13)$$

One can derive from H_{true} Hamilton's equations of motion for ϕ^r and p_s. These variables can, however, only be interpreted as describing embeddings in a space–time *after* these equations (together with the equations for lapse and shift) have been solved.

The constraint (5.10) can be quantized in a straightforward manner by introducing wave functionals $\Psi[\phi^r(\mathbf{x})]$, with the result

$$i\hbar \frac{\delta \Psi[\phi^r(\mathbf{x})]}{\delta X^A(\mathbf{x})} = h_A \left(\mathbf{x}; X^B, \hat{\phi}^r, \hat{p}^s \right] \Psi[\phi^r(\mathbf{x})] \; , \qquad (5.14)$$

in which the X^A have not been turned into an operator. In this respect, the quantization is of a hybrid nature: momenta occurring linearly in the constraints are formally turned into derivatives, although the corresponding configuration variables stay classical—like the t in the Schrödinger equation. Equation (5.14) has the form of a local Schrödinger equation. Such an equation is usually called a 'Tomonaga–Schwinger equation'; strictly speaking, it consists of infinitely many equations with respect to the local 'bubble time' $X^A(\mathbf{x})$. We shall say more about such equations in Section 5.4. The main advantages of this approach to quantization are:

1. One has isolated already at the classical level a time variable (here: 'embedding variables') that is external to the quantum system described by $\hat{\phi}^r, \hat{p}_s$. The formalism thus looks similar to ordinary quantum field theory.
2. Together with such a distinguished notion of time comes a natural Hilbert-space structure and its ensuing probability interpretation.
3. One would consider observables to be any function of the 'genuine' operators $\hat{\phi}^r$ and $\hat{p}_s$. As in the linearized approximation (Chapter 2), the gravitational field would have two degrees of freedom.

On the other hand, one faces many problems:

1. 'Multiple-choice problem': The canonical transformation (5.9) is certainly non-unique and the question arises which choice should be made. One would expect that different choices of 'time' lead to non-unitarily connected quantum theories.

2. 'Global-time problem': One cannot find a canonical transformation (5.9) to find a global time variable (Torre 1993).

3. 'h_A-problem': The 'true' Hamiltonian (5.13) depends on 'time', that is, on the embedding variables X^A. This dependence is expected to be very complicated (leading to square roots of operators, etc.), prohibiting in general a rigorous definition.

4. 'X^A-problem': In the classical theory, the 'bubble time' X^A describes a hypersurface in space–time *only after* the classical equations have been solved. Since no classical equations and therefore no space–time are available in the quantum theory, (5.14) has no obvious space–time interpretation. In particular, an operational treatment of time is unknown.

5. 'Anomalies': Quantum anomalies may spoil the consistency of this approach; cf. Section 5.3.

6. 'Space–time problem': Writing $X^A = (T, X^i)$, the 'time' T must be a space–time scalar. This means that, although it is constructed from the canonical data h_{ab} and p^{cd} on Σ, it must weakly vanish with the Hamiltonian constraint,

$$\left\{ T(\mathbf{x}), \int_\Sigma \mathrm{d}^3 y \, \mathcal{H}_\perp(\mathbf{y}) N(\mathbf{y}) \right\} \approx 0 \qquad (5.15)$$

for all lapse functions $N(\mathbf{y})$ with $N(\mathbf{x}) = 0$. Otherwise, one would get from two hypersurfaces Σ and Σ' crossing at $\mathbf{x}$, two different values of T, depending on whether the canonical data of Σ or of Σ' are used. The variable T would in this case have no use as a time variable. A possible solution to the space–time problem can be obtained by using matter variables, for example, the 'reference fluid' used by Brown and Kuchař (1995). The 'space–time problem' already anticipates a space–time picture which, however, is absent in quantum gravity.

7. 'Problem of construction': The actual transformation (5.9) has been performed only in very special cases, for example, linearized gravity, cylindrical gravitational waves, black holes, and dust shells (Chapter 7), and homogeneous cosmological models (Chapter 8).

In the full theory, concrete proposals for the canonical transformation (5.9) are rare. A possibility that was developed to a certain extent makes use of 'York's time' or 'extrinsic time', which is defined by

$$T\left(\mathbf{x}; h_{ab}, p^{cd}\right] = \frac{2}{3\sqrt{h}} p^{cd} h_{cd} \, , \quad P_T = -\sqrt{h} \, , \qquad (5.16)$$

cf. Altshuler and Barvinsky (1996) and the references therein. Since $p^{cd} h_{cd} = -\sqrt{h} K / 8\pi G$, cf. (4.63), T is proportional to the trace of the extrinsic curvature K. It is canonically conjugated to P_T. Note that T does not obey (5.15) and is thus not a space–time scalar. It has been shown[3] that the Hamiltonian constraint can be written in the form (5.10), that is, written as $P_T + h_T \approx 0$, where h_A is known to exist, but not known in explicit form, that is, not known as an explicit function of T and the remaining variables. From (5.16) it is clear that the 'true' Hamiltonian contains the three-dimensional volume as its dynamical part, that is,

$$H_{\text{true}} = \int d^3x \sqrt{h} + \int d^3x \, N^a \mathcal{H}_a \ . \tag{5.17}$$

The main problem with this approach towards the issue of time in quantum gravity is perhaps its closeness to a classical space–time picture. From various equations, such as (5.14), one gets the illusion that a space–time exists even in the quantum theory, although this cannot be the case, see (5.3). One can, therefore, conclude that attempting to identify time before quantization does not solve the problem of time in the general case, although it might help in special cases (Section 7.4).

5.2.2   *Time after quantization*

Using directly the commutation rules (5.3) and their formal implementation (5.4) and (5.5) one arrives at wave functionals $\Psi[h_{ab}(\mathbf{x})]$ defined on Riem Σ, the space of all three-metrics. This is the central kinematical quantity. The 'dynamics' must be implemented through the quantization of the constraints (4.69) and (4.70)—this is all that remains in the quantum theory. One, then, gets the following equations for the wave functional,

$$\hat{\mathcal{H}}_{\perp}^g \Psi \equiv \left(-16\pi G\hbar^2 G_{abcd} \frac{\delta^2}{\delta h_{ab} \delta h_{cd}} - \frac{\sqrt{h}}{16\pi G}(\,^{(3)}R - 2\Lambda) \right) \Psi = 0 \ , \tag{5.18}$$

$$\hat{\mathcal{H}}_a^g \Psi \equiv -2D_b h_{ac} \frac{\hbar}{i} \frac{\delta \Psi}{\delta h_{bc}} = 0 \ . \tag{5.19}$$

Equation (5.18) is called the *Wheeler–DeWitt equation* in honour of the work by DeWitt (1967*a*) and Wheeler (1968). In fact, these are again infinitely many equations. The constraints (5.19) are called the *quantum diffeomorphism (or momentum) constraints*. Occasionally, both (5.18) and (5.19) are referred to as Wheeler–DeWitt equations. In the presence of non-gravitational fields, these equations are augmented by the corresponding terms.

There are many problems associated with these equations. An obvious problem is the 'factor-ordering problem': the precise form of the kinetic term is open—there could be additional terms proportional to $\hbar$ containing at most first derivatives in the metric. Since second functional derivatives at the same space point usually lead to undefined expressions such as $\delta(0)$, a regularization

---

[3]This involved a detailed study of the 'Lichnerowicz equation', a non-linear (but quasi-linear) elliptical equation for P_T, which under appropriate conditions possesses a unique solution (cf. Choquet-Bruhat and York 1980).

(and perhaps renormalization) scheme has to be employed. Connected with this is the potential presence of anomalies. The general discussion of these problems is continued in Section 5.3. Here we shall address again the problem of time and the related Hilbert-space problem. Since (5.18) does not have the structure of a local Schrödinger equation (5.14), the choice of Hilbert space is not clear a priori.

The first option for an appropriate Hilbert space is the use of a *Schrödinger-type inner product*, that is, the standard quantum-mechanical inner product as generalized to quantum field theory,

$$\langle \Psi_1 | \Psi_2 \rangle = \int_{\mathrm{Riem}\ \Sigma} \mathcal{D}\mu[h]\ \Psi_1^*[h]\Psi_2[h]\ , \qquad (5.20)$$

where h is here a shorthand for h_{ab}. It is known that such a construction is at best formal, since the measure $\mathcal{D}\mu[h]$ cannot be rigorously defined, that is, there is no Lebesgue measure in the functional case. The elementary operators $\hat{h}_{ab}$ and $\hat{p}^{cd}$ are formally self-adjoint with respect to this inner product.

Besides the lack of mathematical rigour, this inner product has other problems. The integration runs over all metric components, including potential unphysical ones (the constraints have not yet been imposed at this stage). This could lead to divergences—similar to an integration over t in the quantum-mechanical case—which one would have to cure by changing the measure, similar to the introduction of the Faddeev–Popov determinant into path integrals (Section 2.2); cf. Woodard (1993) in this context. This is related to the fact that the product (5.20) is defined on the full space $\mathcal{F}$, not the solution space $\mathcal{F}_0 \subset \mathcal{F}$. It could be possible to turn these problems into a virtue by imposing as a 'boundary condition' that physical states solve the constraints *and* lead to a finite inner product (5.20). Such a proposal can at least be implemented within simple models (cf. Chapter 8), but one would face the danger that in the full theory no such solutions would exist at all. An open problem is also the implementation of a probability interpretation in this context (recall that this is the major motivation for using this inner product in quantum mechanics). What does the probability to find a certain three-metric mean? The answer is unclear. It is possible that the Schrödinger inner product only makes sense in the semiclassical approximation (Section 5.4).

We have seen in Section 4.2 that the kinetic term of the Hamiltonian constraint is *indefinite*, due to the indefinite structure of the DeWitt metric G_{abcd}; see the discussion following (4.92). Consequently, the Wheeler–DeWitt equation (5.18), too, possesses an indefinite kinetic term. From this point of view (5.18) resembles a Klein–Gordon equation; strictly speaking, infinitely many Klein–Gordon equations with a non-trivial potential term. This can be made more explicit. Using instead of h_{ab} and $\sqrt{h}$ the variables $\tilde{h}_{ab} = h^{-1/3}h_{ab}$, cf. (4.96), and the local volume element $\sqrt{h}$, the Wheeler–DeWitt equation can explicitly be written as

$$\left(6\pi G\hbar^2\sqrt{h}\frac{\delta^2}{\delta(\sqrt{h})^2} - \frac{16\pi G\hbar^2}{\sqrt{h}}\tilde{h}_{ac}\tilde{h}_{bd}\frac{\delta^2}{\delta\tilde{h}_{ab}\delta\tilde{h}_{cd}}\right.$$
$$\left. -\frac{\sqrt{h}}{16\pi G}(\,^{(3)}R - 2\Lambda)\right)\Psi[\sqrt{h},\tilde{h}_{ab}] = 0 . \tag{5.21}$$

It might therefore be more appropriate to use a *Klein–Gordon-type inner product.* If Ψ_1 and Ψ_2 are solutions of (5.18), the functional version of this inner product would read (DeWitt 1967a)

$$\langle\Psi_1|\Psi_2\rangle = i\int\prod_{\mathbf{x}}d\Sigma^{ab}(\mathbf{x})\Psi_1^*[h_{ab}]\cdot$$
$$\left(G_{abcd}\overrightarrow{\frac{\delta}{\delta h_{cd}}} - \overleftarrow{\frac{\delta}{\delta h_{cd}}}G_{abcd}\right)\Psi_2[h_{ab}] = \langle\Psi_2|\Psi_1\rangle^* . \tag{5.22}$$

Here, the integration is over a $5 \times \infty^3$-dimensional surface in the $6 \times \infty^3$-dimensional space Riem Σ, and $d\Sigma^{ab}$ denotes the corresponding surface element. In view of (5.21), the integration can be taken over the variables $\tilde{h}_{ab}$, referring to 'constant time $\sqrt{h}$=constant'. Of course, the lack of mathematical rigour is the same as with (5.20).

The inner product (5.22) has the advantage that it is invariant under deformations of the $5 \times \infty^3$-dimensional surface. This expresses its 'time independence'. However, this inner product is—like the usual inner product for the Klein–Gordon equation—not positive definite. In particular, one has $\langle\Psi|\Psi\rangle = 0$ for real solutions of (5.18). Since the Wheeler–DeWitt equation is a real equation (unlike the Schrödinger equation), real solutions should possess some significance.

For the standard Klein–Gordon equation in Minkowski space, one can make a separation between 'positive' frequencies and 'negative' frequencies. As long as one can stay within the one-particle picture, it is consistent to make a restriction to the positive-frequency sector. For such solutions, the inner product is positive. On curved backgrounds, a separation into positive and negative frequencies can be made if both the space–time metric and the potential are stationary, that is, if there is a time-like Killing field and if the potential is constant along its orbits. The Killing field can also be a conformal Killing field, but then the potential must obey certain scaling properties. Moreover, the potential must be positive. If these conditions are violated, particles are produced and the one-particle picture breaks down.

Can such a separation into positive and negative frequencies be made for the Wheeler–DeWitt equation? The clear answer is *no* (Kuchař 1992). There exists a conformal Killing field for the DeWitt metric, namely the three-metric h_{ab}. The potential is, however, neither positive definite nor scales in the correct way. Therefore, no Klein–Gordon inner product can be constructed, which is positive definite for the generic case (although this might be achievable for special models). For the standard Klein–Gordon equation, the failure of the one-particle picture leads to 'second quantization' and quantum field theory. The Wheeler–DeWitt equation, however, corresponds already to a field-theoretic situation. It has, therefore, been suggested to proceed with a 'third quantization' and to turn the wave function $\Psi[h]$ into an operator (see Kuchař 1992 for review and references). No final progress, however, has been achieved with such attempts.

One might wonder whether the failure of the above attempts is an indication of the absence of time at the most fundamental level. As will be discussed in Section 5.4, the usual concept of time emerges as an approximate notion on a semiclassical level. This is, in fact, all that is needed to have accordance with experience. Also, the notion of a Hilbert space may be a semiclassical artefact only. It is, however, not yet clear what kind of mathematical structure could replace it at the fundamental level, in order to select physically reasonable solutions from $\mathcal{F}_0$. We shall, therefore, proceed pragmatically and treat in the following the Wheeler–DeWitt equation (5.18) just as a (functional) differential equation.

5.3   The geometrodynamical wave function

5.3.1   *The diffeomorphism constraints*

The general equations (5.18) and (5.19) are very complicated and need a mathematical elaboration. Still, some general features can be studied directly. For the diffeomorphism constraints (5.19), this is much easier to achieve and the present subsection is therefore devoted to them. Since we have seen in Section 4.2 that the classical constraints $\mathcal{H}_a^{\mathrm{g}}$ generate three-dimensional coordinate transformations, the presence of the quantum constraints (5.19) expresses the invariance of the wave functional Ψ under such transformations, more precisely: under infinitesimal coordinate transformations.

This can be seen as follows (Higgs 1958). Under the infinitesimal transformation,

$$x^a \mapsto \bar{x}^a = x^a + \delta N^a(\mathbf{x}) \; ,$$

the three-metric transforms as

$$h_{ab}(\mathbf{x}) \mapsto \bar{h}_{ab}(\bar{\mathbf{x}}) = h_{ab}(\mathbf{x}) - D_a \delta N_b(\mathbf{x}) - D_b \delta N_a(\mathbf{x}) \; .$$

The wave functional then transforms according to

$$\Psi[h_{ab}] \mapsto \Psi[h_{ab}] - 2 \int \mathrm{d}^3 x \, \frac{\delta \Psi}{\delta h_{ab}(\mathbf{x})} D_a \delta N_b(\mathbf{x}) \; .$$

Assuming that $\delta N_b(\mathbf{x})$ vanishes at infinity, one can make a partial integration and conclude from the arbitrariness of $\delta N_b(\mathbf{x})$ that

$$D_a \frac{\delta \Psi}{\delta h_{ab}} = 0 \; ,$$

that is, (5.19) is fulfilled. Therefore, Ψ depends only on the three-dimensional geometry, not on the particular form of the metric, that is, it is defined on superspace (Section 4.2). This is sometimes expressed by the notation $\Psi[\,^3\mathcal{G}]$ (Wheeler 1968). Such a representation is, however, at best pictorial, since one cannot construct a derivative operator of the form $\delta/\delta(\,^3\mathcal{G})$ on superspace; one must work with the equations (5.18) and (5.19) for $\Psi[h_{ab}]$. This is similar to gauge theories (Section 4.1) where one has to work with the connection and where an explicit transition to gauge-invariant variables is in general impossible. Note that the above demonstration of coordinate invariance for Ψ is completely analogous to (3.86)—for parametrized field theories—and (3.49)—for the bosonic string.

A simple analogy to (5.19) is Gauß' law in QED (or its generalizations to the non-Abelian case, see Section 4.1). The quantized version of the constraint $\nabla \mathbf{E} \approx 0$ reads

$$\frac{\hbar}{i}\nabla\frac{\delta\Psi[\mathbf{A}]}{\delta\mathbf{A}} = 0 \, , \tag{5.23}$$

from which invariance of Ψ with respect to gauge transformations $\mathbf{A} \to \mathbf{A} + \nabla\lambda$ follows.

We have seen that the wave functional $\Psi[h_{ab}]$ is invariant under infinitesimal coordinate transformations ('small diffeomorphisms'). There may, however, exist 'large diffeomorphisms', that is, diffeomorphisms which are not connected with the identity, under which Ψ might not be invariant.

This situation is familiar from Yang–Mills theories (see e.g. Huang 1992). The quantized form of the Gauß law (4.30) demands that $\Psi[A_a^i]$ be invariant under infinitesimal ('small') gauge transformations; cf. the QED-example (5.23). We take the Yang–Mills gauge group $\mathcal{G}$ as the map

$$S^3 \longrightarrow \mathrm{SU}(N) \equiv G \, , \tag{5.24}$$

where $\mathbb{R}^3$ has been compactified to the three-sphere S^3; this is possible since it is assumed that gauge transformations approach a constant at spatial infinity. The key role in the study of 'large gauge transformations' is played by

$$\pi_0(\mathcal{G}) \equiv \mathcal{G}/\mathcal{G}_0 \, , \tag{5.25}$$

where $\mathcal{G}_0$ denotes the component of $\mathcal{G}$ connected with the identity. Thus, π_0 counts the number of components of the gauge group. One can also write

$$\pi_0(\mathcal{G}) = [S^3, G] \equiv \pi_3(G) = \mathbb{Z} \, , \tag{5.26}$$

where $[S^3, G]$ denotes the set of homotopy classes of continuous maps from S^3 to G.[4] The 'winding numbers' $n \in \mathbb{Z}$ denote the number of times that the spatial S^3 is covered by the SU(2)-manifold S^3.[5] This, then, leads to a vacuum state for each connected component of $\mathcal{G}$, called 'k-vacua' $|k\rangle$, $k \in \mathbb{Z}$. A state $|k\rangle$ is

[4]Two maps are called homotopic if they can be continuously deformed into each other. All homotopic maps yield a homotopy class.

[5]The SU(N)-case can be reduced to the SU(2)-case.

invariant under small gauge transformations, but transforms as $|k\rangle \to |k+n\rangle$ under large gauge transformations. Defining the central concept of a 'θ-vacuum' by

$$|\theta\rangle = \sum_{k=-\infty}^{\infty} e^{-ik\theta}|k\rangle \; , \tag{5.27}$$

with a real parameter θ, the transformation of this state under a large gauge transformation reads

$$\sum_{k=-\infty}^{\infty} e^{-ik\theta}|k+n\rangle = e^{in\theta}|\theta\rangle \; .$$

The θ-states are thus labelled by Hom $(\mathbb{Z}, U(1))$, the homomorphisms from $\mathbb{Z}$ to $U(1)$. Different values of θ characterize different 'worlds' (compare the ambiguity related with the Barbero–Immirzi parameter in Section 4.3); θ is in principle a measurable quantity and one has, for example, from the limit on the neutron dipole element, the constraint $|\theta| < 10^{-9}$ on the θ-parameter of QCD. Instead of the gauge-dependent wave functions (5.27), one can work with gauge-independent wave functions, but with an additional term in the action, the 'θ-action' (Ashtekar 1991; Huang 1992). A state of the form (5.27) is also, of course, well known from solid state physics ('Bloch wave function').

One can envisage the states $|k\rangle$ as being 'peaked' around a particular minimum in a periodic potential. Therefore, tunnelling is possible between different minima. In fact, tunnelling is described by 'instantons', that is, solutions of the classical Euclidean field equations for which the initial and final value of the gauge potential differ by a large gauge transformation (Huang 1992).

One does not have to restrict oneself to S^3, but can generalize this notion to an arbitrary compact orientable three-space Σ (Isham 1981),

$$|\theta\rangle = \sum_{(k,g)} \theta(k,g)|k,g\rangle \; , \tag{5.28}$$

where $\theta(k,g) \in \text{Hom}([\Sigma, G], U(1))$ appears instead of the $e^{-ik\theta}$ of (5.27). As it turns out, $g \in \text{Hom}(\pi_1(\Sigma), \pi_1(G))$.

Instead of taking the gauge group as the starting point, one can alternatively focus on the physical configuration space of the theory. This is more suitable for the comparison with gravity. For Yang–Mills fields, one has the configuration space $Q = \mathcal{A}/\mathcal{G}$, where $\mathcal{A}$ denotes the set of connections. In gravity, $Q = \mathcal{S}(\Sigma) = \text{Riem } \Sigma/\text{Diff } \Sigma$, see Section 4.2. If the group acts freely on $\mathcal{A}$ (or Riem Σ), that is, if it has no fixed points, then

$$\pi_1(Q) = \pi_0(\mathcal{G}) \; ,$$

and the θ-structure as obtained from $\pi_0(\mathcal{G})$ can be connected directly with the topological structure of the configuration space, that is, with $\pi_1(Q)$. As we have seen in Section 4.2.5, Diff Σ does not act freely on Riem Σ, so $\mathcal{S}(\Sigma)$ had to be transformed into the 'resolution space' $\mathcal{S}_\mathrm{R}(\Sigma)$. Everything is fine if we restrict Diff Σ to $\mathcal{D}_\mathrm{F}(\bar{\Sigma})$ (this is relevant in the open case) and take into account that $\mathcal{S}_\mathrm{R}(\bar{\Sigma}) \cong \mathcal{S}(\Sigma)$. Then,

$$\pi_1(\mathcal{S}(\Sigma)) = \pi_0(\mathcal{D}_\mathrm{F}(\bar{\Sigma})) \ ,$$

and one can classify θ-states by elements of Hom $(\pi_0(\mathcal{D}_\mathrm{F}(\bar{\Sigma})), U(1))$. Isham (1981) has investigated the question as to which three-manifolds Σ can yield a non-trivial θ-structure. He has found that

$$\pi_0(\mathcal{D}_\mathrm{F}(S^3)) = 0 \ ,$$

so no θ-structure is available in the cosmologically interesting case S^3. A θ-structure is present, for example, in the case of 'Wheeler's wormhole', that is, for $\Sigma = S^1 \times S^2$. In that case,

$$\pi_0(\mathcal{D}_\mathrm{F}(S^1 \times S^2)) = \mathbb{Z}_2 \oplus \mathbb{Z}_2 \ ,$$

where $\mathbb{Z}_2 = \{-1, 1\}$. Also for the three-torus $S^1 \times S^1 \times S^1$, one has a non-vanishing π_0, but the expression is more complicated. Therefore, θ-sectors in quantum gravity are associated with the disconnectedness of Diff Σ. In the asymptotically flat case, something interesting may occur in addition (Friedman and Sorkin 1980). If one allows for rotations at infinity, one can get half-integer spin states in case that a 2π-rotation acts non-trivially, that is, if one cannot communicate a rotation by 2π at ∞ to the whole interior of space. An example for a manifold that allows such states is $\Sigma = \mathbb{R}^3 \# T^3$.

5.3.2 WKB approximation

An important approximation in quantum mechanics is the WKB approximation. On a formal level, this can also be performed for equations (5.18) and (5.19). For this purpose, one makes the ansatz

$$\Psi[h_{ab}] = C[h_{ab}] \exp\left(\frac{\mathrm{i}}{\hbar} S[h_{ab}]\right) \ , \tag{5.29}$$

where $C[h_{ab}]$ is a 'slowly varying amplitude' and $S[h_{ab}]$ is a 'rapidly varying phase' (an 'eikonal' like in geometrical optics). This corresponds to

$$p^{ab} \longrightarrow \frac{\delta S}{\delta h_{ab}} \ ,$$

and from (5.18) and (5.19) one finds the approximate equations

$$16\pi G\, G_{abcd}\frac{\delta S}{\delta h_{ab}}\frac{\delta S}{\delta h_{cd}} - \frac{\sqrt{h}}{16\pi G}(\,^{(3)}R - 2\Lambda) = 0 \ , \tag{5.30}$$

$$D_a\frac{\delta S}{\delta h_{ab}} = 0 \ . \tag{5.31}$$

In the presence of matter one has additional terms. Equation (5.30) is the Hamilton–Jacobi equation for the gravitational field (Peres 1962). Equation

(5.31) expresses again the fact that $S[h_{ab}]$ is invariant under coordinate transformations. One can show that (5.30) and (5.31) are fully equivalent to the classical Einstein field equations (Gerlach 1969)—this is one of the 'six routes to geometrodynamics' (Misner *et al.* 1973). This route again shows how the dynamical laws follow from the laws of the instant (Kuchař 1993).

The 'interconnection theorems' mentioned in Section 4.2 have their counterparts on the level of the eikonal $S[h_{ab}]$. For example, if S satisfies (5.30), it must automatically satisfy (5.31). These relations have their counterpart in the full quantum theory, provided there are no anomalies (Section 5.3.5).

More useful than a WKB approximation for all degrees of freedom is a 'mixed' approximation scheme in which gravity is treated differently from other fields. It is then possible to recover the limit of quantum field theory in an external space–time. This method will be presented in Section 5.4.

5.3.3 *Remarks on the functional Schrödinger picture*

The central kinematical object in quantum geometrodynamics is the wave functional. It obeys the quantum constraint equations (5.18) and (5.19), which are functional differential equations. In non-gravitational quantum field theory, this 'Schrödinger picture' is used only infrequently, mainly because the focus there is on perturbative approaches for which other formulations are more appropriate, for example, the Fock-space picture. Still, even there a Schrödinger picture is sometimes used in the form of 'Tomonaga–Schwinger equation (TS)' (although first formulated in Stueckelberg 1938),

$$i\hbar\frac{\delta\Psi}{\delta\tau(\mathbf{x})} = \mathcal{H}\Psi \ , \tag{5.32}$$

where $\tau(\mathbf{x})$ is the local 'bubble time' parameter, and $\mathcal{H}$ is the Hamiltonian density; for instance, in the case of a scalar field, one has

$$\mathcal{H} = -\frac{\hbar^2}{2}\frac{\delta^2}{\delta\phi^2} + \frac{1}{2}(\nabla\phi)^2 + V(\phi) \ . \tag{5.33}$$

In the approach 'choice of time before quantization', the gravitational constraints are directly cast into TS form, see (5.14). Equation (5.32) is at best only of formal significance. First, the bubble time cannot be a scalar; see Giulini and Kiefer (1995) and Section 5.4. Second, although the TS equation describes in principle the evolution along all possible foliations of space–time into space-like hypersurfaces, this evolution cannot be unitarily implemented on Fock space (Giulini and Kiefer 1995; Helfer 1996; Torre and Varadarajan 1999). The only sensible approach is the use of a genuine Schrödinger equation, that is, an integrated version of (5.32) along a privileged foliation of space–time.

The Schrödinger equation can be applied successfully to some non-perturbative aspects of quantum field theory; see Jackiw (1995), section IV.4, for a detailed discussion and references. Among the applications are the θ-structure of QCD (cf. Section 5.3.1), chiral anomalies, and confinement. From a more principal point of view one can show that, at least for the ϕ^4-theory in Minkowski space, the Schrödinger picture exists at each order of perturbation theory, that is, in each order one has an integrated version of (5.32) with renormalized quantities,

$$i\hbar \frac{\partial \Psi_{\text{ren}}}{\partial t} = \int \mathrm{d}^3 x \; \mathcal{H}_{\text{ren}} \Psi_{\text{ren}} \; , \tag{5.34}$$

where one additional renormalization constant is needed in comparison to the Fock-space formulation; see Symanzik (1981).[6]

The simplest example for the Schrödinger picture is the free bosonic field. The implementation of the commutation relations

$$[\hat{\phi}(\mathbf{x}), \hat{p}_\phi(\mathbf{y})] = i\hbar \delta(\mathbf{x} - \mathbf{y}) \tag{5.35}$$

leads to

$$\hat{\phi}(\mathbf{x})\Psi[\phi(\mathbf{x})] = \phi(\mathbf{x})\Psi[\phi(\mathbf{x})] \; , \tag{5.36}$$

$$\hat{p}_\phi \Psi[\phi(\mathbf{x})] = \frac{\hbar}{i} \frac{\delta}{\delta \phi(\mathbf{x})} \Psi[\phi(\mathbf{x})] \; , \tag{5.37}$$

where $\Psi[\phi(\mathbf{x})]$ is a wave functional on the space of all fields $\phi(\mathbf{x})$, which includes not only smooth classical configurations, but also distributional ones. The Hamilton operator for a free massive scalar field reads (from now on again $\hbar = 1$)

$$\hat{H} = 1/2 \int \mathrm{d}^3 x \; \left(\hat{p}_\phi^2(\mathbf{x}) + \hat{\phi}(\mathbf{x})(-\nabla^2 + m^2)\hat{\phi}(\mathbf{x}) \right)$$

$$\equiv 1/2 \int \mathrm{d}^3 x \; \hat{p}_\phi^2(\mathbf{x}) + \frac{1}{2} \int \mathrm{d}^3 x \mathrm{d}^3 x' \; \hat{\phi}(\mathbf{x})\omega^2(\mathbf{x}, \mathbf{x}')\hat{\phi}(\mathbf{x}') \; , \tag{5.38}$$

where

$$\omega^2(\mathbf{x}, \mathbf{x}') \equiv (-\nabla^2 + m^2)\delta(\mathbf{x} - \mathbf{x}') \tag{5.39}$$

is not diagonal in three-dimensional space, but is diagonal in momentum space (due to translation invariance),

$$\omega^2(\mathbf{p}, \mathbf{p}') \equiv \int \mathrm{d}^3 p \; \omega(\mathbf{p}, \mathbf{p}'')\omega(\mathbf{p}'', \mathbf{p}')$$

$$= \frac{1}{(2\pi)^3} \int \mathrm{d}^3 x \mathrm{d}^3 x' \; e^{i\mathbf{p}\mathbf{x}}\omega^2(\mathbf{x}, \mathbf{x}')e^{-i\mathbf{p}'\mathbf{x}'}$$

---

[6]This analysis was generalized by McAvity and Osborn (1993) to quantum field theory on manifolds with arbitrarily smoothly curved boundaries. Non-Abelian fields are treated, for example, in Lüscher et al. (1992).

$$= (p^2 + m^2)\delta(\mathbf{p} - \mathbf{p}') \,, \tag{5.40}$$

with $p \equiv |\mathbf{p}|$. Therefore,

$$\omega(\mathbf{p}, \mathbf{p}') = \sqrt{p^2 + m^2}\delta(\mathbf{p} - \mathbf{p}') \equiv \omega(p)\delta(\mathbf{p} - \mathbf{p}') \,. \tag{5.41}$$

The stationary Schrödinger equation then reads

$$\hat{H}\Psi_n[\phi] \equiv \left(-\frac{1}{2}\int \mathrm{d}^3x \, \frac{\delta^2}{\delta\phi^2} + \frac{1}{2}\int \mathrm{d}^3x \mathrm{d}^3x' \, \phi\omega^2\phi\right)\Psi_n[\phi] = E_n\Psi_n[\phi] \,. \tag{5.42}$$

In analogy to the ground-state wave function of the quantum-mechanical harmonic oscillator, the ground-state solution for (5.42) reads

$$\Psi_0[\phi] = \det{}^{1/4}\left(\frac{\omega}{\pi}\right)\exp\left(-\frac{1}{2}\int \mathrm{d}^3x \mathrm{d}^3x' \, \phi(\mathbf{x})\omega(\mathbf{x}, \mathbf{x}')\phi(\mathbf{x}')\right) \,, \tag{5.43}$$

with the ground-state energy given by

$$E_0 = \frac{1}{2}\mathrm{tr}\omega = \frac{1}{2}\int \mathrm{d}^3x \mathrm{d}^3x' \, \omega(\mathbf{x}, \mathbf{x}')\delta(\mathbf{x} - \mathbf{x}') = \frac{1}{2}\frac{V}{(2\pi)^3}\int \mathrm{d}^3p \, \omega(p) \,. \tag{5.44}$$

This is just the sum of the ground-state energies for infinitely many harmonic oscillators. Not surprisingly, it contains divergences: the infrared (IR) divergence connected with the spatial volume V (due to translational invariance) and the ultraviolet (UV) divergence connected with the sum over all oscillators. This is the usual field theoretic divergence of the ground-state energy and can be dealt with by standard methods (e.g. normal ordering). Note that the normalization factor in (5.43) is also divergent:

$$\det{}^{1/4}\left(\frac{\omega}{\pi}\right) = \exp\left(\frac{1}{4}\mathrm{tr}\ln\frac{\omega}{\pi}\right) = \exp\left(\frac{V}{32\pi^3}\int \mathrm{d}^3p \, \ln\frac{\sqrt{p^2 + m^2}}{\pi}\right) \,. \tag{5.45}$$

One can define the many-particle states (the Fock space) in the usual manner through the application of creation operators on (5.43). The divergence (5.45) cancels in matrix elements between states in Fock space. However, the space of wave functionals is much bigger than Fock space. In fact, because there is no unique ground state in the case of time-dependent external fields, any Gaussian functional is called a 'vacuum state', independent of whether it is the ground state of some Hamiltonian or not. A general Gaussian is of the form

$$\Psi_\Omega[\phi] = \det{}^{1/4}\left(\frac{\Omega_R}{\pi}\right)\exp\left(-\frac{1}{2}\int \mathrm{d}^3x \mathrm{d}^3x' \, \phi(\mathbf{x})\Omega(\mathbf{x}, \mathbf{x}')\phi(\mathbf{x}')\right) \,, \tag{5.46}$$

where $\Omega \equiv \Omega_R + i\Omega_I$ is in general complex and time-dependent. One can define in the usual manner an annihilation operator (skipping the integration variables for simplicity)

$$A = \frac{1}{\sqrt{2}} \int \Omega_R^{-1/2} \left(\Omega \phi + \frac{\delta}{\delta \phi} \right) \tag{5.47}$$

and a creation operator

$$A^\dagger = \frac{1}{\sqrt{2}} \int \Omega_R^{-1/2} \left(\Omega^* \phi - \frac{\delta}{\delta \phi} \right) . \tag{5.48}$$

One has the usual commutation relation $[A(\mathbf{x}), A^\dagger(\mathbf{y})] = \delta(\mathbf{x} - \mathbf{y})$, and the vacuum state (5.46) is annihilated by A, $A\Psi_\Omega = 0$.

Gaussian functionals are used frequently in quantum field theory with external fields. Examples are an external electric field in QED or an external De Sitter-space background in a gravitational context (Jackiw 1995). The functional Schrödinger picture can also be formulated for fermions (see e.g. Kiefer and Wipf 1994; Jackiw 1995; Barvinsky *et al.* 1999*b*). In the context of linearized gravity, we have already encountered the Gaussian functional describing the graviton ground state; see the end of Section 2.1. Many discussions of the geometrodynamical wave functional take their inspiration from the above discussed properties of the Schrödinger picture.

5.3.4 *Connection with path integrals*

We have discussed in Section 2.2 the formulation of a quantum-gravitational path integral. In quantum mechanics, the path integral can be shown to satisfy the Schrödinger equation (Feynman and Hibbs 1965). It is, therefore, of interest to see if a similar property holds in quantum gravity, that is, if the quantum-gravitational path integral (2.71) obeys the quantum constraints (5.18) and (5.19). This is not straightforward since there are two major differences to ordinary quantum theory: first, one has constraints instead of the usual Schrödinger equation. Second, the path integral (2.71) contains an integration over the whole four-metric, that is, including 'time' (in the form of the lapse function). Since the ordinary path integral in quantum mechanics is a propagator, denoted by $\langle q'', T | q', 0 \rangle$, the quantum-gravitational path integral corresponds to an expression of the form

$$\int dT \, \langle q'', T | q', 0 \rangle \equiv G(q'', q'; E)|_{E=0} \, ,$$

where the 'energy Green function'

$$G(q'', q'; E) = \int dT \, e^{iET} \langle q'', T | q', 0 \rangle \tag{5.49}$$

has been introduced. The quantum-gravitational path integral thus resembles an energy Green function instead of a propagator, and due to the T-integration no composition law holds in the ordinary sense (Kiefer 1991, 2001). All this is, of course, true already for the models with reparametrization invariance discussed in Section 3.1. In general, an integration over T yields a divergence. One therefore has to choose appropriate contours in a complex T-plane in order to get a sensible result.

A formal derivation of the constraints from (2.71) is straightforward (Hartle and Hawking 1983). Taking a matter field ϕ into account, the path integral reads

$$Z = \int \mathcal{D}g\mathcal{D}\phi \; e^{iS[g,\phi]} \;, \tag{5.50}$$

where the integration over $\mathcal{D}g$ includes an integration over the three-metric as well as lapse function N and shift vector N^a. From the demand that Z be independent of N and N^a at the three-dimensional boundaries, one gets

$$\frac{\delta Z}{\delta N} = 0 = \int \mathcal{D}g\mathcal{D}\phi \; \frac{\delta S}{\delta N} \Big|_{\text{3-bound.}} \; e^{iS[g,\phi]} \;,$$

and an analogous expression for N^a. The conditions that the path integrals containing $\delta S/\delta N$ and $\delta S/\delta N^a$ vanish, respectively, immediately yield the constraints (5.18) and (5.19).

A more careful derivation has to take care for the definition of the measure in the path integral. This was first attempted by Leutwyler (1964) without yet taking into account ghost terms. Regarding the correct gauge-fixing procedure, this was achieved by Barvinsky and collaborators; see Barvinsky (1993a) for a review and references. Halliwell and Hartle (1991) address general reparametrization-invariant systems and demand that the 'sum over histories' in the path integral respect the invariance generated by the constraints.[7] They assume a set of constraints $H_\alpha \approx 0$ obeying the Poisson-bracket relations

$$\{H_\alpha, H_\beta\} = U^\gamma_{\alpha\beta}H_\gamma \;, \tag{5.51}$$

where the $U^\gamma_{\alpha\beta}$ may depend on the canonical variables p_i and q^i (as happens in GR, where a dependence on the three-metric is present, see (3.90)). The corresponding action is written in the form

$$S[p_i, q^i, N^\alpha] = \int_{t_1}^{t_2} dt \; (p_i\dot{q}^i - N^\alpha H_\alpha) \;, \tag{5.52}$$

where N^α are Lagrange parameters. As in Section 3.1, one considers

$$\delta p_i = \{p_i, \epsilon^\alpha H_\alpha\} \;, \quad \delta q^i = \{q^i, \epsilon^\alpha H_\alpha\} \;.$$

If

$$\delta N^\alpha = \dot{\epsilon}^\alpha - U^\alpha_{\beta\gamma}N^\beta\epsilon^\gamma \;,$$

the action transforms as

$$\delta S = \left[\epsilon^\alpha F_\alpha(p_i, q^i)\right]_{t_1}^{t_2} \;, \quad F_\alpha = p_i\frac{\partial H_\alpha}{\partial \dot{p}_i} - H_\alpha \;.$$

Except for constraints linear in the momenta, the action is only invariant if $\epsilon^\alpha(t_2) = 0 = \epsilon^\alpha(t_1)$; cf. (3.34). Halliwell and Hartle (1991) have shown that for

---

[7]Restriction is made to quantum-mechanical systems, so issues such as field-theoretic anomalies are not discussed.

such systems—together with five natural assumptions for the path integral—the quantum constraints

$$\hat{H}_\alpha \psi(q^i) = 0$$

follow from the path integral. At least on the formal level (neglecting anomalies etc.) GR is included as a special case and the derivation applies. The constraints follow only if the integration range $-\infty < N < \infty$ holds for the lapse function. A *direct* check that the path integral solves the quantum constraints was achieved in Barvinsky (1998) for generic (first class) constrained systems on the one-loop of the semiclassical approximation; see also Barvinsky (1993a) and the references therein.

Quantum gravitational path integrals play also a crucial role for the formulation of boundary conditions in quantum cosmology; cf. Section 8.3.

5.3.5   Anomalies and factor ordering

If classical constraints $\mathcal{G}_a \approx 0$ are quantized à la Dirac, one gets restriction on wave functions according to $\hat{\mathcal{G}}_a \psi = 0$. Also, it is evident that the commutator between two constraints must vanish if applied on wave functions, that is,

$$[\hat{\mathcal{G}}_a, \hat{\mathcal{G}}_b]\psi = 0 \ . \tag{5.53}$$

This requirement is known as 'Dirac consistency'. It only holds if the commutator has the form

$$[\hat{\mathcal{G}}_a, \hat{\mathcal{G}}_b]\psi = C^c_{ab}(\hat{p}, \hat{q})\hat{\mathcal{G}}_c\psi \ , \tag{5.54}$$

with the coefficients $C^c_{ab}(\hat{p}, \hat{q})$ standing to the *left* of the constraints. If this is not the case, additional terms proportional to a power of $\hbar$ appear. They are called 'central terms' (if they are c-numbers), 'Schwinger terms', or simply *anomalies*. We have encountered anomalies already in our discussion of the bosonic string; cf. (3.62). If there are anomalies, it is not possible to implement all constraints in the quantum theory via the equations $\hat{\mathcal{G}}_a \psi = 0$.

The question whether there are anomalies in quantum gravity has not yet been answered. One may hope that the demand for an absence of anomalies may fix the factor ordering of the theory and perhaps other issues such as the allowed number of space–time dimensions or the value of fundamental parameters. The question is: can the structure of the Poisson algebra (3.90)–(3.92) be preserved for the corresponding commutators? Or are there necessarily anomalous terms? In spite of much literature, no definite result has arisen. A warning has been pronounced by Tsamis and Woodard (1987): in order to establish Dirac consistency, one must first properly regularize singular operator products. Otherwise, one can get any result. This is because identities such as

$$f(y)g(x)\delta(x - y) = f(x)g(x)\delta(x - y)$$

are justified for test functions f and g, but not for distributions (and field operators are distributions!).

It might be possible to get some insight by looking at anomalies in ordinary quantum field theories (see e.g. Jackiw 1995; Bertlmann 1996). In Yang–Mills theories, one has the generalized Gauß law, see (4.30) and (4.122),

$$\mathcal{G}_i \equiv \mathcal{D}_a E_i^a - \rho_i \approx 0 \ , \tag{5.55}$$

where ρ_i denotes the (non-Abelian) charge density for the sources. Quantization yields $\hat{\mathcal{G}}_i \Psi = 0$, where Ψ depends on the gauge fields $A_a^i(x)$ and the charged fields. One also uses instead of A_a^i the variable $\mathcal{A}_a = A_a^i T_i$, where T_i denote the generators of the gauge group,

$$[T_i, T_j] = C_{ij}^k T_k \ .$$

Classically,

$$\{\mathcal{G}_i(x), \mathcal{G}_j(y)\} = C_{ij}^k \mathcal{G}_k(x)\delta(x, y) \ . \tag{5.56}$$

Dirac consistency would then be implemented in the quantum theory if one had

$$[\hat{\mathcal{G}}_i(x), \hat{\mathcal{G}}_j(y)] = i\hbar C_{ij}^k \hat{\mathcal{G}}_k(x)\delta(x, y) \ . \tag{5.57}$$

It is, however, known that anomalies may occur in the presence of fermions with definite chirality; cf. Bertlmann (1996) and the references therein,

$$[\hat{\mathcal{G}}_i(x), \hat{\mathcal{G}}_j(y)] = i\hbar C_{ij}^k \hat{\mathcal{G}}_k(x)\delta(x, y) \pm \frac{i\hbar}{24\pi^2}\epsilon^{abc}\mathrm{tr}\{T_i, T_j\}\partial_a \mathcal{A}_b \partial_c \delta(x, y) \ , \tag{5.58}$$

where the sign in front of the second term on the right-hand side depends on the chirality. In perturbation theory, the occurrence of such anomalies arises through triangle graphs. An anomaly is harmless as long as it describes only the breakdown of an external symmetry in the presence of gauge fields. It can then even be responsible for particle decays: the decay $\pi^0 \to \gamma\gamma$, for instance, is fully generated by the axial anomaly (the non-conservation of the axial current). An analogous anomaly, which is of relevance for gravity, is the 'Weyl anomaly' or 'trace anomaly' (see e.g. DeWitt 1979, 2003; Birrell and Davies 1982); cf. also Section 2.2.4: the invariance of a classical action under Weyl transformations (multiplication of the metric with a function) leads to a traceless energy–momentum tensor. Upon quantization, however, the trace can pick up a non-vanishing term proportional to $\hbar$.

An anomaly becomes problematic if the gauge fields are treated as quantized internal fields because this would lead to a violation of gauge invariance with all its consequences such as the destruction of renormalizability. This is the situation described by (5.58). In the standard model of strong and electroweak interactions, such harmful anomalies could in principle emerge from the electroweak (SU(2) × U(1)) sector. However, the respective anomalies of quarks and leptons cancel each other to render the standard model anomaly-free. This is only possible because there is an equal number of quarks and leptons and because quarks have three possible colours. In superstring theory (Chapter 9), anomaly cancellation occurs (for the heterotic string) only if the possible gauge groups are strongly constrained (either SO(32) or E8 × E8). Chiral fermions in external gravitational fields lead to (Lorentz) anomalies only in space–time dimensions $2, 6, 10, \ldots$; see for example, Leutwyler (1986).

It is possible to quantize an anomalous theory, but not through the equations $\hat{\mathcal{G}}_i \Psi = 0$ (Bertlmann 1996). The chiral Schwinger model (chiral QED in 1+1 dimensions), for example, allows a consistent quantum theory with a massive boson.

Another instructive example for the discussion of anomalies is dilaton gravity in 1+1 dimensions. It is well known that GR in 1+1 dimensions possesses no dynamics (see e.g. Brown (1988) for a review), since

$$\int \mathrm{d}^2 x \, \sqrt{-g} \, ^{(2)}R = 4\pi\chi \, , \tag{5.59}$$

where $^{(2)}R$ is the two-dimensional Ricci scalar, and χ is the Euler characteristic of the two-dimensional manifold; if the manifold were a closed compact Riemann surface with genus g, one would have $\chi = 2(1-g)$. Although (5.59) plays a role in string perturbation theory (see Chapter 9), it is of no use in a direct quantization of GR. One can, however, construct non-trivial models in two dimensions if there are degrees of freedom in the gravitational sector in addition to the metric. A particular example is the presence of a *dilaton field*. Such a field occurs, for example, in the 'CGHS model' presented in Callan *et al.* (1992). This model is defined by the action

$$4\pi G S_{\mathrm{CGHS}} = \int \mathrm{d}^2 x \sqrt{-g} \, \mathrm{e}^{-2\phi} \left(\, ^{(2)}R + 4g^{\mu\nu}\partial_\mu\phi\partial_\nu\phi - \lambda \right) + S_{\mathrm{m}} \, , \tag{5.60}$$

where ϕ is the dilaton field, and λ is a parameter ('cosmological constant') with dimension L^{-2}.[8] Note that the gravitational constant G is dimensionless in two dimensions. The name 'dilaton' comes from the fact that ϕ occurs in the combination $\mathrm{d}^2 x \sqrt{-g} \, \mathrm{e}^{-2\phi}$ and can thus be interpreted as describing an effective change of integration measure ('change of volume'). It is commonly found in string perturbation theory (Chapter 9), and its value there determines the string coupling constant.

The simplest choice for the matter action S_{m} is an ordinary scalar-field action,

$$S_{\mathrm{m}} = 1/2 \int \mathrm{d}^2 x \, \sqrt{-g} g^{\mu\nu}\partial_\mu\varphi\partial_\nu\varphi \, . \tag{5.61}$$

Cangemi *et al.* (1996) make a series of redefinitions and canonical transformations (partly non-local) to simplify this action. The result is then defined as providing

---

[8]One can exhaust all dilaton models by choosing instead of λ any potential $V(\phi)$, cf. Louis-Martinez and Kunstatter (1994). A particular example is the dimensional reduction of spherically symmetric gravity to two dimensions; see Grumiller *et al.* (2002) for a general review of dilaton gravity in two dimensions.

the starting point for quantization (independent of whether equivalence to the old variables holds or not). In the Hamiltonian version, one finds again constraints: one Hamiltonian constraint and one momentum constraint. They read (after a rescaling $\lambda \to \lambda/8\pi G$)

$$\mathcal{H}_\perp = \frac{(\pi_1)^2 - (\pi_0)^2}{2\lambda} - \frac{\lambda}{2}([r^0]')^2 + \frac{\lambda}{2}([r^1]')^2 + \frac{1}{2}(\pi_\varphi^2 + [\varphi']^2) , \quad (5.62)$$

$$\mathcal{H}_1 = -[r^0]'\pi_0 - [r^1]'\pi_1 - \varphi'\pi_\varphi , \quad (5.63)$$

where r^0 and r^1 denote the new gravitational variables (found from the metric—the only dynamical part being its conformal part—and the dilaton), and π_0 and π_1 are their respective momenta. The form of $\mathcal{H}_1$ is similar to the case of parametrized field theory and string theory, cf. (3.80) and (3.49). One recognizes explicitly that the kinetic term in $\mathcal{H}_\perp$ is indefinite. In fact, the Hamiltonian constraint describes an 'indefinite harmonic oscillator' (Zeh 1988)—the sum of two ordinary oscillators where one comes with the opposite sign (see also Section 7.3).

According to our general prescription one has in the quantum theory

$$\hat{\mathcal{H}}_\perp \Psi(r^0, r^1, \varphi) = 0 , \quad \hat{\mathcal{H}}_1 \Psi(r^0, r^1, \varphi) = 0 . \quad (5.64)$$

Although both $\mathcal{H}_\perp$ and $\mathcal{H}_1$ are a sum of independent terms, one cannot expect to find a product state as a common solution ('correlation interaction'). All physical states are probably entangled among all degrees of freedom. The algebra of constraints (3.90)–(3.92) then reads in the quantum theory,[9]

$$i[\hat{\mathcal{H}}_\perp(x), \hat{\mathcal{H}}_\perp(y)] = \hbar(\hat{\mathcal{H}}_1(x) + \hat{\mathcal{H}}_1(y))\delta'(x - y) , \quad (5.65)$$

$$i[\hat{\mathcal{H}}_\perp(x), \hat{\mathcal{H}}_\perp(y)] = \hbar(\hat{\mathcal{H}}_\perp(x) + \hat{\mathcal{H}}_\perp(y))\delta'(x - y) - \frac{c\hbar^2}{12\pi}\delta'''(x - y) , \quad (5.66)$$

$$i[\hat{\mathcal{H}}_1(x), \hat{\mathcal{H}}_1(y)] = \hbar(\hat{\mathcal{H}}_1(x) + \hat{\mathcal{H}}_1(y))\delta'(x - y) . \quad (5.67)$$

Note the absence of the metric on the right-hand side of these equations. This is different from the (3+1)-dimensional case. The reason is that $hh^{ab} = 1$ in one spatial dimension and that the constraint generators have been rescaled by a factor $\sqrt{h}$.

In (5.66) an additional 'Schwinger term' with central charge c has been added. The reason is a theorem by Boulware and Deser (1967) stating that there must necessarily be a Schwinger term in the commutator

$$[\hat{\mathcal{H}}_\perp(x), \hat{\mathcal{H}}_1(y)] .$$

This theorem was proven, however, within standard Poincaré-invariant local field theory, with the additional assumption that there be a ground state of the Hamiltonian. This is certainly not a framework that is applicable in a gravitational

[9]The Virasoro form (3.62) of the algebra follows for the combinations $\theta_\pm = (\mathcal{H}_\perp \mp \mathcal{H}_1)$.

context. But since the equations (5.62) and (5.63) have the form of equations in flat space–time, one can tentatively apply this theorem. The central charge is then a sum of three contributions (Cangemi *et al.* 1996),

$$c = c^{\mathrm{g}} + c^{\mathrm{m}} \equiv c_0^{\mathrm{g}} + c_1^{\mathrm{g}} + c^{\mathrm{m}} , \tag{5.68}$$

where c_0^{g} and c_1^{g} are the central charges connected with the gravitational variables r^0 and r^1, respectively, and c^{m} is the central charge connected with the field φ. The result for c depends on the notion of vacuum (if there is one). Standard methods (decomposition into creation and annihilation operators) yield $c_1^{\mathrm{g}} = 1$. What about c_0^{g}? If the sign in front of the $(\pi_0)^2$-term in (5.62) were positive, one would have $c_0^{\mathrm{g}} = 1$, too. But with the minus sign one cannot simultaneously demand positive energy and positive norm. Demanding positive norm one must combine positive frequency with the creation operator (instead, as usual, the annihilation operator). This would yield $c_0^{\mathrm{g}} = -1$. Then, $c = 0$ in the absence of the φ-field and one would have no anomaly. The constraints (5.64) can then be consistently imposed, and one can find the following two solutions in the pure gravitational case,

$$\Psi_{\mathrm{g}}(r^0, r^1) = \exp\left(\pm \frac{\mathrm{i}\lambda}{2\hbar} \int \mathrm{d}x \; (r^0[r^1]' - r^1[r^0]') \right) . \tag{5.69}$$

Exact states describing black holes in generic dilaton models were discussed in Barvinsky and Kunstatter (1996).

The presence of the φ-field would, however, yield $c^{\mathrm{m}} = 1$ and the anomaly would not vanish, $c = 1$. Cangemi *et al.* (1996) have shown that the anomaly can be cancelled by adding an appropriate counterterm, but this leads to a complicated form of the quantum constraints for which no solution is in sight.[10] Albeit obtained within an unrealistically simple model, the above discussion demonstrates what kind of problems can be expected to occur. The presence of anomalies might prevent one to impose all constraints in GR à la Dirac, but one could also imagine that in the full theory a cancellation of the various central charges might occur. The latter is suggested by the indefinite kinetic term in quantum gravity, but an explicit demonstration is far from reach.

The above discussion of anomalies refers to a field-theoretic context. However, even for finite-dimensional models with constraints $\mathcal{H}_\perp \approx 0$, $\mathcal{H}_a \approx 0$ the demand for closure of the quantum algebra leads to restrictions on the possible factor ordering. This was studied by Barvinsky and Krykhtin (1993) for general constrained systems and applied by Barvinsky (1993b) to the gravitational case. The classical constraints are again collectively written as $\mathcal{H}_\alpha \approx 0$. Their

---

[10] These authors also present a proposal for BRST-quantization, $Q_{\mathrm{BRST}}\Psi = 0$, where Q_{BRST} is the BRST-charge; cf. Section 9.1. They find many solutions which, however, cannot be properly interpreted.

Poisson-bracket algebra—the analogue to (3.90)–(3.92)—is given by the shorthand notation (5.51). Demanding equivalence of 'Dirac quantization' and 'BRST quantization', Barvinsky and Krykhtin (1993) find the relation

$$\hat{\mathcal{H}}_\alpha - \hat{\mathcal{H}}_\alpha^\dagger = i\hbar \left(\hat{U}^\lambda_{\alpha\lambda} \right)^\dagger + \mathcal{O}(\hbar^2) \,, \tag{5.70}$$

where the adjoint is defined with respect to the standard Schrödinger inner product. If one demands that the constraints be covariant with respect to redefinitions in configuration space, their quantum form is fixed to read (with $32\pi G = 1$)

$$\hat{\mathcal{H}}_\perp = -\frac{\hbar^2}{2} G_{abcd} \mathcal{D}^{ab} \mathcal{D}^{cd} + V \,, \tag{5.71}$$

$$\hat{\mathcal{H}}_a = -\frac{2\hbar}{i} D_b h_{ac} \mathcal{D}^{bc} + \frac{i\hbar}{2} U^\lambda_{a\lambda} \,, \tag{5.72}$$

where $\mathcal{D}^{ab}$ is the covariant derivative with respect to the DeWitt metric,

$$\mathcal{D}^{ab} \equiv \frac{\mathcal{D}\Psi}{\mathcal{D}h_{ab}} \,.$$

The big open problem is of course to see whether this result survives the transition to the field-theoretic case, that is, whether no anomalies are present after a consistent regularization has been performed.

5.3.6   Canonical quantum supergravity

We have seen in Section 2.3 that the quantization of *super*gravity instead of GR exhibits interesting features. Thus, it seems worthwhile to discuss the canonical quantization of supergravity (SUGRA). This will be briefly reviewed here; more details can be found in D'Eath (1984, 1996) and Moniz (1996). Here, we will only consider the case of $N = 1$ SUGRA given by the action (2.130).

The classical canonical formalism was developed by Fradkin and Vasiliev (1977), Pilati (1977), and Teitelboim (1977). Working again with tetrads, cf. Section 1.1, one has

$$g_{\mu\nu} = \eta_{nm} e^n_\mu e^m_\nu \,. \tag{5.73}$$

For the quantization it is more convenient to use two-component spinors according to

$$e^{AA'}_\mu = e^n_\mu \sigma^{AA'}_n \,, \tag{5.74}$$

where A runs from 1 to 2, A' from $1'$ to $2'$, and the van der Waerden symbols $\sigma^{AA'}_n$ denote the components of the matrices

$$\sigma_0 = -\frac{1}{\sqrt{2}} \mathbb{I} \,, \quad \sigma_a = \frac{1}{\sqrt{2}} \times \text{Pauli matrix} \,, \tag{5.75}$$

with raising and lowering of indices by ϵ^{AB}, ϵ_{AB}, $\epsilon^{A'B'}$, $\epsilon_{A'B'}$, which are all given in matrix form by

$$\begin{pmatrix} 0 & 1 \\ -1 & 0 \end{pmatrix} ,$$

see for example, Wess and Bagger (1992), Sexl and Urbantke (2001) for more details on this formalism. The inverse of (5.74) is given by

$$e_\mu^n = -\sigma_{AA'}^n e_\mu^{AA'} , \qquad (5.76)$$

where $\sigma_{AA'}^n$ is obtained from $\sigma_n^{AA'}$ by raising and lowering indices. One can go from tensors to spinors via $e_\mu^{AA'}$ and from spinors to tensors via $e_{AA'}^\mu$.

One can now rewrite the action (2.130) in two-component language. Instead of e_μ^n (vierbein) and ψ_μ^α (gravitino), one works with the spinor-valued one-form $e_\mu^{AA'}$ and the spinor-valued one-form ψ_μ^A plus its Hermitian conjugate $\bar\psi_\mu^{A'}$. The latter two are odd Grassmann variables, that is, they are anticommuting among themselves. The action then reads (for $\Lambda = 0$)

$$S = \frac{1}{16\pi G} \int \mathrm{d}^4 x \ (\det e_\mu^n) R + \frac{1}{2} \int \mathrm{d}^4 x \ \epsilon^{\mu\nu\rho\sigma} \left(\bar\psi_\mu^{A'} e_{AA'\nu} D_\rho \psi_\sigma^A + \text{h.c.} \right) . \quad (5.77)$$

The derivative D_ρ acts on spinor-valued forms (i.e. acts on their spinor indices only),

$$D_\mu \psi_\nu^A = \partial_\mu \psi_\nu^A + \omega_{B\mu}^A \psi_\nu^B , \qquad (5.78)$$

where $\omega_{B\mu}^A$ denotes the spinorial version of ω_μ^{nm} (see D'Eath 1984). We remark that the presence of gravitinos leads to torsion,

$$D_{[\mu} e_{\nu]}^{AA'} = S_{\mu\nu}^{AA'} = -4\pi \mathrm{i} G \bar\psi_{[\mu}^{A'} \psi_{\nu]}^A , \qquad (5.79)$$

where $S_{\mu\nu}^{AA'}$ denotes the torsion, and the last step follows from variation of the action with respect to the connection forms; see van Nieuwenhuizen (1981). The action (5.77) is invariant under the following infinitesimal local symmetry transformations:

1. *Supersymmetry (SUSY) transformations*:

$$\delta e_\mu^{AA'} = -\mathrm{i}\sqrt{8\pi G}(\epsilon^A \bar\psi_\mu^{A'} + \bar\epsilon^{A'} \psi_\mu^A) , \qquad (5.80)$$

$$\delta \psi_\mu^A = \frac{D_\mu \epsilon^A}{\sqrt{2\pi G}} , \quad \delta \bar\psi_\mu^{A'} = \frac{D_\mu \bar\epsilon^{A'}}{\sqrt{2\pi G}} , \qquad (5.81)$$

where ϵ^A and $\bar\epsilon^{A'}$ denote anticommuting fields.

2. *Local Lorentz transformations*:

$$\delta e_\mu^{AA'} = N_B^A e_\mu^{BA'} + \bar N_{B'}^{A'} e^{AB'\mu} , \qquad (5.82)$$

$$\delta \psi_\mu^A = N_B^A \psi_\mu^B , \quad \delta \bar\psi_\mu^{A'} = \bar N_{B'}^{A'} \bar\psi_\mu^{B'} , \qquad (5.83)$$

with $N^{AB} = N^{(AB)}$.

3. *Local coordinate transformations*:

$$\delta e_\mu^{AA'} = \xi^\nu \partial_\nu e_\mu^{AA'} + e_\nu^{AA'} \partial_\mu \xi^\nu , \tag{5.84}$$

$$\delta \psi_\mu^A = \xi^\nu \partial_\nu \psi_\mu^A + \psi_\nu^A \partial_\mu \xi^\nu , \tag{5.85}$$

where ξ^ν are the parameters defining the (infinitesimal) coordinate transformation.

In analogy to Chapter 4 for GR, one can develop a Hamiltonian formalism for SUGRA. For this purpose, one splits $e_\mu^{AA'}$ into $e_0^{AA'}$ and $e_a^{AA'}$ to get the spatial metric

$$h_{ab} = -e_{AA'a} e_b^{AA'} = g_{ab} , \tag{5.86}$$

where $e_a^{AA'} = e_a^n \sigma_n^{AA'}$ in analogy to (5.74). The spinorial version of the normal vector n^μ reads

$$n^{AA'} = e_\mu^{AA'} n^\mu , \tag{5.87}$$

obeying

$$n^{AA'} e_a^{AA'} = 0 , \quad n_{AA'} n^{AA'} = 1 . \tag{5.88}$$

Analogous to (4.40) one can expand the remaining components of the spinorial tetrad as

$$e_0^{AA'} = N n^{AA'} + N^a e_a^{AA'} , \tag{5.89}$$

with lapse function N and shift vector N^a. The canonical formalism starts with the definition of the momenta. The momenta conjugate to N, N^a, ψ_0^A, and $\bar{\psi}_0^{A'}$ are all zero, since these variables are Lagrange multipliers. The momenta conjugate to the gravitino fields are[11]

$$\pi_A^a = \frac{\delta S}{\delta \dot{\psi}_a^A} = -\frac{1}{2} \epsilon^{abc} \bar{\psi}_b^{A'} e_{AA'c} , \tag{5.90}$$

$$\tilde{\pi}_{A'}^a = \frac{\delta S}{\delta \dot{\bar{\psi}}_a^{A'}} = \frac{1}{2} \epsilon^{abc} \psi_b^A e_{AA'c} . \tag{5.91}$$

Since the action is linear in $D\psi$ and $D\bar{\psi}$, the time derivatives $\dot{\psi}$ and $\dot{\bar{\psi}}$ do not occur on the right-hand sides. Therefore, these equations are in fact constraints. It turns out that these constraints are of *second class*, that is, the Poisson brackets of the constraints do not close on the constraints again; cf. Section 3.1.2. As a consequence, one can eliminate the momenta π_A^a and $\tilde{\pi}_{A'}^a$ from the canonical action by using these constraints. Finally, the momentum conjugate to the spinorial tetrad can be found from

$$p_{AA'}^a = \frac{\delta S}{\delta \dot{e}_a^{AA'}} , \tag{5.92}$$

---

[11] The Grassmann-odd variables must be brought to the left before the functional differentiation is carried out. The momentum $\tilde{\pi}_{A'}^a$ is minus the Hermitian conjugate of π_A^a.

from which the ordinary spatial components follow via $p^{ab} = -e^{AA'a}p^b_{AA'}$. The symmetric part of p^{ab} can be expressed exactly as in (4.63) in terms of the second fundamental form K_{ab} on $t = $ constant,

$$p^{(ab)} = \frac{\sqrt{h}}{16\pi G}\left(K^{(ab)} - Kh^{ab}\right) . \tag{5.93}$$

However, due to the presence of torsion, K_{ab} now possesses also an antisymmetric part,

$$K_{[ab]} = S_{0ab} = n^\mu S_{\mu ab} . \tag{5.94}$$

If second-class constraints are present, one has to use *Dirac brackets* instead of Poisson brackets for the canonically conjugate variables (Dirac 1964; Sundermeyer 1982; Henneaux and Teitelboim 1992). Dirac brackets coincide with Poisson brackets on the constraint hypersurface but have the advantage that the variables of the original phase space can be used. They are denoted by $\{...\}_*$. In the present case, one has

$$\{e_a^{AA'}(x), e_b^{BB'}(y)\}_* = 0 ,$$
$$\{e_a^{AA'}(x), p^b_{BB'}(y)\}_* = \epsilon_B^A \epsilon_{B'}^{A'}\delta_a^b\delta(x,y) ,$$
$$\{p^a_{AA'}(x), p^b_{BB'}(y)\}_* = 1/4\ \epsilon^{bcd}\psi_{Bd}D_{AB'ec}\epsilon^{aef}\bar\psi_{A'f}\delta(x,y) + \text{h.c.} ,$$
$$\{\psi_a^A(x), \psi_b^B(y)\}_* = 0 , \tag{5.95}$$
$$\{\psi_a^A(x), \bar\psi_b^{A'}(y)\}_* = -D_{ab}^{AA'}\delta(x,y) ,$$
$$\{e_a^{AA'}(x), \psi_b^B(y)\}_* = 0 ,$$
$$\{p^a_{AA'}(x), \psi_b^B(y)\}_* = 1/2\ \epsilon^{acd}\psi_{Ad}D^B_{A'bc}\delta(x,y) ,$$

where

$$D_{ab}^{AA'} = -\frac{2\mathrm{i}}{\sqrt{h}}e_b^{AB'}e_{BB'a}n^{BA'} . \tag{5.96}$$

In addition, one has the conjugate relations.

The invariance of the action under local Lorentz transformations yields the primary constraints

$$J_{AB} \approx 0 , \quad \bar J_{A'B'} \approx 0 , \tag{5.97}$$

where

$$J_{AB} = e_{(A}^{A'a}p_{B)A'a} + \psi_{(A}^a\pi_{B)a} . \tag{5.98}$$

In addition one finds the secondary constraints

$$\mathcal{H}_\perp \approx 0 , \quad \mathcal{H}_a \approx 0 , \quad S_A \approx 0 , \quad \bar S_{A'} \approx 0 , \tag{5.99}$$

where S_A and $\bar S_{A'}$ denote the generators of SUSY transformations. One also likes to use the combination

$$\mathcal{H}_{AA'} = -n_{AA'}\mathcal{H}_\perp + e_{AA'}^a\mathcal{H}_a . \tag{5.100}$$

With the definition $\pi^{ab} = -p^{(ab)}/2$, one finds for the Hamiltonian constraint

$$\mathcal{H}_\perp = 16\pi G\ G_{abcd}\pi^{ab}\pi^{cd} - \frac{\sqrt{h}\ ^{(3)}R}{16\pi G}$$

$$+\pi G\sqrt{h}n_{AA'}\bar{\psi}^{A'}_{[a}\psi^A_{b]}n^{BB'}\bar{\psi}^{[a}_{B'}\psi^{b]}_B$$

$$+\frac{1}{2}\epsilon^{abc}\bar{\psi}^{A'}_a n_{AA'}\mathcal{D}_b\psi^A_c + \text{h.c.} \qquad (5.101)$$

plus terms proportional to the Lorentz constraints. $\mathcal{D}_b$ denotes the three-dimensional version of (5.78),

$$\mathcal{D}_b\psi^A_a = \partial_b\psi^A_a + {}^{(3)}\omega^A_{Bb}\psi^B_a , \qquad (5.102)$$

where ${}^{(3)}\omega^A_{Bb}$ are the spatial connection forms. For the explicit expressions of the other constraints, we refer to D'Eath (1984, 1996) and Moniz (1996); see below for the quantum versions of the Lorentz and the SUSY constraints.

The time evolution of a dynamical variable is given by

$$\frac{\mathrm{d}A}{\mathrm{d}t} = \{A, H\}_* , \qquad (5.103)$$

where the Hamiltonian is given by the expression

$$H = \int \mathrm{d}^3x \left(N\mathcal{H}_\perp + N^a\mathcal{H}_a + \psi^A_0 S_A + \bar{S}_{A'}\bar{\psi}^{A'}_0 - \omega_{AB0}J^{AB} - \bar{\omega}_{A'B'0}\bar{J}^{A'B'}\right) , \qquad (5.104)$$

from which the constraints follow after variation with respect to the Lagrange multipliers. This expression holds for the spatially compact case. In the asymptotically flat case one has again terms at spatial infinity—the original ones (Section 4.2.4) plus supercharges at infinity arising from the global SUSY algebra (Section 2.3).

Of particular interest are the Dirac brackets among the SUSY generators, for which one finds

$$\{S_A(x), S_B(y)\}_* = 0 , \quad \{\bar{S}_{A'}(x), \bar{S}_{B'}(y)\}_* = 0 , \qquad (5.105)$$

and

$$\{S_A(x), \bar{S}_{A'}(y)\}_* = 4\pi iG\mathcal{H}_{AA'}(x)\delta(x,y) \qquad (5.106)$$

plus terms proportional to the constraints (5.97). One recognizes from (5.106) that the constraints $\mathcal{H}_{AA'} \approx 0$ already follow from the validity of the remaining constraints. Since the SUSY constraints appear quadratically on the left and the Hamiltonian constraint linearly on the right, one can refer to $N = 1$ SUGRA as the 'square root of gravity' (Teitelboim 1977).

Quantization proceeds by turning Dirac brackets into commutators or anticommutators. Grassmann-even variables are quantized using commutators (omitting hats on operators),

$$\{E_1, E_2\}_* \longrightarrow -\frac{\mathrm{i}}{\hbar}[E_1, E_2] , \qquad (5.107)$$

while Grassmann-odd variables are quantized using anticommutators,

$$\{O_1, O_2\}_* \longrightarrow -\frac{i}{\hbar}[O_1, O_2]_+ \; . \tag{5.108}$$

Mixed variables are quantized via commutators,

$$\{O, E\}_* \longrightarrow -\frac{i}{\hbar}[O, E] \; . \tag{5.109}$$

Proceeding as in Section 5.2.2, one can implement the Dirac brackets (5.95) via wave functionals

$$\Psi[e_a^{AA'}(x), \psi_a^A(x)] \; , \tag{5.110}$$

which can depend either on $\psi_a^A(x)$ (as written here) or on $\bar{\psi}_a^{A'}$, but not both. This is because of the non-trivial anticommutator

$$[\psi_a^A(x), \bar{\psi}_b^{A'}(y)]_+ = -i\hbar D_{ab}^{AA'}\delta(x, y) \; , \tag{5.111}$$

which can be represented on the wave functional as[12]

$$\bar{\psi}_a^{A'}(x) = -i\hbar D_{ba}^{AA'}\frac{\delta}{\delta\psi_b^A(x)} \; . \tag{5.112}$$

The momenta $p_{AA'}^a(x)$ are represented as

$$\begin{aligned}
p_{AA'}^a(x) &= -i\hbar\frac{\delta}{\delta e_a^{AA'}(x)} + \frac{1}{2}\epsilon^{abc}\psi_{Ab}(x)\bar{\psi}_{A'c}(x) \\
&= -i\hbar\frac{\delta}{\delta e_a^{AA'}(x)} - \frac{i\hbar}{2}\epsilon^{abc}\psi_{Ab}(x)D_{A'dc}^B\frac{\delta}{\delta\psi_d^B(x)} \; .
\end{aligned} \tag{5.113}$$

The factor ordering has been chosen such that all derivatives stand on the right. One can choose a formal Schrödinger-type inner product with respect to which $p_{AA'}^a$ is Hermitian and $\bar{\psi}_a^{A'}$, ψ_a^A are Hermitian adjoints (D'Eath 1996).[13]

The quantum constraints then read

$$J_{AB}\Psi = 0 \; , \quad \bar{J}_{A'B'}\Psi = 0 \; , \quad \mathcal{H}_{AA'}\Psi = 0 \; , \quad S_A\Psi = 0 \; , \quad \bar{S}_{A'}\Psi = 0 \; . \tag{5.114}$$

The first two constraints express the invariance of the wave functional under Lorentz transformations, while the last two constraints express its invariance under SUSY transformations. From the quantum version of (5.106), it becomes clear that a solution of the Lorentz and the SUSY constraints is *also* a solution to $\mathcal{H}_{AA'}\Psi = 0$, provided, of course, that there are no anomalies and the quantum algebra closes. The issue of anomalies is here as unsolved as it is in the case of quantum GR.

---

[12]One can also employ other representations; cf. the remarks in Moniz (2003).

[13]One can go via a functional Fourier transformation from Ψ to $\tilde{\Psi}[e_a^{AA'}(x), \bar{\psi}_a^{A'}(x)]$.

The explicit form of the quantum Lorentz constraint operators reads

$$J_{AB} = -\frac{i\hbar}{2}\left(e^{A'}_{Ba}\frac{\delta}{\delta e^{AA'}_a} + e^{A'}_{Aa}\frac{\delta}{\delta e^{BA'}_a} + \psi_{Ba}\frac{\delta}{\delta\psi^A_a} + \psi_{Aa}\frac{\delta}{\delta\psi^B_a}\right) , \quad (5.115)$$

$$\bar{J}_{A'B'} = -\frac{i\hbar}{2}\left(e^{A}_{B'a}\frac{\delta}{\delta e^{AA'}_a} + e^{A}_{A'a}\frac{\delta}{\delta e^{AB'}_a}\right) , \quad (5.116)$$

while the quantum SUSY constraint operators read

$$\bar{S}_{A'} = \epsilon^{abc}e_{AA'a}\,{}^sD_b\psi^A_c + 4\pi G\hbar\psi^A_a\frac{\delta}{\delta e^{AA'}_a} , \quad (5.117)$$

$$S_A = i\hbar\,{}^sD_a\left(\frac{\delta}{\delta\psi^A_a}\right) + 4\pi iG\hbar\frac{\delta}{\delta e^{AA'}_a}\left(D^{BA'}_{ba}\frac{\delta}{\delta\psi^B_b}\right) , \quad (5.118)$$

where ${}^sD_a$ is the 'torsion-free derivative'. In the anomaly-free case, one has only to solve the constraints $J_{AB}\Psi = 0$, $S_A\Psi = 0$ (and their conjugates), since $\mathcal{H}_{AA'}\Psi = 0$ must hold automatically. This could lead to considerable simplification because (5.115)–(5.118) involve at most first-order derivatives.

Canonical quantum SUGRA can be applied, for example, in the context of quantum cosmology (Section 8.1). One can also study some general properties. One of them is the fact that pure bosonic states cannot exist (see e.g. Moniz 1996 for discussion and references). This can easily be shown. Considering a bosonic state $\Psi[e^{AA'}_a]$, one has $\delta\Psi/\delta\psi^A_a = 0$ and recognizes immediately that (5.118) is solved, that is, $S_A\Psi = 0$. Assuming that $\Psi[e^{AA'}_a]$ is Lorentz invariant, that is, that the Lorentz constraints are already fulfilled, it is clear that a state with $\bar{S}^{A'}\Psi = 0$ satisfies all constraints. However, such a state cannot exist. This can be seen as follows (Carroll et al. 1994). One multiplies $\bar{S}^{A'}\Psi = 0$ by $[\Psi]^{-1}$ and integrates over space with an arbitrary spinorial test function $\bar{\epsilon}^{A'}(x)$ to get

$$I \equiv \int d^3x\, \bar{\epsilon}^{A'}(x)\left(\epsilon^{abc}e_{AA'a}\,{}^sD_b\psi^A_c + 4\pi G\hbar\psi^A_a\frac{\delta(\ln\Psi)}{\delta e^{AA'}_a}\right) = 0 . \quad (5.119)$$

This must hold for all fields and all $\bar{\epsilon}^{A'}(x)$. Now replacing $\bar{\epsilon}^{A'}(x)$ by $\bar{\epsilon}^{A'}(x)\exp(-\phi(x))$ and $\psi^A_a(x)$ by $\psi^A_a(x)\exp(\phi(x))$, where $\phi(x)$ is some arbitrary function, the second term in (5.119) cancels out in the difference ΔI between the old and the new integral, and one is left with

$$\Delta I = -\int d^3x\, \epsilon^{abc}e_{AA'a}\bar{\epsilon}^{A'}\psi^A_c\partial_b\phi = 0 , \quad (5.120)$$

which is independent of the state Ψ. It is obvious that one can choose the fields as well as $\bar{\epsilon}^{A'}(x)$ and $\phi(x)$ in such a way that the integral is non-vanishing, leading to a contradiction. Therefore, no physical bosonic states can exist, and a solution of the quantum constraints can be represented in the form

$$\Psi[e^{AA'}_a(x), \psi^A_a(x)] = \sum_{n=1}^{\infty}\Psi^{(n)}[e^{AA'}_a(x), \psi^A_a(x)] , \quad (5.121)$$

where the expansion is into states with fermion number n. In fact, one can show with similar arguments that any solution of the quantum constraints must have infinite fermion number. An explicit solution of a peculiar type was found (without any regularization) by Csordás and Graham (1995).

5.4 The semiclassical approximation

5.4.1 *Analogies from quantum mechanics*

The semiclassical approximation to quantum geometrodynamics discussed here uses, in fact, a mixture of two different approximation schemes. The full system is divided into two parts with very different scales. One part is called the 'heavy part'—for it the standard semiclassical (WKB) approximation is used. The other part is called the 'light part'—it is treated fully quantum and follows adiabatically the dynamics of the heavy part. A mixed scheme of this kind is called a 'Born–Oppenheimer' type of approximation scheme. It is successfully applied in molecular physics, where the division is into the heavy nuclei (moving slowly) and the light electrons (following adiabatically the motion of the nuclei). Many molecular spectra can be explained in this way. In quantum gravity, the 'heavy' part is often taken to be the full gravitational field (motivated by the large value of the Planck mass), while the 'light' part are all non-gravitational degrees of freedom (see e.g. Kiefer 1994). This has the formal advantage that an expansion with respect to the Planck mass can be performed. It is, however, fully consistent to consider part of the gravitational part as fully quantum and therefore include it in the 'light' part (see e.g. Halliwell and Hawking 1985; Vilenkin 1989). Physically, these are gravitons and quantum density fluctuations. It depends on the actual situation one is interested in whether this 'light' gravitational part has to be taken into account or not.

For full quantum geometrodynamics, this semiclassical expansion exists only on a formal level. Therefore, it will be appropriate to discuss quantum-mechanical analogies in this subsection. Albeit formal, the expansion scheme is of the utmost conceptual importance. As we shall see, it enables one to recover the usual time as an approximate concept from 'timeless' quantum gravity (Section 5.4.2). This is the relevant approach for observers within the Universe ('intrinsic viewpont'). Moreover, the scheme allows to go to higher orders and calculate quantum-gravitational correction terms to the functional Schrödinger equation, which could have observational significance (Section 5.4.3).

Let us now consider in some detail a simple quantum-mechanical model (see e.g. Kiefer 1994; Bertoni *et al.* 1996; Briggs and Rost 2001). The total system consists of the 'heavy part' described by the variable Q, while the 'light-part' variable is called q.[14] It is assumed that the full system is described by a stationary Schrödinger equation,

$$H\Psi(q, Q) = E\Psi(q, Q) , \qquad (5.122)$$

---

[14]For simplicity, the total system is considered to be two dimensional. The extension to more dimensions is straightforward.

with the Hamilton operator to be of the form

$$H = -\frac{\hbar^2}{2M}\frac{\partial^2}{\partial Q^2} + V(Q) + h(q,Q) \; , \tag{5.123}$$

where $h(q,Q)$ contains the pure q-part and the interaction between q and Q. In the case of the Wheeler–DeWitt equation, the total energy is zero, $E = 0$. One now makes an expansion of the form

$$\Psi(q,Q) = \sum_n \chi_n(Q)\psi_n(q,Q) \tag{5.124}$$

and assumes that $\langle \psi_n | \psi_m \rangle = \delta_{nm}$ for each value of Q. The inner product here is the ordinary scalar product with respect to q only (only this part will be naturally available in quantum gravity). Inserting the ansatz (5.124) into the Schrödinger equation and projecting on $\psi_m(q,Q)$ yields

$$\sum_n \left\langle \psi_m \left| -\frac{\hbar^2}{2M}\frac{\partial^2}{\partial Q^2} \right| \psi_n \right\rangle \chi_n + V(Q)\chi_m + \sum_n \langle \psi_m | h | \psi_n \rangle \chi_n = E\chi_m \; , \tag{5.125}$$

where the Q-derivative acts on everything to the right. We now introduce the 'Born–Oppenheimer potentials'

$$\epsilon_{mn}(Q) \equiv \langle \psi_m | h | \psi_n \rangle \; , \tag{5.126}$$

which for eigenstates of the 'light' variable, $h|\psi_n\rangle = \epsilon_n|\psi_n\rangle$, just read: $\epsilon_{mn}(Q) = \epsilon_n(Q)\delta_{mn}$. In molecular physics, this is often the case of interest. We shall keep the formalism more general. We also introduce the 'connection'

$$A_{mn}(Q) \equiv i\hbar \left\langle \psi_m \left| \frac{\partial \psi_n}{\partial Q} \right. \right\rangle \tag{5.127}$$

and a corresponding momentum

$$\mathcal{P}_{mn} \equiv \frac{\hbar}{i} \left(\delta_{mn}\frac{\partial}{\partial Q} - \frac{i}{\hbar}A_{mn} \right) \; . \tag{5.128}$$

Making use of (5.126) and (5.128), one can write (5.125) in the form

$$\left(\sum_n \frac{\mathcal{P}_{mn}^2}{2M} + \epsilon_{mn}(Q) \right) \chi_n(Q) + V(Q)\chi_m = E\chi_m(Q) \; . \tag{5.129}$$

The modification in the momentum $\mathcal{P}_{mn}$ and the 'Born–Oppenheimer potential' ϵ_{mn} express the 'back reaction' from the 'light' part onto the 'heavy part'. Inserting the ansatz (5.124) into the Schrödinger equation *without* projection on ψ_m one gets

$$\sum_n \chi_n(Q) \left[h(q,Q) - \left(E - V(Q) + \frac{\hbar^2}{2M\chi_n}\frac{\partial^2\chi_n}{\partial Q^2} \right) \right.$$

$$\left. - \frac{\hbar^2}{2M} \frac{\partial^2}{\partial Q^2} - \frac{\hbar^2}{M\chi_n} \frac{\partial \chi_n}{\partial Q} \frac{\partial}{\partial Q} \right] \psi_n(q, Q) = 0 \ . \qquad (5.130)$$

We emphasize that (5.129) and (5.130) are still *exact* equations, describing the coupling between the 'heavy' and the 'light' part.

Various approximations can now be performed. In a first step, one can assume that the 'heavy' part is approximately insensitive to changes in the 'light' part. This enables one to neglect the off-diagonal parts in (5.129), leading to

$$\left[\frac{1}{2M} \left(\frac{\hbar}{i} \frac{\partial}{\partial Q} - A_{nn}(Q) \right)^2 + V(Q) + E_n(Q) \right] \chi_n(Q) = E\chi_n(Q) \ , \qquad (5.131)$$

where $E_n(Q) \equiv \epsilon_{nn}(Q)$. For real ψ_n, the connection vanishes, $A_{nn} = 0$. Otherwise, it leads to a geometric phase ('Berry phase'); cf. Berry (1984). We shall neglect the connection in the following.[15]

In a second step, one can perform a standard semiclassical (WKB) approximation for the heavy part through the ansatz

$$\chi_n(Q) = C_n(Q) e^{iMS_n(Q)/\hbar} \ . \qquad (5.132)$$

This is inserted into (5.131). For the Q-derivative, one gets

$$\frac{\partial^2 \chi_n}{\partial Q^2} = \frac{\partial^2 C_n}{\partial Q^2} \frac{\chi_n}{C_n} + \frac{2iM}{\hbar} \frac{\partial C_n}{\partial Q} \frac{\partial S_n}{\partial Q} \frac{\chi_n}{C_n}$$
$$- \left(\frac{M}{\hbar} \right)^2 \left(\frac{\partial S_n}{\partial Q} \right)^2 \chi_n + \frac{iM}{\hbar} \frac{\partial^2 S_n}{\partial Q^2} \chi_n \ . \qquad (5.133)$$

Assuming M to be large corresponds to neglecting derivatives of C_n and second derivatives of S_n (the usual assumptions for WKB). One then has

$$\frac{\partial^2 \chi_n}{\partial Q^2} \approx - \left(\frac{M}{\hbar} \right)^2 \left(\frac{\partial S_n}{\partial Q} \right)^2 \chi_n \ . \qquad (5.134)$$

The classical momentum is then given by

$$P_n = M \frac{\partial S_n}{\partial Q} \approx \frac{\hbar}{i\chi_n} \frac{\partial \chi_n}{\partial Q} \ , \qquad (5.135)$$

and (5.131) becomes the Hamilton–Jacobi equation,

$$H_{cl} \equiv \frac{P_n^2}{2M} + V(Q) + E_n(Q) = E \ . \qquad (5.136)$$

Since $E_n(Q) = \langle \psi_n | h | \psi_n \rangle$, this corresponds, in the gravitational context, to the semiclassical Einstein equations discussed in Section 1.2, where the expectation value of the energy–momentum tensor appears.

[15]An intriguing idea would be to derive the connection in gauge theories along these lines.

One can now introduce a time coordinate t via the Hamilton equations of motion,

$$\frac{\mathrm{d}}{\mathrm{d}t}P_n = -\frac{\partial}{\partial Q}H_{\mathrm{cl}} = -\frac{\partial}{\partial Q}(V(Q) + E_n(Q)) ,$$

$$\frac{\mathrm{d}}{\mathrm{d}t}Q = \frac{\partial}{\partial P_n}H_{\mathrm{cl}} = \frac{P_n}{M} . \tag{5.137}$$

In fact, the very definition of t depends on n, and we call it therefore t_n in the following. Since it arises from the WKB approximation (5.132), it is called *WKB time* (Zeh 1988). The last term in (5.130) can then be written as

$$-\frac{\hbar^2}{M\chi_n}\frac{\partial\chi_n}{\partial Q}\frac{\partial\psi_n}{\partial Q} \approx -\mathrm{i}\hbar\frac{\partial S_n}{\partial Q}\frac{\partial\psi_n}{\partial Q} \equiv -\mathrm{i}\hbar\frac{\partial\psi_n}{\partial t_n} . \tag{5.138}$$

This means that ψ_n is evaluated along a particular classical trajectory of the 'heavy' variable, $\psi_n(Q(t_n), q) \equiv \psi_n(t_n, q)$. Assuming slow variation of ψ_n with respect to Q, one can neglect the term proportional to $\partial^2\psi_n/\partial Q^2$ in (5.130). Also using (5.136), one is then left with

$$\sum_n \chi_n \left[h(q, t_n) - E_n(t_n) - \mathrm{i}\hbar\frac{\partial}{\partial t_n} \right] \psi_n(t_n, q) = 0 . \tag{5.139}$$

This equation still describes a coupling between 'heavy' and 'light' part.

In a third and last step one can assume that instead of the whole sum (5.124) only one component is available, that is, one has—up to an (adiabatic) dependence of ψ on Q—a factorizing state,

$$\chi(Q)\psi(q, Q) .$$

This lack of entanglement can of course only arise in certain situations and must be dynamically justified (through decoherence; cf. Chapter 10). If it happens, and after absorbing $E_n(t)$ into a redefinition of ψ (yielding only a phase), one gets from (5.139):

$$\mathrm{i}\hbar\frac{\partial\psi}{\partial t} = h\psi , \tag{5.140}$$

that is, just the Schrödinger equation. The 'heavy' system acts as a 'clock' and defines the time with respect to which the 'light' system evolves. Therefore, a time-dependent Schrödinger equation has arisen for one of the subsystems, although the full Schrödinger equation is of a stationary form; cf. Mott (1931).

Considering the terms with order M in (5.133), one finds an equation for the C_n,

$$2\frac{\partial C_n}{\partial Q}\frac{\partial S_n}{\partial Q} + \frac{\partial^2 S_n}{\partial Q^2}C_n = 0 , \tag{5.141}$$

or in the case of one component only,

$$2\frac{\partial S}{\partial Q}\frac{\partial C}{\partial Q} + \frac{\partial^2 S}{\partial Q^2}C \equiv 2\frac{\partial C}{\partial t} + \frac{\partial^2 S}{\partial Q^2}C = 0 \ . \tag{5.142}$$

This can be written in the form of a continuity equation,

$$\frac{\partial}{\partial Q}\left(C^2\frac{\partial S}{\partial Q}\right) = 0 \ . \tag{5.143}$$

A systematic derivation makes use of an M-expansion,

$$\Psi \equiv \exp\left(\mathrm{i}\mathcal{S}/\hbar\right) \ , \quad \mathcal{S} = MS_0 + S_1 + M^{-1}S_2 + \cdots \tag{5.144}$$

The Hamilton–Jacobi equation (5.136) then appears at order M, and both the Schrödinger equation (5.140) and the prefactor equation (5.143) appear at order M^0. The next order, M^{-1}, then yields corrections to the Schrödinger equation (discussed in Section 5.4.3 for the quantum-gravitational case).

Another example, which is closer to the situation with the Wheeler–DeWitt equation (because of the indefinite kinetic term), is the non-relativistic expansion of the Klein–Gordon equation; cf. Kiefer and Singh (1991). In the absence of an external potential, the latter reads

$$\left(\frac{\hbar^2}{c^2}\frac{\partial^2}{\partial t^2} - \hbar^2\Delta + m^2c^2\right)\varphi(\mathbf{x}, t) = 0 \ . \tag{5.145}$$

One writes

$$\varphi(\mathbf{x}, t) \equiv \exp\left(\mathrm{i}S(\mathbf{x}, t)/\hbar\right) \tag{5.146}$$

and expands the exponent in powers of c,

$$S = c^2 S_0 + S_1 + c^{-2}S_2 + \cdots \tag{5.147}$$

This ansatz is inserted into (5.145), which is then solved at consecutive orders of c^2. The highest order, c^2, yields solutions $S_0 = \pm mt$, from which we choose $S_0 = -mt$. This choice corresponds to

$$\varphi = \exp\left(-\mathrm{i}mc^2 t/\hbar\right) \ ,$$

which is the 'positive-energy solution'. The approximation works as long as 'negative-energy solutions' can be consistently neglected, that is, as long as field-theoretic effects such as particle creation do not play any role. Writing $\psi \equiv \exp(\mathrm{i}S_1/\hbar)$ one gets at order c^0, the Schrödinger equation,

$$\mathrm{i}\hbar\dot{\psi} = -\frac{\hbar^2}{2m}\Delta\psi \ . \tag{5.148}$$

The Schrödinger equation has thus been derived as a non-relativistic approximation to the Klein–Gordon equation. Writing $\Psi \equiv \psi\exp(\mathrm{i}S_2/\hbar c^2)$ one arrives at order c^{-2} at the equation

$$\mathrm{i}\hbar\dot{\Psi} = -\frac{\hbar^2}{2m}\Delta\Psi - \frac{\hbar^4}{8m^3c^2}\Delta\Delta\Psi \ . \tag{5.149}$$

The last term is the first relativistic correction term and can be used to derive testable predictions, for example, for spectra of pionic atoms (for ordinary atoms an expansion of the Dirac equation must be employed). A more general case is the Klein–Gordon equation coupled to gravity and the electromagnetic field. This leads to additional relativistic correction terms (Lämmerzahl 1995).

A major difference of the Klein–Gordon example to the first example is the indefinite structure of the kinetic term (d'Alembertian instead of Laplacian). Therefore, on the full level, the conserved inner product is the Klein–Gordon one (cf. Section 5.2.2). In order c^0 of the approximation, one obtains from this inner product the standard Schrödinger inner product as an approximation. The next order yields corrections to the Schrödinger inner product proportional to c^{-2}. Does this mean that unitarity is violated at this order? Not necessarily. In the case of the Klein–Gordon equation in external gravitational and electromagnetic fields, one can make a (t-dependent!) redefinition of wave functions and Hamiltonian to arrive at a conserved Schrödinger inner product with respect to which the Hamiltonian is Hermitian (Lämmerzahl 1995).

5.4.2 *Derivation of the Schrödinger equation*

Similar to the discussion of the examples in the last subsection one can perform a semiclassical ('Born–Oppenheimer') approximation for the Wheeler–DeWitt equation and the momentum constraints. In this way one can recover approximately the limit of ordinary quantum field theory in an external gravitational background. This is done in the Schrödinger picture, so this limit emerges through the *functional* Schrödinger equation, not the quantum-mechanical Schrödinger equation as in the last subsection. In the following, we shall mainly follow, with elaborations, the presentation in Barvinsky and Kiefer (1998); see also Kiefer (1994) and references therein.

Starting point is the Wheeler–DeWitt equation (5.18) and the momentum constraint (5.19). Taking into account non-gravitational degrees of freedom, these equations can be written in the following form,

$$\left\{ -\frac{1}{2m_{\mathrm{P}}^2} G_{abcd} \frac{\delta^2}{\delta h_{ab} \delta h_{cd}} - 2m_{\mathrm{P}}^2 \sqrt{h}\,^{(3)}R + \hat{\mathcal{H}}_{\perp}^{\mathrm{m}} \right\} |\Psi[h_{ab}]\rangle = 0 \; , \quad (5.150)$$

$$\left\{ -\frac{2}{\mathrm{i}} h_{ab} D_c \frac{\delta}{\delta h_{bc}} + \hat{\mathcal{H}}_a^{\mathrm{m}} \right\} |\Psi[h_{ab}]\rangle = 0 \; . \quad (5.151)$$

Here, $m_{\mathrm{P}}^2 = (32\pi G)^{-1}$, $\hbar = 1$, $\Lambda = 0$, and $\hat{\mathcal{H}}_{\perp}^{\mathrm{m}}$ and $\hat{\mathcal{H}}_a^{\mathrm{m}}$ denote the contributions from non-gravitational fields. To be concrete, we think about the presence of a scalar field. The notation $|\Psi[h_{ab}]\rangle$ means: Ψ is a wave functional with respect to the three-metric h_{ab} and a state in the standard Hilbert space referring to the scalar field (bra- and ket-notation).

The situation is now formally similar to the example discussed in the previous subsection. One of the main differences is the presence of the momentum constraints (5.151), which have no analogue in the quantum-mechanical example. Comparing (5.150) with (5.123), one notes the following formal correspondence between the terms:

$$-\frac{1}{2M}\frac{\partial^2}{\partial Q^2} \quad \leftrightarrow \quad -\frac{1}{2m_{\rm P}^2}G_{abcd}\frac{\delta^2}{\delta h_{ab}\delta h_{cd}} \; ,$$

$$V(Q) \quad \leftrightarrow \quad -2m_{\rm P}^2\sqrt{h}\,^{(3)}R \; ,$$

$$h(q,Q) \quad \leftrightarrow \quad \hat{\mathcal{H}}_\perp^{\rm m} \; ,$$

$$\Psi(q,Q) \quad \leftrightarrow \quad |\Psi[h_{ab}]\rangle \; . \tag{5.152}$$

The same steps as in the quantum-mechanical example can now be performed. We already assume at this stage that we have one component instead of a sum like (5.124) and that this component is written in the form

$$|\Psi[h_{ab}]\rangle = C[h_{ab}]e^{im_{\rm P}^2 S[h_{ab}]}|\psi[h_{ab}]\rangle \; . \tag{5.153}$$

In the highest order of a WKB approximation for the gravitational part, $S[h_{ab}]$ obeys a Hamilton–Jacobi equation similar to (5.136). In addition, it obeys a Hamilton–Jacobi version of the momentum constraints. Therefore,

$$\frac{m_{\rm P}^2}{2}G_{abcd}\frac{\delta S}{\delta h_{ab}}\frac{\delta S}{\delta h_{cd}} - 2m_{\rm P}^2\sqrt{h}\,^{(3)}R + \langle\psi|\hat{\mathcal{H}}_\perp^{\rm m}|\psi\rangle = 0 \; , \tag{5.154}$$

$$-2m_{\rm P}^2 h_{ab}D_c\frac{\delta S}{\delta h_{bc}} + \langle\psi|\hat{\mathcal{H}}_a^{\rm m}|\psi\rangle = 0 \; . \tag{5.155}$$

These equations correspond—in the usual four-dimensional notation—to the semiclassical Einstein equations (1.34). Note that the 'back reaction' terms in these equations are formally suppressed by a factor $m_{\rm P}^{-2}$ compared to the remaining terms. For this reason they are often neglected in this order of approximation and only considered in the next order (Kiefer 1994).

Instead of the single-component version of (5.139), one gets here

$$\left(\hat{\mathcal{H}}_\perp^{\rm m} - \langle\psi|\hat{\mathcal{H}}_\perp^{\rm m}|\psi\rangle - iG_{abcd}\frac{\delta S}{\delta h_{ab}}\frac{\delta}{\delta h_{cd}}\right)|\psi[h_{ab}]\rangle = 0 \; , \tag{5.156}$$

$$\left(\hat{\mathcal{H}}_a^{\rm m} - \langle\psi|\hat{\mathcal{H}}_a^{\rm m}|\psi\rangle - \frac{2}{i}h_{ab}D_c\frac{\delta}{\delta h_{bc}}\right)|\psi[h_{ab}]\rangle = 0 \; . \tag{5.157}$$

One now evaluates $|\psi[h_{ab}]\rangle$ along a solution of the classical Einstein equations, $h_{ab}(\mathbf{x},t)$, corresponding to a solution, $S[h_{ab}]$, of the Hamilton–Jacobi equations (5.154) and (5.155),

$$|\psi(t)\rangle = |\psi[h_{ab}(\mathbf{x},t)]\rangle \; . \tag{5.158}$$

After a certain *choice* of lapse and shift functions, N and N^a, has been made, this solution is obtained from

$$\dot{h}_{ab} = NG_{abcd}\frac{\delta S}{\delta h_{cd}} + 2D_{(a}N_{b)} \; . \tag{5.159}$$

To get an evolutionary equation for the quantum state (5.158), one defines

$$\frac{\partial}{\partial t}|\psi(t)\rangle = \int \mathrm{d}^3x\, \dot{h}_{ab}(\mathbf{x})\, \frac{\delta}{\delta h_{ab}(\mathbf{x})}|\psi[h_{ab}]\rangle \; . \tag{5.160}$$

This, then, leads to the functional Schrödinger equation for quantized matter fields in the chosen external classical gravitational field,

$$\mathrm{i}\frac{\partial}{\partial t}|\psi(t)\rangle = \hat{H}^{\mathrm{m}}|\psi(t)\rangle \; ,$$

$$\hat{H}^{\mathrm{m}} \equiv \int \mathrm{d}^3x\, \Big\{ N(\mathbf{x})\hat{\mathcal{H}}_{\perp}^{\mathrm{m}}(\mathbf{x}) + N^a(\mathbf{x})\hat{\mathcal{H}}_a^{\mathrm{m}}(\mathbf{x}) \Big\} \; . \tag{5.161}$$

Here, $\hat{H}^{\mathrm{m}}$ is the matter-field Hamiltonian in the Schrödinger picture, parametrically depending on (generally non-static) metric coefficients of the curved space–time background. This equation is the analogue of (5.140) in the quantum-mechanical example. (The back-reaction terms have again been absorbed into the phase of $|\psi(t)\rangle$.) The standard concept of time in quantum theory thus emerges only in a semiclassical approximation—the Wheeler–DeWitt equation itself is 'timeless'.[16]

Such a derivation of quantum field theory from the Wheeler–DeWitt equations dates back, on the level of cosmological models, to DeWitt (1967a). It was later performed by Lapchinsky and Rubakov (1979) for generic gravitational systems and discussed in various contexts in Banks (1985), Halliwell and Hawking (1985), Hartle (1987), Kiefer (1987), Barvinsky (1989), Brout and Venturi (1989), Singh and Padmanabhan (1989), Parentani (2000), and others. Even if on a formal level, this derivation yields an important bridge connecting the full theory of quantum gravity with the limit of quantum field theory in an external space–time. It lies behind many cosmological applications.

This 'Born–Oppenheimer type of approach' is also well suited for the calculation of quantum-gravitational corrections terms to the Schrödinger equation (5.161). This will be discussed in the next subsection. As a preparation it is, however, most appropriate to introduce again a condensed, so-called 'DeWitt', notation; cf. Section 2.2. We introduce the notation

$$q^i = h_{ab}(\mathbf{x}) \; , \quad p_i = p^{ab}(\mathbf{x}) \; , \tag{5.162}$$

in which the condensed index $i = (ab, \mathbf{x})$ includes both discrete tensor indices and three-dimensional spatial coordinates $\mathbf{x}$. In this way, the situation is formally the same as for a finite-dimensional model. A similar notation can be introduced for the constraints,

$$H_\mu^{\mathrm{g}}(q, p) = (\mathcal{H}_{\perp}^{\mathrm{g}}(\mathbf{x}), \mathcal{H}_a^{\mathrm{g}}(\mathbf{x})) \; , \quad H_\mu^{\mathrm{m}}(q, \varphi, p_\varphi) = (\mathcal{H}_{\perp}^{\mathrm{m}}(\mathbf{x}), \mathcal{H}_a^{\mathrm{m}}(\mathbf{x})) \; . \tag{5.163}$$

The index μ enumerates the superhamiltonian and supermomenta of the theory as well as their spatial coordinates, $\mu \to (\mu, \mathbf{x})$. In this notation, the functional

---

[16] An attempt to extrapolate the standard interpretational framework of quantum theory into the 'timeless realm' is the use of 'evolving constants' in the Heisenberg picture by Rovelli (1991b).

dependence on phase-space variables is represented in the form of functions of (q^i, p_i), and the contraction of condensed indices includes integration over $\mathbf{x}$ along with discrete summation. In the condensed notation, the gravitational part of the canonical action (4.68) acquires the simple form

$$S^{\mathrm{g}}[q, p, N] = \int \mathrm{d}t \, \left(p_i \dot{q}^i - N^\mu H_\mu^{\mathrm{g}}(q, p)\right) \equiv \int \mathrm{d}t \, L^{\mathrm{g}} , \qquad (5.164)$$

with the superhamiltonian and supermomenta given by expressions which are quadratic and linear in the momenta, respectively,

$$H_\perp^{\mathrm{g}} = \frac{1}{2m_{\mathrm{P}}^2} G_\perp^{ik} p_i p_k + V_\perp , \qquad H_a^{\mathrm{g}} = D_a^i p_i , \qquad (5.165)$$

with $V_\perp = -2m_{\mathrm{P}}^2 \sqrt{h}\,^{(3)}R$. Here the indices $\perp \to (\perp, \mathbf{x})$ and $a \to (a, \mathbf{x})$ are also condensed, $G_\perp^{ik}$ is the ultralocal three-point object containing the matrix of the DeWitt metric, and D_a^i is the generator of the spatial diffeomorphisms (see below). The objects $G_\perp^{ik}$ and D_a^i have the form of the following delta-function type kernels (Barvinsky 1993b),

$$G_\perp^{ik} = G_{abcd} \, \delta(\mathbf{x}_i, \mathbf{x}_k)\delta(\mathbf{x}_\perp, \mathbf{x}_k) ,$$
$$i = (ab, \mathbf{x}_i), \quad k = (cd, \mathbf{x}_k), \quad \perp = (\perp, \mathbf{x}_\perp) , \qquad (5.166)$$
$$D_a^i = -2h_{a(b}D_{c)}\delta(\mathbf{x}_a, \mathbf{x}_i) , \quad i = (bc, \mathbf{x}_i), \quad a = (a, \mathbf{x}_a) . \qquad (5.167)$$

Note that the object $G_\perp^{ik}$ itself is not yet the DeWitt metric because it contains two delta-functions. Only the functional contraction of $G_\perp^{ik}$ with the constant lapse function $N = 1$ converts this object into the distinguished ultralocal metric on the functional space of three-metric coefficients,

$$G^{ik} = G_\perp^{ik} N \big|_{N=1} \equiv \int \mathrm{d}^3 \mathbf{x}_\perp \, G_\perp^{ik} = G_{abcd} \, \delta(\mathbf{x}_i, \mathbf{x}_k) . \qquad (5.168)$$

The Poisson-bracket algebra for the gravitational constraints in condensed notations then reads as in (5.51).

Note that the transformations of the q^i-part of phase space, cf. Section 5.3.4,

$$\delta q^i = D_\mu^i \epsilon^\mu , \qquad D_\mu^i \equiv \frac{\partial H_\mu^{\mathrm{g}}}{\partial p_i} \qquad (5.169)$$

have as generators the vectors D_μ^i which are momentum-independent for space-like diffeomorphisms $\mu = a$ (and, therefore, coincide with the coefficients of the momenta in the supermomentum constraints (5.167)), but involve momenta for normal deformations, $D_\perp^i = G_\perp^{ik} p_k / m_{\mathrm{P}}^2$.

With these condensed notations, one can formulate the operator realization of the gravitational constraints $H_\mu^{\mathrm{g}}(q, p) \to \hat{H}_\mu^{\mathrm{g}}$, closing the commutator version of the Poisson-bracket algebra (5.51),

$$\left[\hat{H}^{\mathrm{g}}_{\mu}, \hat{H}^{\mathrm{g}}_{\nu}\right] = \mathrm{i}\hat{U}^{\lambda}_{\mu\nu}\hat{H}^{\mathrm{g}}_{\lambda} . \tag{5.170}$$

As shown in Barvinsky (1993b), the fact that (5.170) holds follows from the classical gravitational constraints (5.165) by replacing the momenta p_k with the functional covariant derivatives $\mathcal{D}_k/\mathrm{i}$—covariant with respect to the Riemann connection based on the DeWitt metric (5.168)—and by adding a purely imaginary part (anti-Hermitian with respect to the L^2 inner product): the functional trace of structure functions, $\mathrm{i}U^{\nu}_{\mu\nu}/2$. With this definition of covariant derivatives, it is understood that the space of three-metrics q is regarded as a functional differentiable manifold, and that the quantum states $|\Psi(q)\rangle$ are scalar densities of weight $1/2$. Thus, the operator realization for the full constraints including the matter parts has the form

$$\hat{H}_{\perp} = -\frac{1}{2m_{\mathrm{P}}^2}G^{ik}_{\perp}\mathcal{D}_i\mathcal{D}_k + V_{\perp} + \frac{\mathrm{i}}{2}U^{\nu}_{\perp\nu} + \hat{H}^{\mathrm{m}}_{\perp} , \tag{5.171}$$

$$\hat{H}_a = \frac{1}{\mathrm{i}}\nabla^i_a\mathcal{D}_i + \frac{\mathrm{i}}{2}U^{\nu}_{a\nu} + \hat{H}^{\mathrm{m}}_a . \tag{5.172}$$

The imaginary parts of these operators are either formally divergent (being the coincidence limits of delta-function type kernels) or formally zero (as in (5.171) because of vanishing structure-function components). We shall, however, keep them in a general form, expecting that a rigorous operator regularization will exist that can consistently handle these infinities as well as the corresponding quantum anomalies (see Section 5.3.5).

The highest order of the semiclassical approximation leads to a wave functional of the form (5.153). It is sometimes convenient to consider a two-point object ('propagator') instead of a wave functional—one can, for example, easily translate the results into a language using Feynman diagrams. This will be done in the next subsection. We shall, therefore, consider a two-point solution $\boldsymbol{K}(q, q')$ of the Wheeler–DeWitt equation. One can construct a closed expression for the one-loop pre-exponential factor of a solution which is of the semiclassical form corresponding to (5.153); see Barvinsky and Krykhtin (1993),

$$\boldsymbol{K}(q, q') = \boldsymbol{P}(q, q')\mathrm{e}^{\mathrm{i}m_{\mathrm{P}}^2\boldsymbol{S}(q, q')} . \tag{5.173}$$

Here, $\boldsymbol{S}(q, q')$ is a particular solution of the Hamilton–Jacobi equations with respect to both arguments—the classical action calculated at the extremal of equations of motion joining the points q and q',

$$H^{\mathrm{g}}_{\mu}(q, \partial\boldsymbol{S}/\partial q) = 0 . \tag{5.174}$$

The one-loop $(O(m_{\mathrm{P}}^0)$ part of the $1/m_{\mathrm{P}}^2$-expansion) order of the pre-exponential factor $\boldsymbol{P}(q, q')$ here satisfies a set of quasi-continuity equations which follow from the Wheeler–DeWitt equations at one loop and which are analogous to (5.143) in the quantum-mechanical example,

$$\mathcal{D}_i(D^i_{\mu}\boldsymbol{P}^2) = U^{\lambda}_{\mu\lambda}\boldsymbol{P}^2, \tag{5.175}$$

$$D^i_\mu \equiv \left.\frac{\partial H^g_\mu}{\partial p_i}\right|_{p \,=\, \partial \boldsymbol{S}/\partial q} , \qquad (5.176)$$

with the generators D^i_μ *here* evaluated at the Hamilton–Jacobi values of the canonical momenta. The solution of this equation turns out to be a particular generalization of the Pauli–van Vleck–Morette formula—the determinant calculated on the subspace of non-degeneracy for the matrix

$$\boldsymbol{S}_{ik'} = \frac{\partial^2 \boldsymbol{S}(q, q')}{\partial q^i \, \partial q^{k'}} . \qquad (5.177)$$

This matrix has the generators D^i_μ as zero-eigenvalue eigenvectors (Barvinsky and Krykhtin 1993). An invariant algorithm of calculating this determinant is equivalent to the Faddeev–Popov gauge-fixing procedure; cf. Section 2.2.3. It consists in introducing a 'gauge-breaking' term to the matrix (5.177),

$$\boldsymbol{F}_{ik'} = \boldsymbol{S}_{ik'} + \phi^\mu_i c_{\mu\nu} \phi^\nu_{k'} , \qquad (5.178)$$

formed with the aid of the gauge-fixing matrix $c_{\mu\nu}$ and two sets of arbitrary covectors (of 'gauge conditions') ϕ^μ_i and $\phi^\nu_{k'}$ at the points q and q', respectively. They satisfy invertibility conditions for 'Faddeev–Popov operators' at these two points,

$$J^\mu_\nu = \phi^\mu_i D^i_\nu , \ \ J \equiv \det J^\mu_\nu \neq 0 , \ \ J'^\mu_\nu = \phi^\mu_{i'} D^{i'}_\nu , \ \ J' \equiv \det J'^\mu_\nu \neq 0 . \quad (5.179)$$

In terms of these objects, the pre-exponential factor solving the continuity equations (5.175) is given by

$$\boldsymbol{P} = \left[\frac{\det \boldsymbol{F}_{ik'}}{J J' \det c_{\mu\nu}}\right]^{1/2} , \qquad (5.180)$$

which is independent of the gauge fixing. This finishes the discussion of the Born–Oppenheimer scheme at the highest level of approximation.

5.4.3 *Quantum-gravitational correction terms*

We shall now proceed to perform the semiclassical expansion for solutions to the Wheeler–DeWitt equations. Since we are interested in giving an interpretation in terms of Feynman diagrams, we shall not consider wave functionals but—as in the last part of the last subsection— two-point solutions ('propagators'). Due to the absence of an external time parameter in the full theory, such two-point functions play more the role of energy Green functions than ordinary propagators; see Section 5.3.4. In the semiclassical limit, however, a background time parameter is available, with respect to which Feynman 'propagators' can be formulated.

Let us, therefore, look for a two-point solution of the Wheeler–DeWitt equations (i.e. Wheeler–DeWitt equation and momentum constraints) in the form of the ansatz

$$\hat{K}(q_+, q_-) = P(q_+, q_-) e^{im_P^2 S(q_+, q_-)} \hat{U}(q_+, q_-) , \qquad (5.181)$$

where we denote as above by a hat, the operators acting in the Hilbert space of matter fields. Here, $S(q_+, q_-)$ satisfies (5.174), and $P(q_+, q_-)$ is the pre-exponential factor (5.180). Substituting this ansatz into the system of the Wheeler–DeWitt equations and taking into account the Hamilton–Jacobi equations and the continuity equations for $P(q_+, q_-)$, we get for the 'evolution' operator $\hat{U}(q_+, q_-)$ the equations

$$iD_\perp^k \mathcal{D}_k \hat{U} = \hat{H}_\perp^m \hat{U} - \frac{1}{2m_P^2} P^{-1} G_\perp^{mn} \mathcal{D}_m \mathcal{D}_n (P \hat{U}) , \qquad (5.182)$$

$$iD_a^k \mathcal{D}_k \hat{U} = \hat{H}_a^m \hat{U} , \qquad (5.183)$$

where all the derivatives are understood as acting on the argument q_+. Evaluating this operator at the classical extremal $q_+ \to q(t)$,

$$\hat{U}(t) = \hat{U}(q(t), q_-) , \qquad (5.184)$$

where $q(t)$ satisfies the canonical equations of motion corresponding to $S(q_+, q_-)$,

$$\dot{q}^i = N^\mu \nabla_\mu^i , \qquad (5.185)$$

one easily obtains the quasi-evolutionary equation

$$i\frac{\partial}{\partial t} \hat{U}(t) = \hat{H}^{\text{eff}} \hat{U}(t) \qquad (5.186)$$

with the *effective* matter Hamiltonian

$$\hat{H}^{\text{eff}} = \hat{H}^m - \frac{1}{2m_P^2} N G_\perp^{mn} \mathcal{D}_m \mathcal{D}_n [P \hat{U}] P^{-1} \hat{U}^{-1} . \qquad (5.187)$$

(Recall that we use a condensed notation and that this equation is, in fact, an *integral* equation.) The first term on the right-hand side is the Hamiltonian of matter fields at the gravitational background of (q, N)-variables,

$$\hat{H}^m = N^\mu \hat{H}_\mu^m . \qquad (5.188)$$

The second term involves the operator $\hat{U}$ itself in a non-linear way and contributes only at order m_P^{-2} of the expansion. Thus, (5.186) is not a true linear

Schrödinger equation, but semiclassically it can be solved by iterations starting from the lowest order approximation

$$\hat{U}_0 = T \exp \left[-i \int_{t_-}^{t_+} dt \hat{H}^{\mathrm{m}} \right], \tag{5.189}$$

$$\hat{H}_0^{\mathrm{eff}} = \hat{H}^{\mathrm{m}}. \tag{5.190}$$

Here, T denotes the Dyson chronological ordering of the usual unitary evolution operator acting in the Hilbert space of matter fields $(\hat{\varphi}, \hat{p}_\varphi)$. The Hamiltonian $\hat{H}^{\mathrm{m}} = H^{\mathrm{m}}(\hat{\varphi}, \hat{p}_\varphi, q(t), N(t))$ is an operator in the Schrödinger picture of these fields $(\hat{\varphi}, \hat{p}_\varphi)$, parametrically depending on the gravitational background variables $(q(t), N(t))$, that is, evaluated along a *particular* trajectory ('space–time') in configuration space.

The Dyson T-exponent obviously explains the origin of the standard Feynman diagrammatic technique in the matter field sector of the theory which arises in the course of the semiclassical expansion of (5.189). We shall now show that the gravitational part of this diagrammatic technique involving graviton loops naturally arises as a result of iterationally solving (5.186)–(5.187) in powers of $1/m_{\mathrm{P}}^2$.

The effective Hamiltonian in the first-order approximation of such an iterational technique can be obtained by substituting (5.189) into (5.187) to yield

$$\hat{H}_1^{\mathrm{eff}}(t_+) = \hat{H}^{\mathrm{m}} - \frac{1}{2m_{\mathrm{P}}^2} \mathcal{G}^{mn}(\mathcal{D}_m \mathcal{D}_n \hat{U}_0) \hat{U}_0^{-1}$$

$$- \frac{1}{2m_{\mathrm{P}}^2} \mathcal{G}^{mn}(\mathcal{D}_m \mathcal{D}_n \boldsymbol{P}) \boldsymbol{P}^{-1}$$

$$- \frac{1}{m_{\mathrm{P}}^2} \mathcal{G}^{mn}(\mathcal{D}_m \boldsymbol{P}) \boldsymbol{P}^{-1}(\mathcal{D}_n \hat{U}_0) \hat{U}_0^{-1}. \tag{5.191}$$

Here we have used the new notation

$$\mathcal{G}^{mn} = N G_\perp^{mn} \tag{5.192}$$

for another metric on the configuration space (compare with (5.168)) which uses the actual value of the lapse function corresponding to the classical extremal (5.185). This lapse function generally differs from unity. We have also decomposed the first-order corrections in the effective Hamiltonian into three terms corresponding to the contribution of quantum matter (generated by $\hat{U}_0$), a purely quantum gravitational contribution generated by $\boldsymbol{P}$, and their cross-term.

Further evaluation of these terms demands the knowledge of derivatives acting on the configuration space argument q_+ of $\hat{U}_0$ and $\boldsymbol{P}$. Obtaining these derivatives leads to the necessity of considering the special boundary value problem for classical equations of motion, the graviton propagator and vertices—elements of the gravitational Feynman diagrammatic technique. The discussion is long and technical and can be found in Barvinsky and Kiefer (1998). Here we shall only quote the main steps and include a brief discussion of the results.

The following quantities play a role in the discussion. First, we introduce a collective notation for the full set of *Lagrangian* gravitational variables, which includes both the spatial metric as well as lapse and shift functions,

$$g^a \equiv (q^i(t), N^\mu(t)) \ . \tag{5.193}$$

This comprises the space–time metric. (Recall that $q^i(t)$ stands for the three-metric $h_{ab}(\mathbf{x}, t)$.) Next, the second functional derivatives of the gravitational action with respect to the space–time metric is denoted by

$$S_{ab} \equiv \frac{\delta^2 S[g]}{\delta g^a(t) \delta g^b(t')} \ . \tag{5.194}$$

Since S_{ab} is not invertible, one must add gauge-fixing terms similar to (5.178). This leads to an operator F_{ab}. The 'graviton propagator' D^{bc} is then defined as its inverse via

$$F_{ab} D^{bc} = \delta^c_a \ . \tag{5.195}$$

We also need the components of the Wronskian operator obtained from the gravitational Lagrangian L^g,

$$\overrightarrow{W}_{ib} (d/dt) \, \delta g^b(t) = -\delta \frac{\partial L^g(q, \dot{q}, N)}{\partial \dot{q}^i} \ . \tag{5.196}$$

With the help of the 'graviton propagator', one can define

$$\hat{t}^a(t) = -\frac{1}{m_P^2} \int_{t_-}^{t_+} dt' \, D^{ab}(t, t') \hat{T}_b(t') \equiv -\frac{1}{m_P^2} D^{ab} \hat{T}_b \ , \tag{5.197}$$

where $\hat{T}_b$ is the condensed notation for the energy–momentum tensor of the matter field. The quantity $\hat{t}^a(t)$ obeys the linearized Einstein equations with source $\hat{T}_b$ and can thus be interpreted as the gravitational potential generated by the back reaction of quantum matter on the gravitational background.

The first correction term in (5.191)—the contribution of quantum matter—is found to read

$$-\frac{1}{2m_P^2} \mathcal{G}^{mn} (\mathcal{D}_m \mathcal{D}_n \hat{U}_0) \hat{U}_0^{-1}$$

$$= \frac{1}{2} m_P^2 \, \mathcal{G}^{mn} \, T \left(\overrightarrow{W}_{ma} \hat{t}^a \ \overrightarrow{W}_{nb} \hat{t}^b \right)$$

$$+ \frac{i}{2} D^{ab} w_{abc}(t_+) \, \hat{t}^c$$

$$- \frac{i}{2} \mathcal{G}^{mn} (\overrightarrow{W}_{ma} D^{ac}) (\overrightarrow{W}_{nb} D^{bd}) \left(S_{cde} \hat{t}^e + \frac{1}{m_P^2} \hat{S}^{mat}_{cd} \right) \ . \tag{5.198}$$

The resulting three terms can be given a Feynman diagrammatic representation with different structure. Note that because of (5.197) all terms are of the same order m_P^{-2}, despite their appearance. The first term begins with the tree-level structure quadratic in gravitational potential operators $\hat{t}^a$. Note that despite the fact that these operators are taken at one moment of time t_+, their chronological product is non-trivial because it should read as

$$T\left(\hat{t}^a\,\hat{t}^b\right) = \frac{1}{m_P^4}\,D^{ac}D^{bd}\,T\left(\hat{T}_c\,\hat{T}_d\right) \tag{5.199}$$

and, thus, includes all higher order chronological couplings between composite operators of matter stress tensors. The second and third term on the right-hand side of (5.198) are essentially quantum, because their semiclassical expansions start with the one-loop diagrams consisting of one and two 'graviton propagators' D^{ab}, respectively. The quasi-local vertices of these diagrams are built from the Wronskian operators, gravitational three-vertices denoted by $w_{abc}(t_+)$ and S_{cde} (third functional derivatives of the gravitational action) and second-order variation of matter action with respect to gravitational variables, $\hat{S}^m_{cd}$. The corresponding diagrams are shown in Fig. 5.1.

The second and third correction term in (5.191) can be written in a similar way. The second term—the purely quantum gravitational contribution—contains instead of $\hat{t}^a$ a gravitational potential that is generated by the one-loop stress tensor of gravitons which enters as a matter source in the linearized Einstein equations.

Depending on the physical situation, not all of the correction terms are of equal importance. It often happens that the effects of quantum matter dominate over the graviton effects and that from its contribution only the first term on the right-hand side of (5.198) is significant. This means that one has

$$\hat{H}_1^{eff} = \hat{H}^m + 1/2\,m_P^2\,\mathcal{G}^{mn}\,T\left(\vec{W}_{ma}\hat{t}^a\;\vec{W}_{nb}\hat{t}^b\right) + O(1/m_P^2)\,. \tag{5.200}$$

It turns out that this remaining correction term can be interpreted as the kinetic energy of the gravitational radiation produced by the back reaction of quantum matter sources. This term can be decomposed into a component along the semiclassical gravitational trajectory ('longitudinal part') and an orthogonal component. The longitudinal part is ultralocal and basically given by the square of the matter Hamiltonian,

$$\propto \frac{1}{m_P^2}\left(\hat{H}^m_\perp\right)^2\,. \tag{5.201}$$

This is fully analogous to the relativistic correction term to the ordinary Schrödinger equation, as being found from expanding the Klein–Gordon equation; see (5.149).

Can such quantum-gravitational correction terms be observed? If the matter Hamiltonian is dominated by the rest mass of a particle with mass m, the correction term (5.201) is of the order of

$$\propto \left(\frac{m}{m_P}\right)^2\,. \tag{5.202}$$

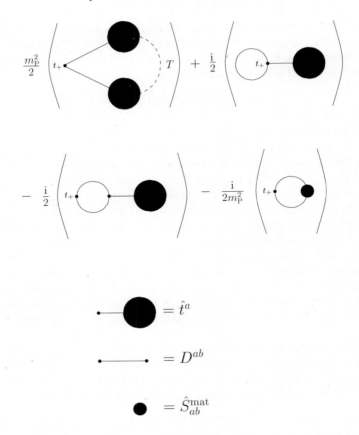

FIG. 5.1. Feynman diagrams illustrating the quantum-gravitational correction terms in (5.198). The time parameter t_+ labels the vertices at the space–time point at which $\hat{H}_1^{\text{eff}}$ is evaluated. Dashed lines labelled by T denote the chronological ordering of matter stress tensors in the bilinear combinations of gravitational potentials $\hat{t}^a$. (From Barvinsky and Kiefer 1998.)

For scalar fields causing an inflationary expansion of the early universe (inflatons), a typical value would be $m \approx 10^{-5} m_\mathrm{P}$. The corrections would thus be of the order 10^{-10}. This is a very small number, but it is nevertheless conceivable that such effects could be seen in future observations of the cosmic microwave anisotropy spectrum.

6

QUANTUM GRAVITY WITH CONNECTIONS AND LOOPS

6.1 The Gauß and diffeomorphism constraints

In Section 4.3, we have encountered a Hamiltonian formulation of GR alternative to geometrodynamics, using the concepts of connections and loops. In the present chapter, we shall review approaches to formulate a consistent quantum theory with such variables, leading to 'quantum connection dynamics' or 'quantum loop dynamics'. More details and references can be found in Rovelli (1998) and Thiemann (2001, 2003). A short overview is in Pullin (1999).

In the present section, the quantization of the Gauß constraint (4.122) and the diffeomorphism constraints (4.127) will be investigated. These constraints already give a picture of the way space might look like on the smallest scales. The main results will be the discrete spectrum of geometric operators representing area or volume in the classical limit. This has a direct bearing on the interpretation of black-hole entropy (see Section 7.1). The more complicated implementation of the Hamiltonian constraint is relegated to Section 6.3. We treat here only pure gravity, but many results can be extended to the case of standard-model matter (Thiemann 2001).

6.1.1 Connection representation

The connection representation is characterized classically by the Poisson bracket (4.120) between the densitized tetrad $E_j^b(\mathbf{y})$ and the SU(2) connection $A_a^i(\mathbf{x})$. These variables are then formally turned into operators obeying the commutation relation

$$\left[\hat{A}_a^i(\mathbf{x}), \hat{E}_j^b(\mathbf{y})\right] = 8\pi\beta \mathrm{i}\hbar \delta_j^i \delta_a^b \delta(\mathbf{x}, \mathbf{y}) \ . \tag{6.1}$$

In the functional Schrödinger representation, one can implement this relation formally through

$$\hat{A}_a^i(\mathbf{x})\Psi[A] = A_a^i(\mathbf{x})\Psi[A] \ , \tag{6.2}$$

$$\hat{E}_j^b(\mathbf{y})\Psi[A] = 8\pi\beta\frac{\hbar}{\mathrm{i}}\frac{\delta}{\delta A_j^b(\mathbf{y})}\Psi[A] \ , \tag{6.3}$$

where the A in the argument of the wave functional is a shorthand for $A_a^i(\mathbf{x})$. As in Chapter 5, the constraints are implemented as conditions on allowed wave functionals. The Gauß constraint (4.122) then becomes

$$\hat{\mathcal{G}}_i\Psi = 0 \quad \longrightarrow \quad \mathcal{D}_a\frac{\delta\Psi}{\delta A_a^i} = 0 \ . \tag{6.4}$$

It expresses the invariance of the wave functional with respect to infinitesimal gauge transformations of the connection. The diffeomorphism constraints (4.127) become

$$\hat{\tilde{\mathcal{H}}}_a \Psi = 0 \quad \longrightarrow \quad F^i_{ab} \frac{\delta \Psi}{\delta A^i_b} = 0 \ . \tag{6.5}$$

Similar to the classical case, it expresses the invariance of the wave functional under a combination of infinitesimal diffeomorphism and gauge transformations; cf. Section 4.3.2.

The Hamiltonian constraint (4.126) cannot be treated directly in this way because the Γ^i_a-terms contain the tetrad in a complicated non-linear fashion. This would lead to similar problems as with the Wheeler–DeWitt equation discussed in Chapter 5, preventing to find any solutions. In Section 6.3, we shall see how a direct treatment of the quantum Hamiltonian constraint can be attempted. Here, we remark only that (4.126) is easy to handle only for the value $\beta = $ i in the Lorentzian case or for the value $\beta = 1$ in the Euclidean case. In the first case, the problem that the resulting formalism uses complex variables and that one has to impose 'reality conditions' at an appropriate stage (which nobody has succeeded in implementing) arises. In the second case, the variables are real but one deals with the unphysical Euclidean case. Nevertheless, for these particular values of β, the second term in (4.126) vanishes, and the quantum Hamiltonian constraint for $\Lambda = 0$ would simply read

$$\epsilon^{ijk} F_{kab} \frac{\delta^2 \Psi}{\delta A^i_a \delta A^j_b} = 0 \ . \tag{6.6}$$

Note that in (6.4)–(6.6), a 'naive' factor ordering has been chosen: all derivatives are put to the right. Formal solutions to these equations have been found; see for example, Brügmann (1994). Many of these solutions are annihilated by the operator corresponding to $\sqrt{h}$ and may therefore be devoid of physical meaning, since matter fields and the cosmological term couple to $\sqrt{h}$. In the case of vacuum gravity with $\Lambda \neq 0$, an exact formal solution in the connection representation was found by Kodama (1990),

$$\Psi_\Lambda \propto \exp\left(-\frac{6}{G\Lambda\hbar} S_{\text{CS}}[A]\right) \ , \tag{6.7}$$

where $S_{\text{CS}}[A]$ denotes the 'Chern–Simons action'

$$S_{\text{CS}}[A] = \int_\Sigma \mathrm{d}^3 x \ \epsilon^{abc} \text{tr}\left(A_a \partial_b A_c + \frac{2}{3} A_a A_b A_c\right) \ . \tag{6.8}$$

Due to the topological nature of the Chern–Simons action, the state (6.7) is both gauge- and diffeomorphism-invariant. It is a solution to the Hamiltonian constraint

$$\left(\epsilon^{ijk} \frac{\delta}{\delta A^i_a} \frac{\delta}{\delta A^j_b} F^k_{ab} + \frac{\Lambda}{6} \epsilon^{ijk} \epsilon_{abc} \frac{\delta}{\delta A^i_a} \frac{\delta}{\delta A^j_b} \frac{\delta}{\delta A^k_c}\right) \Psi[A] = 0 \ .$$

Note that the factor ordering chosen here is different from the one in (6.6). Only with this choice of factor ordering is the state (6.7) a solution of this constraint. In contrast to the states mentioned above, (6.7) is not annihilated by the operator corresponding to $\sqrt{h}$ and may thus have physical content. The Chern–Simons action is also important for GR in 2+1 dimensions; cf. Section 8.1.3.

Since the 'real' quantum Hamiltonian constraint is not given by (6.6), see Section 6.3, we shall not discuss further this type of solutions. What can generally be said about the connection representation? Since (6.4) guarantees that $\Psi[A] = \Psi[A^g]$, where $g \in \mathrm{SU}(2)$, the configuration space after the implementation of the Gauß constraint is actually given by $\mathcal{A}/\mathcal{G}$, where $\mathcal{A}$ denotes the space of connections and $\mathcal{G}$ the local $\mathrm{SU}(2)$ gauge group. Because the remaining constraints have not been considered at this stage, this level corresponds to having in Chapter 5 states $\Psi[h_{ab}]$ before imposing the constraints $\mathcal{H}_a\Psi = 0 = \mathcal{H}_\perp\Psi$. A candidate for an inner product on $\mathcal{A}/\mathcal{G}$ would be

$$\langle\Psi_1|\Psi_2\rangle = \int_{\mathcal{A}/\mathcal{G}} \mathcal{D}\mu[A]\ \Psi_1^*[A]\Psi_2[A]\ , \qquad (6.9)$$

cf. (5.20). The main problem is: can one construct a suitable measure $\mathcal{D}\mu[A]$ in a rigorous way? The obstacles are that the configuration space $\mathcal{A}/\mathcal{G}$ is both non-linear and infinite-dimensional. Such a measure has been constructed; see Ashtekar *et al.* (1994). In the construction process, it was necessary to extend the configuration space to its closure $\overline{\mathcal{A}/\mathcal{G}}$. This space is much bigger than the classical configuration space of smooth field configurations, since it contains distributional analogues of gauge-equivalent connections.

The occurrence of distributional configurations can also be understood from a path-integral point of view where one sums over (mostly) non-differentiable configurations. In field theory, an imprint of this is left on the boundary configuration which shows up as the argument of the wave functional. Although classical configuations form a set of measure zero in the space of all configurations, they nevertheless possess physical significance, since one can construct semiclassical states that are concentrated on them. Moreover, for the measurement of field variables, one would not expect much difference to the case of having smooth field configurations only, since only measurable functions count, and these are integrals of field configurations; see Bohr and Rosenfeld (1933).

6.1.2 *Loop representation*

Instead of considering wave functionals defined on the space of connections, $\Psi[A]$, one can use states defined on the space of *loops* $\alpha^a(s)$, $\Psi[\alpha]$; cf. Section 4.3.3. This is possible due to the availability of the measure on $\overline{\mathcal{A}/\mathcal{G}}$, and the states are obtained by the transformation

$$\Psi[\alpha] = \int_{\overline{\mathcal{A}/\mathcal{G}}} \mathcal{D}\mu[A]\ T[\alpha]\Psi[\alpha]\ , \qquad (6.10)$$

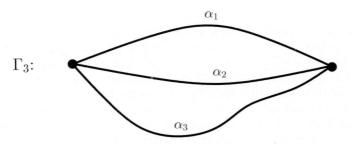

Γ_3:

α_1

α_2

α_3

FIG. 6.1. An example of a graph with three curves.

where $\mathcal{T}[\alpha]$ was defined in (4.143). This corresponds to the usual Fourier transform in quantum mechanics,

$$\tilde{\psi}(p) = \frac{1}{(2\pi\hbar)^{3/2}} \int_{\mathbb{R}^3} \mathrm{d}^3 x \ e^{-i\mathbf{p}\mathbf{x}/\hbar}\psi(\mathbf{x}) \ . \tag{6.11}$$

The plane wave corresponds to $\mathcal{T}[\alpha] \equiv \Psi_\alpha[A]$. We shall refer to the latter as 'loop states'. In the loop approach to quantum gravity they are taken to be the basis states (Rovelli and Smolin 1990). The problem is, however, that this basis is overcomplete. A complete basis can be constructed by a linear combination of this basis; cf. Rovelli and Gaul (2000) for an overview. It is called *spin network basis* and goes back to Penrose (1971); see also, for example, Major (1999) for an introduction. To quote from Penrose (1971):

My basic idea is to try and build up both space-time and quantum mechanics simultaneously—from *combinatorial* principles ... The idea here, then, is to start with the concept of angular momentum—where one has a *discrete* spectrum—and use the rules for combining angular momenta together and see if in some sense one can construct the concept of *space* from this.

How is the spin-network basis defined? Consider first a graph, $\Gamma_n = \{\alpha_1, \ldots, \alpha_n\}$, where the α_i denote curves (also called 'edges' or 'links'), which are oriented and piecewise analytic. If they meet they meet at their endpoints ('vertices'). An example with three curves is depicted in Fig. 6.1. One then associates a holonomy $U[A, \alpha]$, see (4.142), to each link. This leads to the concept of 'cylindrical functions': from a function

$$f_n : [\mathrm{SU}(2)]^n \longrightarrow \mathbb{C}$$

one can define the cylindrical function

$$\Psi_{\Gamma_n, f_n}[A] = f_n(U_1, \ldots, U_n) \ . \tag{6.12}$$

An SU(2)-holonomy has thus been put on each of the n links of the graph.[1] Cylindrical functions are dense in the space of smooth functions on $\mathcal{A}$. One

[1]It has also been suggested to use the group SO(3) instead of SU(2); cf. the discussion in Section 7.3 on black-hole entropy.

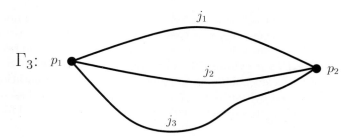

Γ_3:

FIG. 6.2. The graph of Fig. 6.1 with spins attached to the links.

can define a scalar product between two cylindrical functions f and g, which is invariant under gauge transformations and diffeomorphisms,

$$\langle \Psi_{\Gamma,f} | \Psi_{\Gamma,f} \rangle = \int_{[SU(2)]^n} dU_1 \cdots dU_n \, f^*(U_1, \ldots, U_n) g(U_1, \ldots, U_n) , \qquad (6.13)$$

where $dU_1 \cdots dU_n$ denotes the Haar measure.

At the beginning of Section 5.1, we have distinguished between three spaces satisfying $\mathcal{F}_{phys} \subset \mathcal{F}_0 \subset \mathcal{F}$, which do not necessarily have to be Hilbert spaces. Here, the intention is on using the Hilbert-space machinery of ordinary quantum theory as much as possible, and one would like to employ the chain

$$\mathcal{H}_{phys} \subset \mathcal{H}_{diff} \subset \mathcal{H}_g \subset \mathcal{H} \qquad (6.14)$$

of Hilbert spaces in which the three sets of constraints (Gauß, diffeomorphism, and Hamiltonian constraint) are implemented consecutively. However, this would be possible only if the solutions to the constraints were normalizable. Since this is not the case, one has to employ the formalism of Gel'fand tripels, which here will not be elaborated on; cf. Thiemann (2001).

The 'biggest' space $\mathcal{H}$ is obtained by completing the space of all finite linear combinations of cylindrical functions with respect to the norm

$$\| \Psi_{\Gamma,f} \| = \langle \Psi_{\Gamma,f} | \Psi_{\Gamma,f} \rangle^{1/2} .$$

The Hilbert space $\mathcal{H}$ itself is of course infinite-dimensional and carries unitary representations of local SU(2) transformations and diffeomorphisms.

With these preparations, a spin network is defined as follows. One associates with each link α_i a non-trivial irreducible representation of SU(2), that is, attaches a 'spin' j_i to it ('colouring of the link'). The representation acts on a Hilbert space $\mathcal{H}_{j_i}$. An example is shown in Fig. 6.2. Consider now a 'node' p where k links meet and associate to it the Hilbert space

$$\mathcal{H}_p = \mathcal{H}_{j_1} \otimes \ldots \otimes \mathcal{H}_{j_k} . \qquad (6.15)$$

One fixes an orthonormal basis in $\mathcal{H}_p$ and calls an element of this basis a 'colouring' of the node p. A spin network is then a triple $S(\Gamma, \vec{j}, \vec{N})$, where $\vec{j}$ denotes the collection of spins and $\vec{N}$ the collection of basis elements at the nodes, that is, $\vec{N} = (N_{p_1}, N_{p_2}, \ldots)$, where N_{p_1} is a basis element at p_1, N_{p_2} a basis element at p_2, and so on. Note that S is not yet gauge-invariant. A *spin-network state* $\Psi_S[A]$ is then defined as a cylindrical function f_S associated with S. How is it constructed? One takes a holonomy at each link in the representation corresponding to j_i (described by 'matrices' $R^{j_i}(U_i)$) and contracts all these matrices with the chosen vector $\in \mathcal{H}_p$ at each node where these links meet. This gives a complex number. Thus,

$$\Psi_S[A] = f_S(U_1, \ldots, U_n) = \prod_{\text{links } i} R^{j_i}(U_i) \otimes \prod_{\text{nodes } p} N_p . \qquad (6.16)$$

One can prove that any two states Ψ_S are orthonormal,

$$\langle \Psi_S | \Psi_{S'} \rangle = \delta_{\Gamma\Gamma'} \delta_{\vec{j}\vec{j}'} \delta_{\vec{N},\vec{N}'} \equiv \delta_{SS'} . \qquad (6.17)$$

The states Ψ_S form a complete basis in the unconstrained Hilbert space $\mathcal{H}$.

A gauge-invariant spin network can be constructed by imposing the constraint (6.4), leading to the Hilbert space $\mathcal{H}_g$. For this purpose, one first decomposes $\mathcal{H}_p$ into its irreducible parts ('Clebsch–Gordon decomposition'),

$$\mathcal{H}_p = \mathcal{H}_{j_1} \otimes \cdots \otimes \mathcal{H}_{j_k} = \bigoplus_J (\mathcal{H}_J)^{k_J} , \qquad (6.18)$$

where k_J is the multiplicity of the spin-J irreducible representation. One then selects the *singlet* ($J = 0$) subspace, $(\mathcal{H}_0)^{k_0}$, which is gauge-invariant. One chooses at each node p an arbitrary basis and assigns one basis element to the node. The corresponding colouring $\vec{N}$ belongs to a gauge-invariant spin network.[2] At each node p, the spin of the meeting links have to obey the Clebsch–Gordan condition for any two pairs, for example, $|j_1 - j_2| \leq j_3 \leq j_1 + j_2$, etc. for the situation shown in Fig. 6.3. In this example there exists a unique intertwiner given by the Wigner $3j$-coefficient (there exists only one possibility to combine these three representations into a singlet). We only want to mention that spin-network states can be decomposed into loop states; see Rovelli and Gaul (2000) for an illustrative example.

The next step is the implementation of the diffeomorphism constraint (6.5). We shall denote the gauge-invariant spin-network states by

$$\Psi_S[A] \equiv \langle A | S \rangle . \qquad (6.19)$$

Diffeomorphisms move points on Σ around, so that the spin network will be 'smeared' over Σ. This leads to the concept of an 's-knot': two spin networks S

---

[2]The colouring $\vec{N}$ is a collection of invariant tensors N_p, which are also called 'intertwiners' because they couple different representations of SU(2).

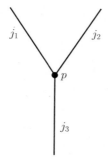

FIG. 6.3. Spins at meeting links have to obey Clebsch–Gordon conditions.

and S' lie in the same s-knot if there exists a diffeomorphism $\phi \in \text{Diff } \Sigma$ such that $S' = \phi \circ S$. One defines

$$\langle s|S \rangle = \begin{cases} 0, & S \notin s \ , \\ 1, & S \in s \ , \end{cases} \tag{6.20}$$

and

$$\langle s|s' \rangle = \begin{cases} 0, & s \neq s' \ , \\ c(s), & s = s' \ , \end{cases} \tag{6.21}$$

where $c(s)$ denotes the number of discrete symmetries of the s-knots under diffeomorphisms. The diffeomorphism-invariant quantum states of the gravitational field are then denoted by $|s\rangle$. The important property is the non-local, 'smeared', character of these states, avoiding problems that such constructions would have, for example, in QCD. For details and references to the original literature, we refer to Thiemann (2001).

6.2 Quantization of area

Up to now we have not considered operators acting on spin-network states. In the following, we shall construct one particular operator of central interest—the 'area operator'. It corresponds in the classical limit to the area of two-dimensional surfaces. Surprisingly, this area operator will turn out to have a discrete spectrum. The discussion can be made both within the connection representation (Ashtekar and Lewandowski 1997) and the loop representation (Rovelli and Gaul 2000). We shall restrict to the latter.

Instead of (6.2) one can consider the operator corresponding to the Wilson loop (4.143) acting on spin-network states,

$$\hat{T}[\alpha]\Psi_S[A] = T[\alpha]\Psi_S[A] \ . \tag{6.22}$$

Instead of the operator acting in (6.3), which is an operator-valued distribution, it turns out to be more appropriate to consider a 'smeared-out' version in which (6.3) is integrated over a two-dimensional manifold $\mathcal{S}$ embedded in Σ,

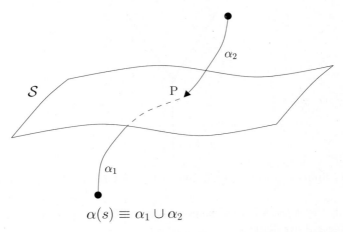

$$\alpha(s) \equiv \alpha_1 \cup \alpha_2$$

FIG. 6.4. Example of an intersection of a link α with a surface $\mathcal{S}$.

$$\hat{E}_i(\mathcal{S}) \equiv -8\pi\beta\hbar i \int_{\mathcal{S}} d\sigma^1 d\sigma^2 \ n_a(\vec{\sigma}) \frac{\delta}{\delta A_a^i[\mathbf{x}(\vec{\sigma})]} \ , \qquad (6.23)$$

where the embedding is given by $(\sigma^1, \sigma^2) \equiv \vec{\sigma} \mapsto x^a(\sigma^1, \sigma^2)$, and

$$n_a(\vec{\sigma}) = \epsilon_{abc} \frac{\partial x^b(\vec{\sigma})}{\partial \sigma^1} \frac{\partial x^c(\vec{\sigma})}{\partial \sigma^2} \qquad (6.24)$$

is the usual vectorial hypersurface element.

We now want to calculate the action of $\hat{E}_i(\mathcal{S})$ on spin-network states $\Psi_{\mathcal{S}}[A]$. For this one needs its action on holonomies $U[A, \alpha]$. This was calculated in detail in Lewandowski *et al.* (1993) by using the differential equation (4.141) for the holonomy. In the simplest case of one intersection of α with $\mathcal{S}$ (cf. Fig. 6.4), one has

$$\frac{\delta U[A, \alpha]}{\delta A_a^i[\mathbf{x}(\vec{\sigma})]} = \frac{\delta}{\delta A_a^i[\mathbf{x}(\vec{\sigma})]} \left(\mathcal{P} \exp\left[G \int_{\alpha} ds \ \dot{\alpha}^a A_a^i(\alpha(s)) \tau_i \right] \right)$$

$$= G \int_{\alpha} ds \ \dot{\alpha}^a \delta^{(3)}(\mathbf{x}(\vec{\sigma}), \alpha(s)) U[A, \alpha_1] \tau_i U[A, \alpha_2] \ , \qquad (6.25)$$

where α_1 refers to the part of α below $\mathcal{S}$ and α_2 to its part above $\mathcal{S}$ (Fig. 6.4). One can now act with the operator $\hat{E}_i(\mathcal{S})$, (6.23), on $U[A, \alpha]$. This yields

$$\hat{E}_i(\mathcal{S})U[A, \alpha] = -8\pi\beta\hbar i \int_{\mathcal{S}} d\sigma^1 d\sigma^2 \ \epsilon_{abc} \frac{\partial x^a(\vec{\sigma})}{\partial \sigma^1} \frac{\partial x^b(\vec{\sigma})}{\partial \sigma^2} \frac{\delta U[A, \alpha]}{\delta A_c^i[\mathbf{x}(\vec{\sigma})]}$$

$$= -8\pi\beta\hbar G i \int_{\mathcal{S}} d\sigma^1 d\sigma^2 \int_{\alpha} ds \ \epsilon_{abc} \frac{\partial x^a(\vec{\sigma})}{\partial \sigma^1} \frac{\partial x^b(\vec{\sigma})}{\partial \sigma^2} \frac{\partial \alpha^c}{\partial s}$$

$$\times \delta^{(3)}(\mathbf{x}(\vec{\sigma}), \alpha(s))U[A, \alpha_1]\tau_i U[A, \alpha_2] \ .$$

The coordinate transformation $(\sigma^1, \sigma^2, s) \mapsto (x^1, x^2, x^3)$ with $\alpha \equiv (0, 0, x^3)$ leads to the Jacobian

$$J \equiv \frac{\partial(\sigma^1, \sigma^2, s)}{\partial(x^1, x^2, x^3)} = \epsilon_{abc} \frac{\partial x^a}{\partial \sigma^1} \frac{\partial x^b}{\partial \sigma^2} \frac{\partial \alpha^c}{\partial s}$$

(note that the right-hand side is zero for curves lying within $\mathcal{S}$). Therefore,

$$\int_{\mathcal{S}} \int_{\alpha} d\sigma^1 d\sigma^2 ds \ \epsilon_{abc} \frac{\partial x^a(\vec{\sigma})}{\partial \sigma^1} \frac{\partial x^b(\vec{\sigma})}{\partial \sigma^2} \frac{\partial x^c(s)}{\partial s} \delta^{(3)}(\mathbf{x}(\vec{\sigma}), \mathbf{x}(s))$$

$$= \int dx^1 dx^2 dx^3 \ \delta^{(3)}(\mathbf{x}(\vec{\sigma}), \mathbf{x}(s)) = \pm 1 \ , \tag{6.26}$$

where the sign depends on the relative orientation of curve and surface. One thus gets

$$\hat{E}_i(\mathcal{S})U[A, \alpha] = \pm 8\pi\beta\hbar G i U[A, \alpha_1]\tau_i U[A, \alpha_2] \ . \tag{6.27}$$

If there is no intersection, the action of this operator will be zero. For more than one point of intersection, one has to sum over all intersections.

What is the action of $\hat{E}_i(\mathcal{S})$ on a spin network? Consider a gauge-invariant spin network S intersecting $\mathcal{S}$ at a single point P. Then, decompose $\Psi_S[A]$, cf. (6.16), as

$$\Psi_S[A] = \Psi^{mn}_{S-\alpha}[A]R^j_{mn}(U[A, \alpha]) \ , \tag{6.28}$$

where $R^j_{mn}(U[A, \alpha])$ is the holonomy along α in the irreducible representation corresponding to spin j, and $\Psi^{mn}_{S-\alpha}[A]$ is the remaining part of (6.16). The action of $\hat{E}_i(\mathcal{S})$ on R^j is similar to (6.27), with $\tau_i \to \tau^{(j)}_i$ according to the representation associated with j. Then,

$$\hat{E}_i(\mathcal{S})\Psi_S[A] = \pm 8\pi\beta l^2_P i \left[R^j(U[A, \alpha_1]) \tau^{(j)}_i R^j(U[A, \alpha_2]) \right]_{mn}$$

$$\times \Psi^{mn}_{S-\alpha}[A] \ . \tag{6.29}$$

This action is not yet gauge-invariant. One can obtain a gauge-invariant operator by 'squaring', that is, by considering

$$\hat{E}^2(\mathcal{S}) \equiv \hat{E}_i(\mathcal{S})\hat{E}_i(\mathcal{S}) \ . \tag{6.30}$$

In order to calculate the action of this operator, consider again a spin network with a single point of interaction, P. We assume that P belongs to the α_1-part of the curve. Therefore, in the action

$$\hat{E}^2(\mathcal{S})\Psi_S[A] = -(8\pi\beta l_P)^2 \hat{E}_i(\mathcal{S}) \left[R^j(U[A, \alpha_1]) \tau^{(j)}_i R^j(U[A, \alpha_2]) \right]_{mn}$$

$$\times \Psi_{S-\alpha}^{mn}[A] \ ,$$

the operator $\hat{E}_i(S)$ on the right-hand side acts only on $R^j \left(U[A, \alpha_1] \right)$ to give $R^j \left(U[A, \alpha_1] \right) \tau_i^{(j)} \mathbb{I}$. Since one has for the 'Casimir operator'

$$\tau_i^{(j)} \tau_i^{(j)} = -j(j+1) \mathbb{I}$$

(recall that we have defined $\tau_i = (i/2)\sigma_i$ for $j = 1/2$ and similarly for higher j), we get

$$\hat{E}^2(S)\Psi_S[A] = (8\pi\beta l_{\mathrm{P}})^2 j(j+1)\Psi_S[A] \ , \qquad (6.31)$$

where (6.16) has been used. If there is more than one intersection of S with S, one will have to consider a partition ρ of S into $n(\rho)$ smaller surfaces S_n such that the points of intersection lie in different S_n (for a given S). Otherwise, the action of $\hat{E}^2(S)$ would not be gauge-invariant (due to 'crossterms' in $\hat{E}^2(S)\Psi_S[A]$).

We now define the 'area operator' (see below for its interpretation)

$$\hat{A}(S) \equiv \lim_{\rho \to \infty} \sum_{n(\rho)} \sqrt{\hat{E}_i(S_n)\hat{E}_i(S_n)} \ , \qquad (6.32)$$

which is independent of ρ. If there are no nodes on S and only a finite number of intersections ('punctures' P), one obtains (Rovelli and Smolin 1995; Ashtekar and Lewandowski 1997)

$$\hat{A}(S)\Psi_S[A] = 8\pi\beta l_{\mathrm{P}}^2 \sum_{P \in S \cap S} \sqrt{j_P(j_P+1)}\Psi_S[A] \equiv A(S)\Psi_S[A] \ . \qquad (6.33)$$

The operator $\hat{A}(S)$ is self-adjoint, that is, it is diagonal on spin-network states and is real on them. If a node lies on S, a more complicated expression is obtained (Frittelli *et al.* 1996; Ashtekar and Lewandowski 1997): denoting the nodes by $\vec{j}_i = (j_i^u, j_i^d, j_i^t)$, $i = 1, \ldots, n$, where j_i^u denotes the colouring of the upper link, j_i^d the colouring of the lower link, and j_i^t the colouring of a link tangential to S, one obtains

$$\hat{A}(S)\Psi_S[A] = 4\pi\beta l_{\mathrm{P}}^2 \sum_{i=1}^n \sqrt{2j_i^u(j_i^u+1) + 2j_i^d(j_i^d+1) - j_i^t(j_i^t+1)}\Psi_S[A] \ . \quad (6.34)$$

For the special case $j_i^t = 0$ and $j_i^u = j_i^d$, we obtain again the earlier result (6.33).

We now show that $\hat{A}(S)$ is indeed an 'area operator', that is, the classical version of (6.32) is just the classical area of S. This classical version is

$$E_i(S_n) = \int_{S_n} \mathrm{d}\sigma^1 \mathrm{d}\sigma^2 \ n_a(\vec{\sigma}) E_i^a(\mathbf{x}(\vec{\sigma})) \approx \Delta\sigma^1 \Delta\sigma^2 n_a(\vec{\sigma}) E_i^a(\mathbf{x}_n(\vec{\sigma})) \ ,$$

where S_n refers to a partition of S, and $\mathbf{x}_n(\vec{\sigma})$ is an arbitrary point in S_n (it is assumed that the partition is sufficiently fine-grained). For the area, that is, the classical version of (6.32), one then obtains

$$A(S) = \lim_{\rho \to \infty} \sum_{n(\rho)} \Delta\sigma^1 \Delta\sigma^2 \sqrt{n_a(\vec{\sigma}) E_i^a(\mathbf{x}_n(\vec{\sigma})) n_b(\vec{\sigma}) E_i^b(\mathbf{x}_n(\vec{\sigma}))}$$

$$= \int_{\mathcal{S}} \mathrm{d}^2\sigma \; \sqrt{n_a(\vec{\sigma}) E_i^a(\mathbf{x}_n(\vec{\sigma})) n_b(\vec{\sigma}) E_i^b(\mathbf{x}_n(\vec{\sigma}))} \; .$$

Adapting coordinates on $\mathcal{S}$ as $x^3(\vec{\sigma}) = 0$, $x^1(\vec{\sigma}) = \sigma^2$, $x^2(\vec{\sigma}) = \sigma^2$, one gets from (6.24) that $n_1 = n_2 = 0$, and $n_3 = 1$. Using in addition (4.102) and (4.104), one obtains for the area

$$A(\mathcal{S}) = \int_{\mathcal{S}} \mathrm{d}^2\sigma \; \sqrt{h(\mathbf{x}) h^{33}(\mathbf{x})} = \int_{\mathcal{S}} \mathrm{d}^2\sigma \; \sqrt{h_{11} h_{22} - h_{12} h_{21}}$$

$$= \int_{\mathcal{S}} \mathrm{d}^2\sigma \; \sqrt{^{(2)}h} \; , \tag{6.35}$$

where $^{(2)}h$ denotes the determinant of the two-dimensional metric on $\mathcal{S}$. Therefore, the results (6.33) and (6.34) demonstrate that *area is quantized* in units proportional to the Planck area l_{P}^2. A similar result holds for volume and length (cf. Thiemann 2001). This could be an indication of the discreteness of space at the Planck scale already mentioned in Chapter 1. Since these geometric quantities refer to three-dimensional space, they only indicate the discrete nature of space, not space–time. In fact, as we have seen in Section 5.4, space–time itself emerges only in a semiclassical limit. The manifold Σ still plays the role of an 'absolute' structure; cf. Section 1.3.

The quantization (6.33) and (6.34) is considered as one of the central results of quantum loop (quantum connection) dynamics. Whether all eigenvalues of (6.34) are indeed realized depends on the topology of $\mathcal{S}$ (Ashtekar and Lewandowski 1997). In the case of an open $\mathcal{S}$ whose closure is contained in Σ, they are all realized. This is not the case for a closed surface.

The smallest eigenvalue of the area operator is zero. Its smallest non-zero eigenvalue is (in the case of open $\mathcal{S}$)

$$A_0 = 2\pi\sqrt{3}\beta l_{\mathrm{P}}^2 \; , \tag{6.36}$$

which is obtained from (6.34) for $j^d = 0$ and $j^u = j^t = j = 1/2$. The area gap (6.36) is referred to as one 'quantum of area'. Ashtekar and Lewandowski (1997) also showed that for $A(\mathcal{S}) \to \infty$, the difference ΔA between an eigenvalue A and its closest eigenvalue obeys

$$\Delta A \leq 4\pi\beta l_{\mathrm{P}}^2 \frac{\sqrt{8\pi\beta}}{\sqrt{A}} + \mathcal{O}\left(\frac{l_{\mathrm{P}}^2}{A}\right) l_{\mathrm{P}}^2 \; . \tag{6.37}$$

Therefore, $\Delta A \to 0$ for large A.

Although the area operator $\hat{A}(\mathcal{S})$ is gauge-invariant (invariant under SU(2)- or SO(3)-transformations), it is *not* invariant under three-dimensional diffeomorphisms. The reason is that it is defined for an abstract surface in terms of coordinates. For the same reason it is not an observable in the sense of Section 3.1.2. The general opinion is that it will become diffeomorphism-invariant (and an observable) if the surface is defined *intrinsically* through curvature invariants of the gravitational field or concrete matter fields; see for example, Rovelli and Gaul (2000). A concrete demonstration is, however, still lacking.

The discrete spectrum of the area operator is also at the heart of the statistical foundation of black-hole entropy, that is, the recovery of the black-hole entropy through a quantitative counting of microscopic states for the gravitational field. This will be discussed in Section 7.3.

6.3   Quantum Hamiltonian constraint

The next task in the quantization process is the treatment of the Hamiltonian constraint. In Section 4.3, we have considered the rescaled Hamiltonian constraint

$$\tilde{\mathcal{H}}_\perp = -8\pi G\beta^2 \mathcal{H}_\perp^{\text{g}} \ ,$$

see (4.126) and (4.129). It can be written as

$$\tilde{\mathcal{H}}_\perp = \frac{1}{\sqrt{h}}\text{tr}\left(\left(F_{ab} + \frac{\beta^2\sigma - 1}{\beta^2}R_{ab}\right)[E^a, E^b]\right)$$

$$\equiv \mathcal{H}_{\text{E}} + \frac{\beta^2\sigma - 1}{\beta^2\sqrt{h}}\text{tr}\left(R_{ab}[E^a, E^b]\right) \ . \tag{6.38}$$

The central idea in the quantization of this constraint is to use the fact that area and volume operators can be rigorously defined. (For the area operator this was discussed in the last section. The volume operator can be treated analogously; cf. Ashtekar and Lewandowski 1998) A direct substitution of E^a by a derivative operator as in (6.3) would not lead very far, since R_{ab} depends on E^a in a complicated way. In this respect the situation has not improved compared to the geometrodynamic approach discussed in Chapter 5.

In a first step one addresses the 'Euclidean' part $\mathcal{H}_{\text{E}}$ of (6.38). One recognizes from (4.135) that this part depends solely on the volume, V, and on F_{ab} and A_c. The volume operator—the quantum equivalent to (4.134)—is well defined and yields a self-adjoint operator on the Hilbert space $\mathcal{H}$ with a finite action on cylindrical functions. There are therefore no problems with factor ordering. The operators corresponding to F_{ab} and A_c can be treated by using holonomies. In a second step one makes use of (4.139), which in the Lorentzian case $\sigma = -1$ reads

$$\tilde{\mathcal{H}}_\perp + \frac{\mathcal{H}_{\text{E}}}{\beta^2} = -\frac{\beta^2 + 1}{2(4\pi\beta)^3}\epsilon^{abc}\text{tr}\left(\{A_a, T\}\{A_b, T\}\{A_c, V\}\right) \ . \tag{6.39}$$

Using (4.137) one can quantize T and then obtain a quantization of the full operator $\tilde{\mathcal{H}}_\perp$.

In order to get a well-defined operator $\mathcal{H}_{\text{E}}$, Thiemann (1996) considered the integral of (4.135) with respect to a lapse function $N(\mathbf{x})$,

$$H_{\text{E}}[N] = -\frac{1}{4\pi\beta}\int_\Sigma \text{d}^3x \ N(\mathbf{x})\epsilon^{abc}\text{tr}\left(F_{ab}\{A_c, V\}\right) \ . \tag{6.40}$$

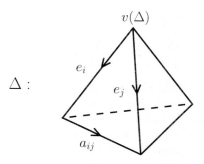

$$v(\Delta)$$

$$\Delta :$$

$$e_i$$

$$e_j$$

$$a_{ij}$$

FIG. 6.5. An elementary tetrahedron Δ.

One can make a triangulation (called 'Tri') of Σ into elementary tetrahedra Δ and choose one vertex, $v(\Delta)$, for each Δ; see Fig. 6.5. Calling $e_i(\Delta)$, $i = 1, 2, 3$, the three edges of Δ meeting at $v(\Delta)$, one can consider the loop

$$\alpha_{ij}(\Delta) \equiv e_i(\Delta) \circ a_{ij}(\Delta) \circ e_j^{-1}(\Delta) \ ,$$

where $a_{ij}(\Delta)$ connects the other vertices different from $v(\Delta)$. Thiemann (1996) could then show that one can get the correct Euclidean Hamiltonian (6.40) in the limit where all tetrahedra Δ shrink to their base points $v(\Delta)$: consider

$$H_{\mathrm{E}}^{\mathrm{Tri}}[N] = \sum_{\Delta \in \mathrm{Tri}} H_{\mathrm{E}}^{\Delta}[N] \ , \tag{6.41}$$

with

$$H_{\mathrm{E}}^{\Delta}[N] \equiv \frac{1}{12\pi\beta} N\left(v(\Delta)\right) \epsilon^{ijk} \mathrm{tr}\left(U_{\alpha_{ij}(\Delta)} U_{e_k(\Delta)} \{U_{e_k(\Delta)}^{-1}, V\}\right) \ , \tag{6.42}$$

where the $U_{...}$ denotes the holonomies along the corresponding loops and edges. Using

$$\lim_{\Delta \to v(\Delta)} U_{\alpha_{ij}(\Delta)} = 1 + 1/2 \ F_{ab} e_i^a(\Delta) e_j^b(\Delta) \ ,$$

$$\lim_{\Delta \to v(\Delta)} U_{e_k(\Delta)} = 1 + A_a e_k^a(\Delta) \ ,$$

the expression (6.41) tends to (6.40) in this limit.

The quantum operator corresponding to (6.41) is then *defined* by replacing $V \to \hat{V}$ and by substituting Poisson brackets with commutators,

$$\hat{H}_{\mathrm{E}}^{\mathrm{Tri}}[N] \equiv \sum_{\Delta \in \mathrm{Tri}} \hat{H}_{\mathrm{E}}^{\Delta}[N] \ , \tag{6.43}$$

with

$$\hat{H}_{\mathrm{E}}^{\Delta}[N] \equiv -\frac{\mathrm{i}}{12\pi\beta\hbar} N\left(v(\Delta)\right) \epsilon^{ijk} \mathrm{tr}\left(\hat{U}_{\alpha_{ij}(\Delta)} \hat{U}_{e_k(\Delta)} \{\hat{U}_{e_k(\Delta)}^{-1}, V\}\right) \ . \tag{6.44}$$

Furthermore, one can show that

$$\hat{H}_{\mathrm{E}}^{\mathrm{Tri}}[N]f_\alpha = \sum_{\Delta \in \mathrm{Tri}; \Delta \cap \alpha \neq \emptyset} \hat{H}_{\mathrm{E}}^{\Delta}[N]f_\alpha \,, \qquad (6.45)$$

where f_α denotes a cylindrical function associated with a graph α. From inspection of the right-hand side, one recognizes that there is a contribution only if Δ intersects α (in fact, as one can show, only if it intersects it at a vertex). This gives only a finite number of such terms, yielding an 'automatic regularization' that survives in the 'continuum limit' $\Delta \to v(\Delta)$. Thiemann (1996) also shows that no anomalies arise and that a Hermitian factor ordering can be chosen.

All this being done, it is clear from (6.39) that a regularization can also be achieved for the full Hamiltonian constraint operator $\hat{\tilde{\mathcal{H}}}_\perp$. It is unclear whether the operator $\hat{H}_{\mathrm{E}}$ has anything to do with the left-hand side of (6.6); the solutions mentioned there will most likely not be annihilated by $\hat{H}_{\mathrm{E}}$. There remain, of course, interesting open question which are at the focus of current research. Among these are:

1. Can physically interesting solutions to all constraints be obtained? This was at least achieved for (2+1)-dimensional gravity and for cosmological models (cf. Thiemann 2001).

2. Does one obtain the correct classical limit of the constraint algebra?

3. How does the semiclassical approximation scheme work?

The latter question has not yet been addressed in the same way as it was done with the geometrodynamical constraints in Section 5.4, since the constraints do not have the simple form in which a 'Born–Oppenheimer-type' of method can be straightforwardly employed. Instead, one has tried to use the methods of coherent states; cf. Thiemann (2001, 2003).

Alternative approaches to the quantum Hamiltonian constraint are the 'spin-foam models', which use a path-integral type of approach employing the evolution of spin networks in 'time'. We shall not discuss this here, but refer the reader to the literature; see Rovelli and Gaul (2000) and the references therein.

7

QUANTIZATION OF BLACK HOLES

7.1 Black-hole thermodynamics and Hawking radiation

In this subsection, we shall briefly review the thermodynamical behaviour of black holes and the Hawking effect. These issues arise at a semiclassical level—the gravitational field is treated as an external classical background—but they are supposed to play a key role in the search for quantum gravity. More details can be found, for example, in Kiefer (1998, 1999) and the references therein, see also the brief review Kiefer (2003a) which we shall partially follow.

7.1.1 *The laws of black-hole mechanics*

It is a most amazing fact that black holes obey *uniqueness theorems* (see Heusler 1996 for a detailed exposition). If an object collapses to form a black hole, a stationary state will be reached asymptotically. One can prove within the Einstein–Maxwell theory that stationary black holes are uniquely characterized by only three parameters: mass M, angular momentum $J \equiv Ma$, and electric charge q.[1] In this sense, black holes are much simpler objects than ordinary stars—given these parameters, they all look the same. All other degrees of freedom that might have been initially present have thus been radiated away, for example, in the form of electromagnetic or gravitational radiation during the collapse, or just disappeared (such as baryon number). Since these other degrees of freedom constitute some form of 'hair' (structure), one calls this theorem the *no-hair theorem*. The three parameters are associated with conservation laws at spatial infinity. In principle, one can thus decide about the nature of a black hole far away from the hole itself, without having to approach it. In astrophysical situations, electrically charged black holes do not play an important role, so the two parameters M and J suffice. The corresponding solution of Einstein's equations is called the Kerr solution (Kerr–Newman in the presence of charge). Stationary black holes are axially symmetric with spherical symmetry being obtained as a special case for $J = 0$. Charged black holes are of interest for theoretical reasons.

In the presence of other fields, the uniqueness theorems do not always hold, see for example, Núñez *et al.* (1998). This is, in particular, the case for non-Abelian gauge fields. In addition to charges at spatial infinity, such 'coloured black holes' have to be characterized by additional variables, and it is necessary to approach the hole to determine them. The physical reason for the occurrence of such solutions is the non-linear character of these gauge fields. Fields in regions

---

[1] Black holes could also have a magnetic-monopole charge, but this possibility will not be considered here.

closer to the black hole (that would otherwise be swallowed by the hole) are tied
to fields far away from the hole (that would otherwise be radiated away) to reach
an equilibrium situation. In most examples this equilibrium is, however, unstable
and the corresponding black-hole solution does not represent a physical solution.
Since classical non-Abelian fields have never been observed (the description of
objects such as quarks necessarily requires quantized gauge fields which, due to
confinement, have no macroscopic limit), they will not be taken into account in
the subsequent discussion.

In 1971, Stephen Hawking proved an important theorem about stationary
black holes: their area can never decrease with time. More precisely, he showed
that for a predictable black hole satisfying $R_{ab}k^a k^b \geq 0$ for all null k^a, the surface
area of the *future* event horizon *never* decreases with time. A 'predictable' black
hole is one for which the cosmic censorship hypothesis holds—this is thus a
major assumption for the area law. Cosmic censorship assumes that all black
holes occurring in nature have an event horizon, so that the singularity cannot be
seen by far-away observers (the singularity is not 'naked'). The time asymmetry
in this theorem comes into play because a statement is made about the future
horizon, not the past horizon; the analogous statement for white holes would then
be that the past event horizon never increases. It is a feature of our universe that
white holes seem to be absent, in contrast to black holes, cf. Section 10.2. It must
be emphasized that the area law only holds in the classical theory, not in the
quantum theory.

The area law seems to exhibit a close formal analogy to the Second Law of
Thermodynamics—there the *entropy* can *never* decrease with time (for a closed
system). However, the conceptual difference could not be more pronounced: while
the Second Law is related to statistical behaviour, the area law is just a theorem
in differential geometry.

Further support for this analogy is given by the existence of analogues to
the other laws of thermodynamics. The Zeroth Law states that there exists a
quantity, the temperature, that is constant on a body in thermal equilibrium.
Does there exist an analogous quantity for a black hole? One can in fact prove
that the surface gravity κ is constant over the event horizon (Wald 1984). For a
Kerr black hole, κ is given by

$$\kappa = \frac{\sqrt{(GM)^2 - a^2}}{2GMr_+} \xrightarrow{a \to 0} \frac{1}{4GM} = \frac{GM}{R_0^2}, \tag{7.1}$$

where r_+ denotes the location of the event horizon. In the Schwarzschild limit,
one recognizes the well-known expression for the Newtonian gravitational acceleration. ($R_S \equiv 2GM$ there denotes the Schwarzschild radius.) One can show for
a static black hole that κ is the limiting force that must be exerted at infinity to
hold a unit test mass in place when approaching the horizon. This justifies the
name surface gravity.

With a tentative formal relation between surface gravity and temperature,
and between area and entropy, the question arises whether a First Law of Ther-

Table 7.1 *Analogies between the laws of thermodynamics and the laws of black-hole mechanics*

Law	Thermodynamics	Stationary black holes
Zeroth	T constant on a body in thermal equilibrium	κ constant on the horizon of a black hole
First	$dE = TdS - pdV + \mu dN$	$dM = \dfrac{\kappa}{8\pi G}dA + \Omega_H dJ + \Phi dq$
Second	$dS \geq 0$	$dA \geq 0$
Third	$T = 0$ cannot be reached	$\kappa = 0$ cannot be reached

modynamics can be proved. This can in fact be achieved and the result for a Kerr–Newman black hole is

$$dM = \frac{\kappa}{8\pi G}dA + \Omega_H dJ + \Phi dq\,, \tag{7.2}$$

where A, Ω_H, Φ denote the area of the event horizon, the angular velocity of the black hole, and the electrostatic potential, respectively. This relation can be obtained by two conceptually different methods: a *physical process version* in which a stationary black hole is altered by infinitesimal physical processes, and an *equilibrium state version* in which the areas of two stationary black-hole solutions of Einstein's equations are compared. Both methods lead to the same result (7.2).

Since M is the energy of the black hole, (7.2) is the analogue of the First Law of Thermodynamics given by

$$dE = TdS - pdV + \mu dN\,. \tag{7.3}$$

'Modern' derivations of (7.2) make use of both Hamiltonian and Lagrangian methods of GR. For example, the First Law follows from an arbitrary diffeomorphism invariant theory of gravity whose field equations can be derived from a Lagrangian; see Wald (2001) and the references therein.

What about the Third Law of Thermodynamics? A 'physical process version' was proved by Israel (1986)—it is impossible to reach $\kappa = 0$ in a finite number of steps, although it is unclear whether this is true under all circumstances (Farrugia and Hajicek 1979). This corresponds to the 'Nernst version' of the Third Law. The stronger 'Planck version', which states that the entropy goes to zero (or a material-dependent constant) if the temperature approaches zero, does not seem to hold. The above analogies are summarized in Table 7.1.

7.1.2   *Hawking and Unruh radiation*

What is the meaning of black-hole temperature and entropy? According to classical GR, a black hole cannot radiate and, therefore, the temperature can only have a formal meaning. Important steps towards the interpretation of black-hole entropy were made by Bekenstein (1973); cf. also Bekenstein (2001). He argued that the Second Law of Thermodynamics would only be valid if a black hole possessed an entropy S_{BH}; otherwise, one could lower the entropy in the universe by just throwing matter possessing a certain amount of entropy into a black hole. Comparing (7.2) with (7.3) one recognizes that black-hole entropy must be a function of the area, $S_{BH} = f(A)$. Since the temperature must be positive, one must demand that $f'(A) > 0$. The simplest case $f(A) \propto \sqrt{A}$, that is, $S_{BH} \propto M$ would violate the Second Law because if two black holes merged, the mass of the resulting hole would be smaller than the sum of the masses of the original holes (due to energy emission through gravitational waves). With some natural assumptions, one can conclude that $S_{BH} \propto A/l_P^2$ (Bekenstein 1973, 2001). Note that Planck's constant $\hbar$ has entered the scene through the Planck length. This has happened since no fundamental length scale can be constructed from G and c alone. A sensible interpretation of black-hole temperature and entropy thus cannot be obtained in pure GR—quantum theory has to be taken into account.

Thus, one can write

$$T_{BH} \propto \frac{\hbar \kappa}{k_B}, \quad S_{BH} \propto \frac{k_B A}{G\hbar}, \tag{7.4}$$

and the important question is how the proportionality factor can be determined. This was achieved in the important paper by Hawking (1975). The key ingredient in Hawking's discussion is the behaviour of *quantum* fields on the background of an object collapsing to form a black hole. Similar to the situation of an external electric field (Schwinger effect), there is no uniquely defined notion of a *vacuum*. This leads to the occurrence of particle creation. The peculiarity of the black-hole case is the *thermal* distribution of the particles created, which is due to the presence of an event horizon.

It is helpful to understand the 'Hawking effect' by first considering an analogous effect in flat space–time: a uniformly accelerated observer experiences the Minkowski vacuum as being filled with thermal particles; cf. Unruh (1976). Whereas all inertial observers in Minkowski space agree on the notion of vacuum (and therefore on particles), this is no longer true for *non-inertial* observers.

Consider an observer who is uniformly accelerating along the X-direction in (1+1)-dimensional Minkowski space–time (Fig. 7.1). The Minkowski cartesian coordinates are labelled here by uppercase letters. The orbit of this observer is the hyperbola shown in Fig. 7.1. One recognizes that, as in the Kruskal diagram for the Schwarzschild metric, the observer encounters a horizon (here called 'acceleration horizon'). There is, however, no singularity behind this horizon. The region I is a globally hyperbolic space–time on its own—called *Rindler space–time*. This space–time can be described by coordinates (τ, ρ) which are connected to the cartesian coordinates via the transformation

$$\begin{pmatrix} T \\ X \end{pmatrix} = \rho \begin{pmatrix} \sinh a\tau \\ \cosh a\tau \end{pmatrix}, \tag{7.5}$$

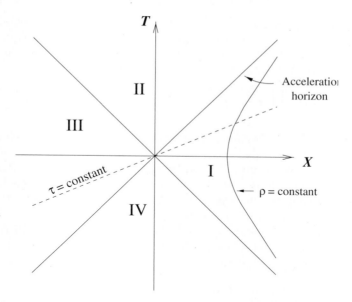

FIG. 7.1. Uniformly accelerated observer in Minkowski space.

where a is a constant (the orbit in Fig. 7.1 describes an observer with acceleration a, who has $\rho = 1/a$). Since

$$ds^2 = dT^2 - dX^2 = a^2\rho^2 d\tau^2 - d\rho^2 \ , \tag{7.6}$$

the orbits $\rho = $ constant are also orbits of a time-like Killing field $\partial/\partial\tau$. It is clear that τ corresponds to the external Schwarzschild coordinate t and that ρ corresponds to r. As in the Kruskal case, $\partial/\partial\tau$ becomes space-like in regions II and IV.

The analogy with Kruskal becomes even more transparent if the Schwarzschild metric is expanded around the horizon at $r = 2GM$. Introducing there a new coordinate ρ via $\rho^2/(8GM) = r - 2GM$ and recalling (7.1), one has

$$ds^2 \approx \kappa^2\rho^2 dt^2 - d\rho^2 - \frac{1}{4\kappa^2}d\Omega^2 \ . \tag{7.7}$$

Comparison with (7.6) shows that the first two terms on the right-hand side of (7.7) correspond exactly to the Rindler space–time (7.6) with the acceleration a replaced by the surface gravity κ. The last term[2] in (7.7) describes a two-sphere with radius $(2\kappa)^{-1}$.

How does the accelerating observer experience the standard Minkowski vacuum $|0\rangle_M$? The key point is that the vacuum is a *global* state correlating regions

---

[2]It is this term that is responsible for the non-vanishing curvature of (7.7) compared to the flat-space metric (7.6) whose extension into the (neglected) other dimensions would be just $-dY^2 - dZ^2$.

I and III in Fig. 7.1 (similar to Einstein–Podolsky–Rosen correlations), but that the accelerated observer is restricted to region I. Considering for simplicity the case of a massless scalar field, the global vacuum state comprising the regions I and III can be written in the form

$$|0\rangle_{\mathrm{M}} = \prod_{\omega} \sqrt{1 - e^{-2\pi\omega a^{-1}}} \sum_{n} e^{-n\pi\omega a^{-1}} |n_{\omega}^{\mathrm{I}}\rangle \otimes |n_{\omega}^{\mathrm{III}}\rangle , \qquad (7.8)$$

where $|n_{\omega}^{\mathrm{I}}\rangle$ and $|n_{\omega}^{\mathrm{III}}\rangle$ are n-particle states with frequency $\omega = |\mathbf{k}|$ in regions I and III, respectively. These n-particle states are defined with respect to the 'Rindler vacuum', which is the vacuum defined by an accelerating observer. The expression (7.8) is an example for the Schmidt expansion of two entangled quantum systems, see for example, Joos $et\ al.$ (2003); note also the analogy of (7.8) with a BCS-state in the theory of superconductivity.

For an observer restricted to region I, the state (7.8) cannot be distinguished, by operators with support in I only, from a density matrix that is found from (7.8) by tracing out all degrees of freedom in region III,

$$\rho_{\mathrm{I}} \equiv \mathrm{tr}_{\mathrm{III}} |0\rangle_{\mathrm{M}} \langle 0|_{\mathrm{M}}$$
$$= \prod_{\omega} \left(1 - e^{-2\pi\omega a^{-1}}\right) \sum_{n} e^{-2\pi n\omega a^{-1}} |n_{\omega}^{\mathrm{I}}\rangle\langle n_{\omega}^{\mathrm{I}}| . \qquad (7.9)$$

Note that the density matrix ρ_{I} has exactly the form corresponding to a thermal canonical ensemble with the Davies–Unruh temperature

$$T_{\mathrm{DU}} = \frac{\hbar a}{2\pi k_{\mathrm{B}}} \approx 4.05 \times 10^{-23}\, a \left[\frac{\mathrm{cm}}{\mathrm{s}^2}\right] \mathrm{K} , \qquad (7.10)$$

cf. (1.33). An observer who is accelerating uniformly through Minkowski space thus sees a $thermal$ distribution of particles. This is an important manifestation of the non-uniqueness of the vacuum state in quantum field theory, even for flat space–time. A more detailed discussion invoking models of particle detectors confirms this result.

We shall now turn to black holes. From the form of the line element near the horizon, (7.7), one can already anticipate that—according to the equivalence principle—a black hole radiates with a temperature as specified in (7.10) with a being replaced by κ. This is, in fact, what Hawking (1975) found. In his approach he considered the situation depicted in Fig. 7.2: the vacuum modes of a quantum field are calculated on the background of a star collapsing to form a black hole.

Due to the dynamical background, an initial vacuum does not stay a vacuum but becomes a thermal state with respect to late-time observers. The 'Hawking temperature' reads

$$T_{\mathrm{BH}} = \frac{\hbar\kappa}{2\pi k_{\mathrm{B}}} , \qquad (7.11)$$

cf. (1.31). For the total luminosity of the black hole, one finds

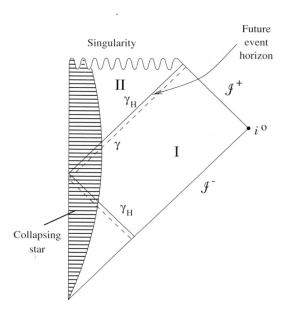

FIG. 7.2. Penrose diagram showing the collapse of a spherically symmetric star to form a black hole; γ denotes a light ray propagating from $\mathcal{J}^-$ through the collapsing star to $\mathcal{J}^+$. The limiting case is the ray γ_{H} which stays on the horizon.

$$L = -\frac{\mathrm{d}M}{\mathrm{d}t} = \frac{1}{2\pi} \sum_{l=0}^{\infty} (2l+1) \int_{0}^{\infty} \mathrm{d}\omega\, \omega \frac{\Gamma_{\omega l}}{e^{2\omega\pi\kappa^{-1}} - 1} \,. \tag{7.12}$$

The term $\Gamma_{\omega l}$—called 'greybody factor' because it encodes a deviation from the black-body spectrum—takes into account the fact that some of the particle modes are back-scattered into the black hole by means of space–time curvature.

For the special case of the Schwarzschild metric, where $\kappa = (4GM)^{-1}$, (7.11) becomes the expression (1.32). One can estimate the life-time of such a black hole by making the plausible assumption that the decrease in mass is equal to the energy radiated to infinity. Using Stefan–Boltzmann's law, one gets

$$\frac{\mathrm{d}M}{\mathrm{d}t} \propto -AT_{\mathrm{BH}}^4 \propto -M^2 \times \left(\frac{1}{M}\right)^4 = -\frac{1}{M^2} \,,$$

which when integrated yields

$$t(M) \propto (M_0^3 - M^3) \approx M_0^3 \,. \tag{7.13}$$

Here M_0 is the initial mass. It has been assumed that after the evaporation $M \ll M_0$. Very roughly, the life-time of a black hole is thus given by

$$\tau_{\text{BH}} \approx \left(\frac{M_0}{m_{\text{P}}}\right)^3 t_{\text{P}} \approx 10^{65} \left(\frac{M_0}{M_\odot}\right)^3 \text{ years} . \tag{7.14}$$

Hawking used the semiclassical approximation in which the non-gravitational fields are quantum but the gravitational field is treated as an external classical field; cf. Section 5.4. This approximation is expected to break down when the black-hole mass approaches the Planck mass, that is, after a time given by (7.14). Only a full theory of quantum gravity can describe the final stage of black-hole evaporation.

As an intermediate step towards full quantum gravity, one might consider the heuristic 'semiclassical' Einstein equations discussed in Section 1.2; see (1.34). The evaluation of $\langle T_{ab} \rangle$—which requires regularization and renormalization—is a difficult subject on its own (Frolov and Novikov 1998). The renormalized value for $\langle T_{ab} \rangle$ is essentially unique (its ambiguities can be absorbed in coupling constants) if certain sensible requirements are imposed. Evaluating the components of the renormalized $\langle T_{ab} \rangle$ near the horizon, one finds that there is a flux of *negative energy* into the hole. Clearly this leads to a decrease of the mass. These negative energies represent a typical quantum effect and are well known from the—accurately measured—Casimir effect. This occurrence of negative energies is also responsible for the breakdown of the classical area law in quantum theory.

The negative flux near the horizon lies also at the heart of the 'pictorial' representation of Hawking radiation that is often used; see for example, Parikh and Wilczek (2000). In vacuum, virtual pairs of 'particles' are created and destroyed. However, close to the horizon one partner of this virtual pair might fall into the black hole, thereby liberating the other partner to become a real particle and escaping to infinity as Hawking radiation. The global quantum field exhibits quantum entanglement between the in- and outside of the black hole, similar to the case of the accelerated observer discussed above.

One can also give explicit expressions for the Hawking temperature (7.11) in the case of rotating and charged black holes. For the Kerr solution, one has

$$k_{\text{B}} T_{\text{BH}} = \frac{\hbar \kappa}{2\pi} = 2 \left(1 + \frac{M}{\sqrt{M^2 - a^2}}\right)^{-1} \frac{\hbar}{8\pi M} < \frac{\hbar}{8\pi M} . \tag{7.15}$$

Rotation thus reduces the Hawking temperature. For the Reissner–Nordström solution (describing a charged spherically symmetric black hole), one has

$$k_{\text{B}} T_{\text{BH}} = \frac{\hbar}{8\pi M} \left(1 - \frac{(Gq)^4}{r_+^4}\right) < \frac{\hbar}{8\pi M} . \tag{7.16}$$

Electric charge thus also reduces the Hawking temperature. For an extremal black hole, $r_+ = GM = \sqrt{G}|q|$, and therefore $T_{\text{BH}} = 0$.

7.1.3 Bekenstein–Hawking entropy

After the Hawking temperature has been calculated, the entropy is also given. From the First Law (7.2) one finds the 'Bekenstein–Hawking entropy'

$$S_{BH} = \frac{k_B A}{4G\hbar} , \qquad (7.17)$$

in which the unknown factor in (7.4) has now been fixed. For the special case of a Schwarzschild black hole, this yields

$$S_{BH} = \frac{k_B \pi R_0^2}{G\hbar} . \qquad (7.18)$$

It can be easily estimated that S_{BH} is much bigger than the entropy of the star that collapsed to form the black hole. The entropy of the sun, for example, is $S_\odot \approx 10^{57} k_B$, whereas the entropy of a solar-mass black hole is about $10^{77} k_B$, which is 20 orders of magnitude larger (recall that the number of states is given by $\exp(S/k_B)$).

Can a physical interpretation of this huge discrepancy be given? In the above discussion, the laws of black-hole mechanics have been treated as phenomenological thermodynamical laws. The central open question therefore is: can S_{BH} be derived from quantum-statistical considerations? This would mean that S_{BH} could be calculated from a Gibbs-type formula according to

$$S_{BH} \stackrel{?}{=} -k_B \text{tr}(\rho \ln \rho) \equiv S_{SM} , \qquad (7.19)$$

where ρ denotes an appropriate density matrix; S_{BH} would then somehow correspond to the number of quantum microstates that are consistent with the macrostate of the black hole. According to the no-hair theorem, the macrostate is uniquely characterized by mass, angular momentum, and charge. Some important questions are:

- Does S_{BH} correspond to states hidden behind the horizon?
- Or does S_{BH} correspond to the number of possible initial states from which the black hole might have formed?
- What are the microscopic degrees of freedom?
- Where are they located (if at all)?
- Can one understand the universality of the result?
- What happens with S_{BH} after the black hole has evaporated?
- Is the entropy a 'one-loop' or a 'tree-level' effect?

The attempts to calculate S_{BH} by state counting are usually done in the 'one-loop limit' of quantum field theory in curved space–time—this is the limit where gravity is classical but non-gravitational fields are fully quantum; cf. Sections 2.2 and 5.4. Equation (7.11) has been derived in this limit. The expression (7.17) can already be calculated from the so-called 'tree level' approximation of the theory, where only the gravitational degrees of freedom are taken into account. Usually a saddle-point approximation for a Euclidean path integral is being performed. Such derivations are, however, equivalent to derivations within classical thermodynamics; cf. Wald (2001).

The emergence of the thermal nature of black-hole radiation has led to the discussion of the *information-loss problem*; cf. Kiefer (2004 *a*) and the references therein. If the black hole evaporates completely and leaves only thermal radiation behind, one would have a conflict with established principles in quantum theory: any initial state (in particular, a pure state) would evolve into a mixed state. In ordinary quantum theory this is forbidden by the unitary evolution of the total system. Formally, $\text{tr}\rho^2$ remains *constant* under the von Neumann equation; the same is true for the entropy $S_{\text{SM}} = -k_B \text{tr}(\rho \ln \rho)$: for a unitarily evolving system, there is no increase in entropy. If these laws were violated during black-hole evaporation, information would be destroyed. This is, indeed, the speculation that Hawking made after his discovery of black-hole radiation. The attitudes towards this information-loss problem can be roughly divided into the following classes:

1. The information is indeed lost during black-hole evaporation, and the quantum-mechanical Liouville equation is replaced by an equation of the form

$$\rho \longrightarrow \$\rho \neq S\rho S^\dagger \,, \tag{7.20}$$

   where $\$$ is Hawking's 'dollar matrix' which generalizes the ordinary S-matrix S.

2. The full evolution is in fact unitary; the black-hole radiation contains subtle quantum correlations that cannot be seen in the semiclassical approximation.

3. The black hole does not evaporate completely, but leaves a 'remnant' with mass in the order of the Planck mass that carries the whole information.

A theory of quantum gravity should give a definite answer to this question. In the following subsections as well as in Chapter 9, we shall discuss attempts to describe black holes and their evolution within quantum gravity.

7.2   Canonical quantization of the Schwarzschild black hole

What can the methods of Chapters 5 and 6 say about the quantum behaviour of Schwarzschild black holes? In the following, we shall present an outline of the canonical quantization procedure for spherically symmetric systems. This was developed in the connection representation by Thiemann and Kastrup (1993) and in the geometrodynamical representation by Kuchař (1994). An extension to charged black holes (Reissner–Nordström solutions) can be found in Louko and Winters-Hilt (1996). In this subsection, we discuss the geometrodynamical approach and follow, with elaborations, the presentation in Kiefer (1998).

7.2.1 Classical formalism

Starting point is the ansatz for a general spherically symmetric metric on $\mathbb{R} \times \mathbb{R} \times S^2$,

$$ds^2 = -N^2(r,t)dt^2 + \Lambda^2(r,t)(dr + N^r(r,t)dt)^2 + R^2(r,t)d\Omega^2 . \qquad (7.21)$$

The lapse function N encodes the possibility to perform arbitrary reparametrizations of the time parameter, while the shift function N^r is responsible for reparametrizations of the radial coordinate (this is the only freedom in performing spatial coordinate transformations that is left after spherical symmetry is imposed). The parameter r is only a label for the spatial hypersurfaces; if the hypersurface extends from the left to the right wedge in the Kruskal diagram, one takes $r \in (-\infty, \infty)$. If the hypersurface originates at the bifurcation point where path and future horizon meet, one has $r \in (0, \infty)$. If one has in addition a spherically symmetric electromagnetic field, one makes the following ansatz for the one-form potential,

$$A = \phi(r,t)dt + \Gamma(r,t)dr . \qquad (7.22)$$

In the Hamiltonian formulation, ϕ as well as N and N^r are Lagrange multipliers whose variations yield the constraints of the theory. Variation of the Einstein–Hilbert action with respect to N yields the Hamiltonian constraint which for the spherically symmetric model is given by

$$\mathcal{H} = \frac{G}{2}\frac{\Lambda P_\Lambda^2}{R^2} - G\frac{P_\Lambda P_R}{R} + \frac{\Lambda P_\Gamma^2}{2R^2} + G^{-1}V^{\mathrm{g}} \approx 0 , \qquad (7.23)$$

where the gravitational potential term reads

$$V^{\mathrm{g}} = \frac{RR''}{\Lambda} - \frac{RR'\Lambda'}{\Lambda^2} + \frac{R'^2}{2\Lambda} - \frac{\Lambda}{2} . \qquad (7.24)$$

(A prime denotes differentiation with respect to r.) Variation with respect to N^r yields one (radial) diffeomorphism constraint,

$$\mathcal{H}_r = P_R R' - \Lambda P_\Lambda' \approx 0 . \qquad (7.25)$$

One recognizes from this constraint that R transforms as a scalar, while Λ transforms as a scalar density.

Variation of the action with respect to ϕ yields as usual the Gauß constraint

$$\mathcal{G} = P_\Gamma' \approx 0 . \qquad (7.26)$$

The constraint (7.25) generates radial diffeomorphisms for the fields R, Λ, and their canonical momenta. It does not generate diffeomorphisms for the electromagnetic variables. This can be taken into account if one uses the multiplier

$\tilde{\phi} = \phi - N^r\Gamma$ instead of ϕ and varies with respect to $\tilde{\phi}$ (Louko and Winters-Hilt 1996), but for the present purpose it is sufficient to stick to the above form (7.25).

The model of spherical symmetric gravity can be embedded into a whole class of models usually referred to as 'two-dimensional dilaton gravity theories'. This terminology comes from effective two-dimensional theories (usually motivated by string theory), which contain in the gravitational sector a scalar field (the 'dilaton') in addition to the two-dimensional metric (of which only the conformal factor is relevant). An example is the 'CGHS model' defined in (5.60) within which one can address the issues of Hawking radiation and back reaction. This model is classically soluble even if another, conformally coupled, scalar field is included. The canonical formulation of this model can be found, for example, in Louis-Martinez *et al.* (1994) and Demers and Kiefer (1996). The dilaton field is analogous to the field R from above, while the conformal factor of the two-dimensional metric is analogous to Λ.

Consider now the boundary conditions for $r \to \infty$. One has in particular

$$\Lambda(r,t) \to 1 + \frac{GM(t)}{r}, \ R(r,t) \to r, \ N \to N(t) , \tag{7.27}$$

as well as

$$P_\Gamma(r,t) \to q(t), \quad \phi(r,t) \to \phi(t) . \tag{7.28}$$

From the variation with respect to Λ, one then finds the boundary term $G \int dt \ N\delta m$. In order to avoid the unwanted conclusion $N = 0$ (no evolution at infinity), one has to compensate this term in advance by adding the boundary term

$$-G \int dt \ NM$$

to the classical action. Note that M is just the ADM mass. The need to include such a boundary term was recognized by Regge and Teitelboim (1974); cf. Section 4.2.4. Similarly, one has to add for charged black holes the term

$$- \int dt \ \phi q$$

to compensate for $\int dt \ \phi\delta q$, which arises from varying P_Γ. If one wished instead to consider q as a given, external parameter, this boundary term would be obsolete.

As long as restriction is made to the eternal hole, appropriate canonical transformations allow one to simplify the classical constraint equations considerably (Kuchař 1994; Louko and Winters-Hilt 1996). One gets

$$(\Lambda, P_\Lambda; R, P_R; \Gamma, P_\Gamma) \longrightarrow (\mathcal{M}, P_\mathcal{M}; \mathcal{R}, P_\mathcal{R}; Q, P_Q) .$$

In particular,

$$\mathcal{M}(r,t) = \frac{P_\Gamma^2 + P_\Lambda^2}{2R} + \frac{R}{2}\left(1 - \frac{R'^2}{\Lambda^2}\right) \overset{r \to \infty}{\longrightarrow} M(t) , \tag{7.29}$$

$$Q(r,t) = P_\Gamma \xrightarrow{r\to\infty} q(t) \ . \tag{7.30}$$

(We only mention that $\mathcal{R} = R$ and that the expression for $P_\mathcal{R}$ is somewhat lengthy and will not be given here.)

The new constraints, which are equivalent to the old ones, read

$$\mathcal{M}' = 0 \quad \Rightarrow \quad \mathcal{M}(r,t) = M(t), \tag{7.31}$$
$$Q' = 0 \quad \Rightarrow \quad Q(r,t) = q(t) \ , \tag{7.32}$$
$$P_\mathcal{R} = 0 \ . \tag{7.33}$$

Note that $N(t)$ and $\phi(t)$ are prescribed functions that must not be varied; otherwise one would be led to the unwanted restriction that $M = 0 = q$. A variation is allowed if the action is being parametrized, bringing in new dynamical variables,

$$N(t) \equiv \dot\tau(t),$$
$$\phi(t) \equiv \dot\lambda(t) \ . \tag{7.34}$$

Here, τ is the proper time that is measured with standard clocks at infinity, and λ is the variable conjugate to charge; λ is therefore connected with the electromagnetic gauge parameter at the boundaries. In the canonical formalism one has to introduce momenta conjugate to these variables, which will be denoted by π_τ and π_λ, respectively. This, in turn, requires the introduction of additional constraints linear in momenta,

$$C_\tau = \pi_\tau + GM \approx 0 \ , \tag{7.35}$$
$$C_\lambda = \pi_\lambda + q \approx 0 \ , \tag{7.36}$$

which have to be added to the action:

$$-G \int \mathrm{d}t \ M\dot\tau \quad \longrightarrow \quad \int \mathrm{d}t \ (\pi_\tau\dot\tau - NC_\tau) \ , \tag{7.37}$$
$$-\int \mathrm{d}t \ q\dot\lambda \quad \longrightarrow \quad \int \mathrm{d}t \ (\pi_\lambda\dot\lambda - \phi C_\lambda) \ . \tag{7.38}$$

The remaining constraints in this model are thus (7.33), and (7.35) and (7.36).

7.2.2 *Quantization*

Quantization then proceeds in the way discussed in Chapter 5 by acting with an operator version of the constraints on wave functionals $\Psi[\mathcal{R}(r); \tau, \lambda]$. Since (7.33) leads to $\delta\Psi/\delta\mathcal{R} = 0$, one is left with a purely quantum *mechanical* wave function $\psi(\tau, \lambda)$. The implementation of the constraints (7.35) and (7.36) then yields

$$\frac{\hbar}{\mathrm{i}} \frac{\partial\psi}{\partial\tau} + M\psi = 0 \ , \tag{7.39}$$

$$\frac{\hbar}{i}\frac{\partial \psi}{\partial \lambda} + q\psi = 0 \; , \tag{7.40}$$

which can be readily solved to give

$$\psi(\tau, \lambda) = \chi(M, q)e^{-i(M\tau + q\lambda)/\hbar} \tag{7.41}$$

with an arbitrary function $\chi(M, q)$. Note that M and q are considered here as being fixed. The reason for this is that up to now we have restricted attention to one semiclassical component of the wave function only (eigenstates of mass and charge).

If the hypersurface goes through the whole Kruskal diagram of the eternal hole, only the boundary term at $r \to \infty$ (and an analogous one for $r \to -\infty$) contributes. Of particular interest in the black-hole case, however, is the case where the surface originates at the 'bifurcation surface' ($r \to 0$) of past and future horizons. This makes sense since data on such a surface suffice to construct the whole right Kruskal wedge, which is all that is accessible to an observer in this region. Moreover, this mimicks the situation where a black hole is formed by collapse, in which the regions III and IV of the Kruskal diagram are absent.

What are the boundary conditions to be adopted at $r \to 0$? They are chosen in such a way that the classical solutions have a non-degenerate horizon and that the hypersurfaces start at $r = 0$ asymptotically to hypersurfaces of constant Killing time (Louko and Whiting 1995). In particular,

$$N(r, t) = N_1(t)r + \mathcal{O}(r^3) \; , \tag{7.42}$$
$$\Lambda(r, t) = \Lambda_0(t) + \mathcal{O}(r^2) \; , \tag{7.43}$$
$$R(r, t) = R_0(t) + R_2(t)r^2 + \mathcal{O}(r^4) \; . \tag{7.44}$$

Variation leads, similarly to the situation at $r \to \infty$, to a boundary term at $r = 0$,

$$-N_1 R_0 (G\Lambda_0)^{-1}\delta R_0 \; .$$

If $N_1 \neq 0$, this term must be subtracted ($N_1 = 0$ corresponds to the case of extremal holes, $|q| = GM$, which is characterized by $\partial N/\partial r(r = 0) = 0$). Introducing the notation $N_0 \equiv N_1/\Lambda_0$, the boundary term to be added to the classical action reads

$$(2G)^{-1}\int dt \; N_0 R_0^2 \; .$$

The quantity

$$\alpha \equiv \int_{t_1}^{t} dt \; N_0(t) \tag{7.45}$$

can be interpreted as a 'rapidity' because it boosts the normal vector to the hypersurfaces t=constant, n^a, in the way described by

$$n^a(t_1)n_a(t) = -\cosh\alpha \; , \tag{7.46}$$

see Hayward (1993). To avoid fixing N_0, one introduces an additional parametrization (Brotz and Kiefer 1997)

$$N_0(t) = \dot{\alpha}(t) . \tag{7.47}$$

Similarly to (7.37) and (7.38) above, one must perform in the action the following replacement:

$$(2G)^{-1} \int dt \; R_0^2 \dot{\alpha} \quad \rightarrow \quad \int dt \; (\pi_\alpha \dot{\alpha} - N_0 \mathcal{C}_\alpha) , \tag{7.48}$$

with the new constraint

$$\mathcal{C} = \pi_\alpha - \frac{A}{8\pi G} \approx 0 , \tag{7.49}$$

where $A = 4\pi R_0^2$ is the surface of the bifurcation sphere. One finds that α and A are canonically conjugate variables; see Carlip and Teitelboim (1995).

The quantum constraints can then be solved, and a plane-wave-like solution reads

$$\Psi(\alpha, \tau, \lambda) = \chi(M, q) \exp\left[\frac{i}{\hbar}\left(\frac{A(M, q)\alpha}{8\pi G} - M\tau - q\lambda\right)\right] , \tag{7.50}$$

where $\chi(M, q)$ is an arbitrary function of M and q; one can construct superpositions of the solutions (7.50) in the standard way by integrating over M and q.

Varying the phase in (7.50) with respect to M and q yields the classical equations

$$\alpha = 8\pi G \left(\frac{\partial A}{\partial M}\right)^{-1} \tau = \kappa\tau , \tag{7.51}$$

$$\lambda = \frac{\kappa}{8\pi G} \frac{\partial A}{\partial q} \tau = \Phi\tau . \tag{7.52}$$

The solution (7.50) holds for non-extremal holes. If one made a similar quantization for extremal holes on their own, the first term in the exponent of (7.50) would be absent.

An interesting analogy with (7.50) is the plane-wave solution for a free non-relativistic particle,

$$\exp(ikx - \omega(k)t) . \tag{7.53}$$

As in (7.50), the number of parameters is one less than the number of arguments, since $\omega(k) = k^2/2m$. A quantization for extremal holes on their own would correspond to choosing a particular value for the momentum at the classical level, say p_0, and demanding that no dynamical variables (x, p) exist for $p = p_0$. This is, however, not the usual way to find classical correspondence—such a correspondence is not gained from the plane-wave solution (7.53) but from *wave packets* which are obtained by superposing different wave numbers k. This then yields quantum states which are sufficiently concentrated around classical trajectories such as $x = p_0 t/m$.

It seems, therefore, appropriate to proceed similarly for black holes: construct wave packets for non-extremal holes that are concentrated around the classical values (7.51) and (7.52) and then *extend* them by hand to the extremal limit. This would correspond to 'extremization after quantization', in contrast to the 'quantization after extremization' made above. Expressing in (7.50) M as a function of A and q and using Gaussian weight functions, one has

$$\Psi(\alpha, \tau, \lambda) = \int_{A>4\pi q^2} \mathrm{d}A\mathrm{d}q \, \exp\left[-\frac{(A - A_0)^2}{2(\Delta A)^2} - \frac{(q - q_0)^2}{2(\Delta q)^2}\right]$$

$$\times \exp\left[\frac{\mathrm{i}}{\hbar}\left(\frac{A\alpha}{8\pi G} - M(A, q)\tau - q\lambda\right)\right]. \tag{7.54}$$

The result of this calculation is given and discussed in Kiefer and Louko (1999). As expected, one finds Gaussian packets that are concentrated around the classical values (7.51) and (7.52). As for the free particle, the wave packets exhibit dispersion with respect to Killing time τ. Using for ΔA the Planck-length squared, $\Delta A \propto G\hbar \approx 2.6 \times 10^{-66} \mathrm{cm}^2$, one finds for the typical dispersion time in the Schwarzschild case

$$\tau_* = \frac{128\pi^2 R_0^3}{G\hbar} \approx 10^{65}\left(\frac{M}{M_\odot}\right)^3 \text{years}. \tag{7.55}$$

Note that this is just of the order of the black-hole evaporation time (7.14). The dispersion of the wave packet gives the time scale after which the semiclassical approximation breaks down.

Coming back to the charged case, and approaching the extremal limit $\sqrt{G}M = |q|$, one finds that the widths of the wave packet (7.54) are *independent* of τ for large τ. This is due to the fact that for the extremal black hole, $\kappa = 0$ and therefore no evaporation takes place. If one takes, for example, $\Delta A \propto G\hbar$ and $\Delta q \propto \sqrt{G\hbar}$, one finds for the α-dependence of (7.54) for $\tau \to \infty$, the factor

$$\exp\left(-\frac{\alpha^2}{128\pi^2}\right), \tag{7.56}$$

which is independent of both τ and $\hbar$. It is clear that this packet, although concentrated at the value $\alpha = 0$ for extremal holes, has support also for $\alpha \neq 0$ and is qualitatively not different from a wave packet that is concentrated at a value $\alpha \neq 0$ close to extremality.

An interesting question is the possible occurrence of a naked singularity for which $\sqrt{G}M < |q|$. Certainly, the above boundary conditions do not comprise the case of a singular three-geometry. However, the wave packets discussed above also contain parameter values that would correspond to the 'naked' case. Such geometries could be avoided if one imposed the boundary condition that the wave function vanishes for such values. But then continuity would enforce the wave function also to vanish on the boundary, that is, at $\sqrt{G}M = |q|$. This would mean that extremal black holes could not exist at all in quantum gravity—an interesting speculation.

A possible thermodynamical interpretation of (7.50) can only be obtained if an appropriate transition into the Euclidean regime is performed. This transition is achieved by the 'Wick rotations' $\tau \to -i\beta\hbar$, $\alpha \to -i\alpha_E$ (from (7.47) it is clear that α is connected to the lapse function and must be treated similar to τ), and $\lambda \to -i\beta\hbar\Phi$. Demanding regularity of the Euclidean line element, one arrives at the conclusion that $\alpha_E = 2\pi$. But this means that the Euclidean version of (7.51) just reads $2\pi = \kappa\beta\hbar$, which with $\beta = (k_B T_{BH})^{-1}$ is just the expression for the Hawking temperature (1.31). Alternatively, one could use (1.31) to derive $\alpha_E = 2\pi$.

The Euclidean version of the state (7.50) then reads

$$\Psi_E(\alpha, \tau, \lambda) = \chi(M, q) \exp\left(\frac{A}{4G\hbar} - \beta M - \beta\Phi q\right). \tag{7.57}$$

One recognizes in the exponent of (7.57), the occurrence of the Bekenstein–Hawking entropy. Of course, (7.57) is still a pure state and should not be confused with a partition sum. But the factor $\exp[A/(4G\hbar)]$ in (7.57) directly gives the enhancement factor for the rate of black-hole pair creation relative to ordinary pair creation (Hawking and Penrose 1996). It must be emphasized that S_{BH} fully arises from a boundary term at the horizon ($r \to 0$).

It is now clear that a quantization scheme that treats extremal black holes as a limiting case gives $S_{BH} = A/(4G\hbar)$ also for the extremal case. This coincides with the result found from string theory; see Section 9.2.5. On the other hand, quantizing extremal holes on their own would yield $S_{BH} = 0$. From this point of view, it is also clear why the extremal (Kerr) black hole that occurs in the transition from the disk-of-dust solution to the Kerr-solution has entropy $A/(4G\hbar)$; see Neugebauer (1998). If $S_{BH} \neq 0$ for the extremal hole (which has temperature zero), the stronger version of the Third Law of Thermodynamics (that would require $S \to 0$ for $T \to 0$) apparently does not hold. This is not particularly disturbing, since many systems in ordinary thermodynamics (such as glasses) violate the strong form of the Third Law; it just means that the system does not approach a unique state for $T \to 0$.

The above discussion has been performed for a pure black hole without inclusion of matter degrees of freedom. In the presence of other variables, it is no longer possible to find simple solutions such as (7.50). One possible treatment is to perform a semiclassical approximation as presented in Section 5.4. In this way, one can recover a functional Schrödinger equation for matter fields on a black-hole background. For simple situations this equation can be solved (Demers and Kiefer 1996). Although the resulting solution is, of course, a pure state, the expectation value of the particle number operator exhibits a Planckian distribution with respect to the Hawking temperature—this is how Hawking radiation is being recovered in this approach.[3] For this reason, one might even speculate that the information-loss problem for black holes is not a real problem, since only pure states appear for the full system. In fact, the mixed nature of Hawking radiation can be understood by the process of decoherence (Kiefer 2004 a). It is even possible that the Bekenstein–Hawking entropy (7.17) could be calculated from the decohering influence of additional degrees of freedom such as the quasi-normal modes of the black hole.

7.3 Black-hole spectroscopy and entropy

The results of the last subsection indicate that black holes are truly quantum objects. In fact, as especially Bekenstein (1999) has emphasized, they might play the same role for the development for quantum gravity that atoms had played for quantum mechanics. In the light of this possible analogy, one may wonder whether black holes possess a discrete spectrum of states similar to atoms. Arguments in favour of this idea were given in Bekenstein (1974, 1999); cf. also Mukhanov (1986). Bekenstein (1974) noticed that the horizon area of a (non-extremal) black hole can be treated in the classical theory as a mechanical adiabatic invariant. This is borne out from gedanken experiments in which one shoots into the hole charged particles (in the Reissner–Nordström case) or scalar waves (in the Kerr case) with appropriate energies. From experience with quantum mechanics, one would expect that the corresponding quantum entity possesses a discrete spectrum. The easiest possibility is certainly to have constant spacing between the eigenvalues, that is, $A_n \propto n$ for $n \in \mathbb{N}$. (In the Schwarzschild case, this would entail for the mass values $M_n \propto \sqrt{n}$.) This can be tentatively concluded, for example, from the 'Euclidean' wave function (7.57): if one imposed an *ad hoc* Bohr–Sommerfeld quantization rule, one would find

$$nh = \oint \pi_{\alpha_{\mathrm{E}}} \mathrm{d}\alpha_{\mathrm{E}} = \int_0^{2\pi} \frac{A \mathrm{d}\alpha_{\mathrm{E}}}{8\pi G} = \frac{A}{4G} \ . \tag{7.58}$$

(Recall $\alpha_{\mathrm{E}} = 2\pi$.) A similar result follows if the range of the time parameter τ in the Lorentzian version is assumed to be compact, similar to momentum quantization on finite spaces (Kastrup 1996).

A different argument to fix the factor in the area spectrum goes as follows (Mukhanov 1986; Bekenstein and Mukhanov 1995). One assumes the quantization condition

$$A_n = \alpha l_{\mathrm{P}}^2 n \ , n \in \mathbb{N} \ , \tag{7.59}$$

with some undetermined constant α. The energy level n will degenerate with multiplicity $g(n)$, so one would expect the identification

$$S_{\mathrm{BH}} = \frac{A}{4l_{\mathrm{P}}^2} + \mathrm{constant} = \ln g(n) \ . \tag{7.60}$$

---

[3]It can be recovered by similar methods also in the quantum collapse of a spherically symmetric dust cloud; cf. Vaz *et al.* (2003).

Demanding $g(1) = 1$ (i.e. assuming that the entropy of the ground state vanishes), this leads with (7.59) to

$$g(n) = e^{\alpha(n-1)/4} .$$

Since this must be an integer, one has the options

$$\alpha = 4 \ln k , \quad k = 2, 3, \dots \tag{7.61}$$

and thus $g(n) = k^{n-1}$. Note that the spectrum would then differ from (7.58). From information-theoretic reasons ('it from bit', cf. e.g. Wheeler (1990)) one would prefer the value $k = 2$, leading to $A_n = (4 \ln 2) l_P^2 n$. Arguments from loop quantum gravity seem to suggest the choice $k = 3$ (see below).

In the Schwarzschild case, the energy spacing between consecutive levels is obtained from

$$\Delta A = 32 \pi G^2 M \Delta M = (4 \ln k) l_P^2$$

to read

$$\Delta M = \Delta E \equiv \hbar \tilde{\omega}_k = \frac{\hbar \ln k}{8 \pi G M} , \tag{7.62}$$

with the fundamental frequency

$$\tilde{\omega}_k = \frac{\ln k}{8 \pi G M} = (\ln k) \frac{T_H}{\hbar} . \tag{7.63}$$

The black-hole emission spectrum would then be concentrated at multiples of this fundamental frequency—unlike the continuous thermal spectrum of Hawking radiation. In fact, one would have a deviation from the Hawking spectrum even for large black holes, that is, black holes with masses $M \gg m_P$. Another consequence would be that quanta with $\omega < \tilde{\omega}_k$ could not be absorbed by the black hole.

A discrete spectrum for the black-hole horizon is found in the context of loop quantum gravity (Chapter 6). This is of course a consequence of the area quantization discussed in Section 6.2. However, this spectrum is not equidistant. Consider the intersection of a spin network with the surface of a black hole. In the case of 'punctures' only, cf. (6.33), the state is characterized by a set of spins $\{j_i\}$. The dimension of the corresponding boundary Hilbert space is thus

$$\prod_{i=1}^{n} (2j_i + 1) .$$

The spectrum is in this case given by (6.33). Denoting the minimal spin by $j_{\min}$, the corresponding area value is

$$A_0 \equiv 8 \pi \beta l_P^2 \sqrt{j_{\min}(j_{\min} + 1)} . \tag{7.64}$$

It has been suggested that the dominating contribution to the entropy comes from the $j_{\min}$-contributions. Following Dreyer (2003), we consider the number of links with $j_{\min}$,

$$N = \frac{A}{A_0} = \frac{A}{8\pi\beta l_{\mathrm{P}}^2 \sqrt{j_{\min}(j_{\min}+1)}} \ .$$

The number of microstates is then

$$N_{\mathrm{ms}} = (2j_{\min}+1)^N \ .$$

This is only equal to the desired result from the Bekenstein–Hawking entropy, that is, equal to $\exp(A/4l_{\mathrm{P}}^2)$, if the Barbero–Immirzi parameter has the special value

$$\beta = \frac{\ln(2j_{\min}+1)}{2\pi\sqrt{j_{\min}(j_{\min}+1)}} \ . \tag{7.65}$$

What is the value of $j_{\min}$? If the underlying group is SU(2), as is usually assumed to be the case, one has $j_{\min} = 1/2$ and thus $\beta = \ln 2/\pi\sqrt{3}$. This result has also been found in a calculation by Ashtekar et al. (1998) where it has been assumed that the degrees of freedom are given by a Chern–Simons theory on the horizon.[4] An important concept in this context is played by the 'isolated horizon' of a black hole; cf. Ashtekar et al. (2000). An isolated horizon is a generalization of an event horizon to non-stationary black holes. It is a local concept and does not need to admit a Killing field.

Taking instead SO(3) as the underlying group, one has $j_{\min} = 1$ and thus $\beta = \ln 3/\pi\sqrt{2}$. If a link with spin $j_{\min}$ is absorbed or created at the black-hole horizon, the area changes by

$$\Delta A = A_0 = 8\pi\beta l_{\mathrm{P}}^2 \sqrt{j_{\min}(j_{\min}+1)} \ , \tag{7.66}$$

which is equal to $4(\ln 2)l_{\mathrm{P}}^2$ in the SU(2)-case and to $4(\ln 3)l_{\mathrm{P}}^2$ in the SO(3)-case. In the SU(2)-case, the result for ΔA corresponds to the one advocated by Bekenstein and Mukhanov (1995), although the spectrum here is not equidistant. This leads to a change in the mass given by

$$\Delta M = \frac{\ln 2 \ m_{\mathrm{P}}^2}{8\pi M} \ , \tag{7.67}$$

cf. (7.62). In the SO(3)-case, one has

$$\Delta M = \frac{\ln 3 \ m_{\mathrm{P}}^2}{8\pi M} \tag{7.68}$$

corresponding to the fundamental frequency

$$\tilde{\omega}_3 = \frac{\ln 3}{8\pi GM} \equiv \frac{\ln 3 \ T_{\mathrm{H}}}{\hbar} \ . \tag{7.69}$$

It was emphasized by Dreyer (2003) that $\tilde{\omega}_3$ coincides with the real part of the asymptotic frequency for the quasi-normal modes of the black hole. These

---

[4]Corrections to the Bekenstein–Hawking entropy have been calculated by Kaul and Majumdar (2000) by going beyond the semiclassical limit. They find a logarithmic correction of the form $-(3/2)\ln(A/4l_{\mathrm{P}}^2)$ where the coefficient is independent of the Barbero–Immirzi parameter β. This result was also obtained for various black-hole models by Carlip (2000) using two-dimensional conformal field theory as a tool.

modes are characteristic oscillations of the black hole before it settles to its stationary state; see for example, Kokkotas and Schmidt (1999) for a review. As was conjectured by Hod (1998) on the basis of numerical evidence and shown by Motl (2003), the frequency of the quasi-normal modes is for $n \to \infty$ given by

$$\omega_n = -\frac{\mathrm{i}(n+\frac{1}{2})}{4GM} + \frac{\ln 3}{8\pi GM} + \mathcal{O}\left(n^{-1/2}\right) = -\mathrm{i}\kappa\left(n+\frac{1}{2}\right) + \frac{\kappa}{2\pi}\ln 3 + \mathcal{O}\left(n^{-1/2}\right) ,$$

(7.70)

see also Neitzke (2003). The imaginary part indicates that one is dealing with damped oscillations. If for some not yet understood reason the frequency $\tilde{\omega}_k$ has to coincide with the real part of ω_n for big n, the group is fixed to be SO(3). It might be that the black-hole entropy arises from the quantum entanglement between the black hole and the quasi-normal modes (Kiefer 2004 a). The quasi-normal modes would then serve as an environment leading to decoherence (Section 10.1). This, however, is still to be shown.

The imaginary part of the frequency (7.70) in this limit is equidistant in n. This could indicate an intricate relation with Euclidean quantum gravity and provide an explanation of why the Euclidean version readily provides expressions for the black-hole temperature and entropy: if one considered in the Euclidean theory a wave function of the form

$$\psi_{\mathrm{E}} \sim \mathrm{e}^{\mathrm{i}n\kappa t_{\mathrm{E}}} ,$$

one would have to demand that the Euclidean time t_{E} be periodic with period $8\pi GM$. This, however, is just the inverse of the Hawking temperature, in accordance with the result that Euclidean time must have this periodicity if the line element is to be regular (see e.g. Hawking and Penrose 1996).

7.4   Quantum theory of collapsing dust shells

In this section, a particular model will be described in some detail, but without too many technicalities. This concerns the collapse of a null dust shell. In the classical theory, the collapse leads to the formation of a black hole. We shall see that it is possible to construct an exact quantum theory of this model in which the dynamical evolution is *unitary* with respect to asymptotic observers (since one has an asymptotically flat space, a semiclassical time exists, which is just the Killing time at asymptotic infinity). As a consequence of the unitary evolution, the classical singularity is fully avoided in the quantum theory: if the collapsing shell is described by a wave packet, the evolution leads to a superposition of black-hole and white-hole horizon yielding a vanishing wave function for zero radial coordinate. At late times only an expanding wave packet (expanding shell) is present. A detailed exposition of this model can be found in Hájíček (2003).

7.4.1   *Covariant gauge fixing*

A problem of principle that arises is the need to represent the expanding and collapsing shell on the same background manifold. What does this mean? Following section 2 of Hájíček and Kiefer (2001a), let $(\mathcal{M}, g)$ be a globally hyperbolic space–time,

$$\mathcal{M} = \Sigma \times \mathbb{R} \ .$$

The manifold $\mathcal{M}$ is called 'background manifold' and is uniquely determined for a given three-manifold Σ (the 'initial-data manifold'). The four-dimensional diffeomorphism group Diff $\mathcal{M}$ is often considered as the 'gauge group' of GR; it pushes the points of $\mathcal{M}$ around, so points are not a gauge-invariant concept. In fact, usually only a subgroup of Diff $\mathcal{M}$ plays the role of a gauge group (describing 'redundancies' in the language of Section 4.2.5). Everything else describes physically relevant symmetries, for example, asymptotic rotations in an asymptotically flat space.

The group Diff $\mathcal{M}$ acts on Riem $\mathcal{M}$, the space of all Lorentzian metrics on $\mathcal{M}$. A particular representative metric for each geometry on $\mathcal{M}$ (in some open set) is chosen by 'covariant gauge fixing', that is, the choice of a section σ,

$$\sigma : \text{Riem}\,\mathcal{M}/\text{Diff}\,\mathcal{M} \mapsto \text{Riem}\,\mathcal{M} \ .$$

Thereby points are defined by coordinates on a fixed manifold. A transformation between two covariant gauge fixings σ and σ' is not a single diffeomorphism, but forms a much larger group (Bergmann and Komar 1972); it corresponds to one coordinate transformation for each solution of the field equations, which is different from a single coordinate transformation on the background manifold. Hájíček and Kijowski (2000) have shown that, given a section σ, one can construct a map from Riem $\mathcal{M}/\text{Diff}\,\mathcal{M} \times \text{Emb}(\Sigma, \mathcal{M})$, where $\text{Emb}(\Sigma, \mathcal{M})$ is the space of embeddings of the initial data surface Σ into $\mathcal{M}$, to the ADM phase space Γ of GR. (This works only if the evolved space–times do not admit an isometry.) It was shown that this map is invertible and extensible to neighbourhoods of Γ and Riem $\mathcal{M}/\text{Diff}\,\mathcal{M} \times \text{Emb}(\Sigma, \mathcal{M})$. In this way, a transformation from ADM variables to embedding variables (see Section 5.2.1) has been performed. The use of embedding variables is called 'Kuchař decomposition' by Hájíček and Kijowski (2000).

The Schwarzschild case may serve as a simple illustration. The transformation between Kruskal coordinates and Eddington–Finkelstein coordinates is *not* a coordinate transformation on a background manifold because this transformation is solution dependent (it depends on the mass M of the chosen Schwarzschild solution). It thus represents a set of coordinate transformations, one for each M, and is thus a transformation between different gauge fixings σ and σ'. A background manifold is obtained if one identifies all points with the same values of the Kruskal coordinates, and another background manifold results if one identifies all points with the same Eddington–Finkelstein coordinates.

7.4.2 *Embedding variables for the classical theory*

Here we consider the dynamics of a (spherically symmetric) null dust shell in GR. In this subsection, we shall first identify appropriate coordinates on a background manifold and then perform the explicit transformation to embedding variables. This will then serve as the natural starting point for the quantization in the next subsection.

Any classical solution describing this system has a simple structure: inside the shell the space–time is flat, whereas outside it is isometric to a part of the Schwarzschild metric with mass M. Both geometries must match along a spherically symmetric null hypersurface describing the shell. All physically distinct solutions can be labelled by three parameters: $\eta \in \{-1, +1\}$, distinguishing between the outgoing ($\eta = +1$) and ingoing ($\eta = -1$) null surfaces; the asymptotic time of the surface, that is, the retarded time $u = T - R \in (-\infty, \infty)$ for $\eta = +1$, and the advanced time $v = T + R \in (-\infty, \infty)$ for $\eta = -1$; and the mass $M \in (0, \infty)$. An ingoing shell creates a black-hole (event) horizon at $R = 2M$ and ends up in the singularity at $R = 0$. The outgoing shell starts from the singularity at $R = 0$ and emerges from a white-hole (particle) horizon at $R = 2M$. We shall follow Hájíček and Kiefer (2001a, b), see also Hájíček (2003).

The Eddington–Finkelstein coordinates do not define a covariant gauge fixing, since it turns out that there are identifications of points in various solutions which do not keep fixed an asymptotic family of observers. Instead, double-null coordinates U and V are chosen on the background manifold $\mathcal{M} = \mathbb{R}_+ \times \mathbb{R}$ (being effectively two-dimensional due to spherical symmetry). In these coordinates (which will play the role of the embedding variables), the metric has the form

$$\mathrm{d}s^2 = -A(U, V)\mathrm{d}U\mathrm{d}V + R^2(U, V)(\mathrm{d}\theta^2 + \sin^2\theta\mathrm{d}\phi^2) . \qquad (7.71)$$

From the demand that the metric be regular at the centre and continuous at the shell, the coefficients A and R are uniquely defined for any physical situation defined by the variables M (the energy of the shell), η, and w (the location of the shell, where $w = u$ for the outgoing and $w = v$ for the ingoing case).

Consider first the case $\eta = 1$. In the Minkowski part, $U > u$ of the solution, one finds

$$A = 1 , \quad R = \frac{V - U}{2} . \qquad (7.72)$$

In the Schwarzschild part, $U < u$, one finds

$$R = 2M\kappa\left(\left(\frac{V - u}{4M} - 1\right)\exp\left(\frac{V - U}{4M}\right)\right) \equiv 2M\kappa(f_+) , \qquad (7.73)$$

where κ is the 'Kruskal function' (not to be confused with the surface gravity κ) defined by its inverse as

$$\kappa^{-1}(y) = (y - 1)e^y , \qquad (7.74)$$

and

$$A = \frac{1}{\kappa(f_+)e^{\kappa(f_+)}}\frac{V - u}{4M}\exp\left(\frac{V - U}{4M}\right) . \qquad (7.75)$$

With these expressions one can verify that A and R are continuous at the shell, as required. We note that these expressions contain u as well as M, which will become conjugate variables in the canonical formalism.

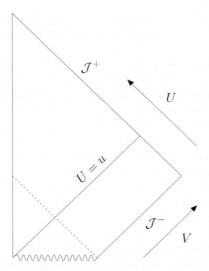

FIG. 7.3. Penrose diagram for the outgoing shell in the classical theory. The shell is at $U = u$.

The Penrose diagram for the outgoing shell is shown in Fig. 7.3. It is important to note that the background manifold possesses a unique asymptotic region with $\mathcal{J}^-$ defined by $U \to -\infty$ and $\mathcal{J}^+$ by $V \to +\infty$.

In the case of ingoing shells ($\eta = -1$), one finds for $V < v$ again (7.72) and for $V > v$,

$$R = 2M\kappa(f_-), \quad A = \frac{1}{\kappa(f_-)e^{\kappa(f_-)}} \frac{v - U}{4M} \exp\left(\frac{V - U}{4M}\right), \tag{7.76}$$

where

$$f_- \equiv \left(\frac{v - U}{4M} - 1\right) \exp\left(\frac{V - U}{4M}\right) .$$

These expressions result from (7.75) by the substitution $V - u \to v - U$.

As the result of the gauge fixing, the set of solutions (η, M, w) can be written as a set of (η, M, w)-dependent metric fields (7.71) and a set of shell trajectories on a *fixed* background manifold $\mathcal{M}$. Here, the corresponding functions A and R have the form

$$A(\eta, M, w; U, V) , \quad R(\eta, M, w; U, V) , \tag{7.77}$$

and the trajectory of the shell on the background manifold is simply $U = u$ for $\eta = +1$ and $V = v$ for $\eta = -1$.

The next step is the explicit transformation to embedding variables (the Kuchař decomposition). The standard (ADM) formulation of the shell was studied in Louko *et al.* (1998); see also Kraus and Wilczek (1995). The spherically symmetric metric is written in the form

$$\mathrm{d}s^2 = -N^2\mathrm{d}\tau^2 + \Lambda^2(\mathrm{d}\rho + N^\rho\mathrm{d}\tau)^2 + R^2\mathrm{d}\Omega^2 \ ,$$

and the shell is described by its radial coordinate $\rho = \mathbf{r}$. The action reads

$$S_0 = \int \mathrm{d}\tau \left[\mathbf{p}\dot{\mathbf{r}} + \int d\rho(P_\Lambda\dot{\Lambda} + P_R\dot{R} - H_0)\right] , \qquad (7.78)$$

and the Hamiltonian is

$$H_0 = N\mathcal{H} + N^\rho\mathcal{H}_\rho + N_\infty E_\infty \ ,$$

where $N_\infty := \lim_{\rho\to\infty} N(\rho)$, E_∞ is the ADM mass, N and N^ρ are the lapse and shift functions, $\mathcal{H}$ and $\mathcal{H}_\rho$ are the constraints,

$$\mathcal{H} = \frac{\Lambda P_\Lambda^2}{2R^2} - \frac{P_\Lambda P_R}{R} + \frac{RR''}{\Lambda} - \frac{RR'\Lambda'}{\Lambda^2} + \frac{R'^2}{2\Lambda} - \frac{\Lambda}{2} + \frac{\eta\mathbf{p}}{\Lambda}\delta(\rho - \mathbf{r}) \approx 0 \quad (7.79)$$

$$\mathcal{H}_\rho = P_R R' - P_\Lambda'\Lambda - \mathbf{p}\delta(\rho - \mathbf{r}) \approx 0 \quad , \qquad (7.80)$$

and the prime (dot) denotes the derivative with respect to ρ (τ). These are the same constraints as in (7.23) and (7.25), except for the contribution of the shell.

The task is to transform the variables in the action S_0. This transformation will be split into two steps. The first step is a transformation of the canonical coordinates $\mathbf{r}$, $\mathbf{p}$, Λ, P_Λ, R, and P_R *at the constraint surface* Γ defined by the constraints (7.79) and (7.80). The new coordinates are u and $p_u = -M$ for $\eta = +1$, v and $p_v = -M$ for $\eta = -1$, and the embedding variables $U(\rho)$ and $V(\rho)$.

The second step is an *extension* of the functions u, v, p_u, p_v, $U(\rho)$, $P_U(\rho)$, $V(\rho)$, and $P_V(\rho)$ *out of the constraint surface*, where the functions u, v, p_u, p_v, $U(\rho)$, and $V(\rho)$ are defined by the above transformation, and $P_U(\rho)$, $P_V(\rho)$ by $P_U(\rho)|_\Gamma = P_V(\rho)|_\Gamma = 0$. The extension must satisfy the condition that the functions form a canonical chart in a neighbourhood of Γ. That such an extension exists was shown in Hájíček and Kijowski (2000). The details of the calculation can be found in Hájíček and Kiefer (2001a). The result is the action

$$S = \int d\tau \, (p_u\dot{u} + p_v\dot{v} - np_up_v) + \int d\tau \int_0^\infty d\rho(P_U\dot{U} + P_V\dot{V} - H) , \qquad (7.81)$$

where $H = N^U P_U + N^V P_V$, and n, $N^U(\rho)$, and $N^V(\rho)$ are Lagrange multipliers. The first term in (7.81) contains the physical variables (observables), while the second term contains the gauge variables. Both classical solutions are contained in the single constraint

$$p_up_v = 0 \ , \qquad (7.82)$$

that is, one has $p_v = 0$ for $\eta = 1$ and $p_u = 0$ for $\eta = -1$. Observe that the Poisson algebra of the chosen set of observables p_u and u for $\eta = +1$ as well as p_v and v for $\eta = -1$ is gauge-invariant in spite of the fact that it has been obtained by a calculation based on a gauge choice (the double-null coordinates U and V). Therefore, the quantum theory will also be gauge-invariant. A crucial point is that the new phase space has non-trivial boundaries,

$$p_u \leq 0, \quad p_v \leq 0 , \quad \frac{-u+v}{2} > 0 . \tag{7.83}$$

The boundary defined by the last inequality is due to the classical singularity. The system has now been brought into a form that can be taken as the starting point for quantization.

7.4.3 *Quantization*

The task is to quantize the physical degrees of freedom defined by the action

$$S_{\text{phys}} = \int d\tau \, (p_u \dot{u} + p_v \dot{v} - n p_u p_v) , \tag{7.84}$$

cf. (7.81). The appropriate method is *group quantization*; see e.g. Isham (1984). This method is suited in particular to implement conditions such as (7.83). It is based on the choice of a set of Dirac observables forming a Lie algebra. This algebra generates a group of transformations respecting all boundaries which insures that information about such boundaries are implemented in the quantum theory. The method automatically leads to self-adjoint operators for the observables. One obtains in particular a self-adjoint Hamiltonian and, consequently, a unitary dynamics.

The application of this method to the null-dust shell was presented in detail in Hájíček (2001, 2003). A complete system of Dirac observables is given by p_u, p_v, as well as $D_u \equiv u p_u$ and $D_v \equiv v p_v$. Thus, they commute with the constraint $p_u p_v$. The only non-vanishing Poisson brackets are

$$\{D_u, p_u\} = p_u , \quad \{D_v, p_v\} = p_v . \tag{7.85}$$

The Hilbert space is constructed from complex functions $\psi_u(p)$ and $\psi_v(p)$, where $p \in [0, \infty)$. The scalar product is defined by

$$(\psi_u, \phi_u) := \int_0^\infty \frac{dp}{p} \, \psi_u^*(p) \phi_u(p) \tag{7.86}$$

(similarly for $\psi_v(p)$). To handle the inequalities (7.83), it is useful to perform the following canonical transformation,

$$t = (u+v)/2, \qquad r = (-u+v)/2, \tag{7.87}$$

$$p_t = p_u + p_v, \qquad p_r = -p_u + p_v . \tag{7.88}$$

The constraint function then assumes the form $p_u p_v = (p_t^2 - p_r^2)/4$. Upon quantization, one obtains the operator $-\hat{p}_t$ which is self-adjoint and has a positive spectrum, $-\hat{p}_t \varphi(p) = p \varphi(p)$, $p \geq 0$. It is the generator of time evolution and corresponds to the energy operator $E \equiv M$. Since r is not a Dirac observable, it cannot directly be transformed into a quantum observable. It turns out that the following construction is useful,

$$\hat{r}^2 := -\sqrt{p}\frac{d^2}{dp^2}\frac{1}{\sqrt{p}} \ . \tag{7.89}$$

This is essentially a Laplacian and corresponds to a concrete choice of factor ordering. It is a symmetric operator which can be extended to a self-adjoint operator. In this process, one is naturally led to the following eigenfunctions of $\hat{r}^2$,

$$\psi(r,p) := \sqrt{\frac{2p}{\pi}}\sin rp \ , \quad r \geq 0 \ . \tag{7.90}$$

One can also construct an operator $\hat{\eta}$ that classically would correspond to the direction of motion of the shell.

The formalism has now reached a stage in which one can start to study concrete physical applications. Of particular interest is the representation of the shell through a narrow wave packet. One takes for $t = 0$, the following family of wave packets,

$$\psi_{\kappa\lambda}(p) \equiv \frac{(2\lambda)^{\kappa+1/2}}{\sqrt{(2\kappa)!}}p^{\kappa+1/2}e^{-\lambda p} \ , \tag{7.91}$$

where κ is a positive integer, and λ is a positive number with dimension of length. By an appropriate choice of these constants, one can prescribe the expectation value of the energy and its variation. A sufficiently narrow wave packet can thus be constructed.

One can show that the wave packets are normalized and that they obey $\psi_{\kappa\lambda}(p) = \psi_{\kappa 1}(\lambda p)$ ('scale invariance'). The expectation value of the energy is calculated as

$$\langle E \rangle_{\kappa\lambda} \equiv \int_0^\infty \frac{dp}{p}\ p\psi_{\kappa\lambda}^2(p) \ , \tag{7.92}$$

with the result

$$\langle E \rangle_{\kappa\lambda} = \frac{\kappa+1/2}{\lambda} \ . \tag{7.93}$$

In a similar way, one finds for the variation

$$\Delta E_{\kappa\lambda} = \frac{\sqrt{2\kappa+1}}{2\lambda} \ . \tag{7.94}$$

Since the time evolution of the packet is generated by $-\hat{p}_t$, one has

$$\psi_{\kappa\lambda}(t,p) = \psi_{\kappa\lambda}(p)e^{-ipt} \ . \tag{7.95}$$

More interesting is the evolution of the wave packet in the r-representation. This is obtained by the integral transform (7.86) of $\psi_{\kappa\lambda}(t,p)$ with respect to the eigenfunctions (7.90). It leads to the exact result.

$$\Psi_{\kappa\lambda}(t,r) = \frac{1}{\sqrt{2\pi}} \frac{\kappa!(2\lambda)^{\kappa+1/2}}{\sqrt{(2\kappa)!}} \left[\frac{i}{(\lambda+it+ir)^{\kappa+1}} - \frac{i}{(\lambda+it-ir)^{\kappa+1}} \right]. \qquad (7.96)$$

One interesting consequence can be immediately drawn:

$$\lim_{r\to0} \Psi_{\kappa\lambda}(t,r) = 0 . \qquad (7.97)$$

This means that the probability to find the shell at vanishing radius is zero! In this sense, the singularity is avoided in the quantum theory. It must be emphasized that this is not a consequence of a certain boundary condition—it is a consequence of the *unitary evolution*. If the wave function vanishes at $r = 0$ for $t \to -\infty$ (asymptotic condition of ingoing shell), it will continue to vanish at $r = 0$ for all times. It follows from (7.96) that the quantum shell bounces and re-expands. Hence, no absolute event horizon can form, in contrast to the classical theory. The resulting object might still be indistinguishable from a black hole due to the huge time delay from the gravitational redshift—the re-expansion would be visible from afar only in the far future. Similar features follow from a model by Frolov and Vilkovisky (1981) who consider a null shell for the case where 'loop effects' in the form of Weyl curvature terms are taken into account.

Of interest is also the expectation value of the shell radius; see Hájíček (2001, 2003) for details. Again one recognizes that the quantum shell always bounces and re-expands. An intriguing feature is that an essential part of the wave packet can even be squeezed below the expectation value of its Schwarzschild radius. The latter is found from (7.92) to read (re-inserting G),

$$\langle R_0 \rangle_{\kappa\lambda} \equiv 2G\langle E \rangle_{\kappa\lambda} = (2\kappa+1)\frac{l_{\mathrm{P}}^2}{\lambda} , \qquad (7.98)$$

while its variation follows from (7.94),

$$\Delta(R_0)_{\kappa\lambda} = 2G\Delta E_{\kappa\lambda} = \sqrt{2\kappa+1}\frac{l_{\mathrm{P}}^2}{\lambda} . \qquad (7.99)$$

The main part of the wave packet is squeezed below the Schwarzschild radius if

$$\langle r \rangle_{\kappa\lambda} + (\Delta r)_{\kappa\lambda} < \langle R_0 \rangle_{\kappa\lambda} - \Delta(R_0)_{\kappa\lambda} .$$

It turns out that this can be achieved if either $\lambda \approx l_{\mathrm{P}}$ (and $\kappa > 2$) or, for bigger λ, if κ is larger by a factor of $(\lambda/l_{\mathrm{P}})^{4/3}$. The wave packet can thus be squeezed below its Schwarzschild radius if its energy is bigger than the Planck energy—a genuine quantum effect.

How can this behaviour be understood? The unitary dynamics ensures that the ingoing quantum shell develops into a *superposition* of ingoing and outgoing shell if the region where in the classical theory a singularity would form is reached. In other words, the singularity is avoided by destructive interference in the quantum theory. This is similar to the quantum-cosmological example of Kiefer and Zeh (1995), where a superposition of a black hole with a white hole leads to a singularity-free quantum universe; cf. Section 10.2. Also here, the horizon becomes a superposition of 'black hole' and 'white hole'—its 'grey' nature can be characterized by the expectation value of the operator $\hat{\eta}$ (a black-hole horizon would correspond to the value -1 and a white-hole horizon to the value $+1$). In this scenario, no information-loss paradox would ever arise if such a behaviour occurred for all collapsing matter (which sounds reasonable). In the same way, the principle of cosmic censorship would be implemented, since no naked singularities (in fact, no singularities at all) would form.

8

QUANTUM COSMOLOGY

8.1 Minisuperspace models

8.1.1 *General introduction*

As we have discussed at length in Chapters 5 and 6, all information about canonical quantum gravity lies in the constraints (apart from possible surface terms). These constraints assume in the Dirac approach the form of conditions for physically allowed wave functionals. In quantum geometrodynamics (Chapter 5), the wave functional depends on the three-metric, while in quantum connection or quantum loop dynamics (Chapter 6), it depends on a non-Abelian connection or a Wilson-loop type variable.

A common feature for all variables is that the quantum constraints are difficult if not impossible to solve. In classical GR, the field equations often become tractable if *symmetry reductions* are performed; one can, for example, impose spherical symmetry, axial symmetry, or homogeneity. This often corresponds to interesting physical situations: stationary black holes are spherically or axially symmetric (Section 7.1), while the Universe as a whole can be approximated by homogeneous and isotropic models. The idea is to apply a similar procedure in the quantum theory. One may wish to make a symmetry reduction at the classical level and to quantize only a restricted set of variables. The quantization of black holes discussed in Chapter 7 is a prominent example. The problem is that such a reduction violates the uncertainty principle, since degrees of freedom are neglected together with the corresponding momenta. Still, the reduction may be an adequate approximation in many circumstances. In quantum mechanics, for example, the model of a 'rigid top' is a good approximation as long as other degrees of freedom remain unexcited due to energy gaps. Such a situation can also hold for quantum gravity; cf. Kuchař and Ryan (1989) for a discussion of this situation in a quantum-cosmological context.

Independent of this question whether the resulting models are realistic or not, additional reasons further support the study of dimensionally reduced models. First, they can play the role of toy models to study conceptual issues which are independent of the number of variables. Examples are the problem of time, the role of observers, and the emergence of a classical world, cf. Chapter 10. Second, they can give the means to study mathematical questions such as the structure of the wave equation and the implementation of boundary conditions. Third, one can compare various quantization schemes in the context of simple models (e.g. reduced versus Dirac quantization).

In quantum geometrodynamics, the wave functional Ψ is defined—apart from non-gravitational degrees of freedom—on Riem Σ, the space of all three-metrics. The presence of the diffeomorphism constraints guarantees that the true configuration space is 'superspace', that is, the space Riem Σ/Diff Σ. Restricting the infinitely many degrees of freedom of superspace to only a finite number, one arrives at a finite-dimensional configuration space called *minisuperspace*. If the number of the restricted variables is still infinite, the term 'midisuperspace' has been coined. The example of spherical symmetry discussed in the last chapter is an example of midisuperspace. Since the most important example in the case of finitely many degrees of freedom is cosmology, the minisuperspace examples are usually applied to *quantum cosmology*, that is, the application of quantum theory to the Universe as a whole. The present chapter deals with quantum cosmology. For simplicity, we shall restrict our attention mostly to quantum geometrodynamics, although quantum cosmology is discussed in detail also in the approaches of Chapter 6, see the remarks below.

The first quantum-cosmological model was presented—together with its semiclassical approximation—in DeWitt (1967*a*). It dealt with the homogeneous and isotropic case. The extension to anisotropic models (in particular, Bianchi models) was performed by Misner; cf. Misner (1972) and Ryan (1972). Kuchař (1971) made the extension to the midisuperspace case and discussed the quantization of cylindrical gravitational waves; see also Ashtekar and Pierri (1996). Classically, different spatial points seem to decouple near a big-bang singularity for general solutions of the Einstein equations (Belinskii *et al.* 1982). Such solutions consist of a collection of homogeneous spaces each of which is like a 'mixmaster solution' (in which the universe behaves like a particle in a time-dependent potential wall, with an infinite sequence of bounces). For this reason the use of minisuperspace models may even provide a realistic description of the universe near its classical singularity.

The usual symmetry reduction proceeds as follows; see for example, Torre (1999). One starts from a classical field theory and specifies the action of a group with respect to which the fields are supposed to be invariant. A prominent example is the rotation group. One then constructs the invariant ('reduced') fields and evaluates the field equations for them. An important question is whether there is a shortcut in the following way. Instead of reducing the field equations one might wish to reduce first the Lagrangian and then derive from it directly the reduced field equations. (Alternatively, this can be attempted at the Hamiltonian level.) This would greatly simplify the procedure, but in general it is not possible: reduction of the Lagrangian is equivalent to reduction of the field equations only in special situations. When do such situations occur? In other words, when do critical points of the reduced action define critical points of the full action? Criteria for this *symmetric criticality principle* were developed by Palais (1979). If restriction is made to local Lagrangian theories, one can specify such criteria more explicitly (Torre 1999; Fels and Torre 2002).

Instead of spelling out the general conditions, we focus on three cases that are relevant for us:

1. The conditions are always satisfied for a compact symmetry group, that is, the important case of *spherical symmetry* obeys the symmetric criticality principle.

2. In the case of homogeneous cosmological models, the conditions are satisfied if the structure constants $c_{ab}{}^{c}$ of the isometry group satisfy $c_{ab}{}^{b} = 0$. Therefore, Bianchi-type A cosmological models and the Kantowski–Sachs universe can be treated via a reduced Lagrangian. For Bianchi-type B models, the situation is more subtle (cf. MacCallum (1979) and Ryan and Waller (1997)).

3. The symmetric criticality principle also applies to cylindrical or toroidal symmetry reductions (which are characterized by two commuting Killing vector fields). The reduced theories can be identified with parametrized field theories on a flat background. With such a formal identification it is easy to find solutions. Quantization can then be understood as quantization on a fixed background with arbitrary foliation into Cauchy surfaces. In two space–time dimensions, where these reduced models are effectively defined, time evolution is unitarily implementable along arbitrary foliations. This ceases to hold in higher dimensions; cf. Giulini and Kiefer (1995), Helfer (1996), and Torre and Varadarajan (1999).

In the case of homogeneous models, the wave function is of the form $\psi(q^{i})$, $i = 1, \ldots, n$, that is, it is of a 'quantum-mechanical' type. In this section, we follow a pragmatic approach and discuss the differential equations for the wave functions. The subtle issue of boundary conditions will be dealt with in Section 8.3.

8.1.2   *Quantization of a Friedmann universe*

As an example, we shall treat in some detail the case of a closed Friedmann universe with a massive scalar field; cf. Kiefer (1988) and Halliwell (1991). Classically, the model is thus characterized by the scale factor $a(t)$ and the (homogeneous) field $\phi(t)$ with mass m. In the quantum theory, the classical time parameter t is absent, and the system is fully characterized by a wave function $\psi(a, \phi)$. The ansatz for the classical line element is

$$\mathrm{d}s^2 = -N^2(t)\mathrm{d}t^2 + a^2(t)\mathrm{d}\Omega_3^2 \,, \tag{8.1}$$

where $\mathrm{d}\Omega_3^2 = \mathrm{d}\chi^2 + \sin^2\chi(\mathrm{d}\theta^2 + \sin^2\theta\mathrm{d}\varphi^2)$ is the standard line element on S^3. A special foliation has thus been chosen in order to capture the symmetries of this model. For this reason no shift vector appears, only the lapse function N. The latter occurs, of course, always in combination with $\mathrm{d}t$, expressing the classical invariance under reparametrizations of the time parameter; see Chapter 3.

The three-metric h_{ab} is fully specified by the scale factor a. The second fundamental form, cf. (4.48), reads here

$$K_{ab} = -\frac{1}{2N}\frac{\partial h_{ab}}{\partial t} = -\frac{\dot{a}}{aN}h_{ab} \,. \tag{8.2}$$

Its trace is

$$K \equiv K_{ab}h^{ab} = -\frac{3\dot{a}}{Na} , \tag{8.3}$$

which is thus proportional to the Hubble parameter $\dot{a}/a$. Its inverse, K^{-1}, is for this reason called 'extrinsic time'; cf. Section 5.2.

Since the model fulfils the symmetric criticality principle, we can insert the ansatz (8.1) directly into the Einstein–Hilbert action (1.1) and derive the Euler–Lagrange equations from the reduced action. For the surface term in (1.1), one obtains

$$2 \int \mathrm{d}^3x \, K \sqrt{h} = -6 \int \mathrm{d}^3x \, \frac{\dot{a}}{Na} \sqrt{h} .$$

Inserting $\sqrt{h} = a^3 \sin^2 \chi \sin \theta$ one finds that this surface term is cancelled by a term that appears after partial integration from the first term in (1.1). This is how the general surface term is constructed. Integrating over d^3x and choosing units such that $2G/3\pi = 1$, one obtains from (1.1), the gravitational part of the 'minisuperspace action',

$$S_{\mathrm{g}} = \frac{1}{2} \int \mathrm{d}t \, N \left(-\frac{a\dot{a}^2}{N^2} + a - \frac{\Lambda a^3}{3} \right) . \tag{8.4}$$

The matter action reads, after a rescaling $\phi \to \phi/\sqrt{2\pi}$,

$$S_{\mathrm{m}} = \frac{1}{2} \int \mathrm{d}t \, Na^3 \left(\frac{\dot{\phi}^2}{N^2} - m^2\phi^2 \right) . \tag{8.5}$$

The full minisuperspace action can then be written in the form

$$S = S_{\mathrm{g}} + S_{\mathrm{m}} \equiv \int \mathrm{d}t \, L(q, \dot{q}) \equiv \int \mathrm{d}t \, N \left(\frac{1}{2}G_{AB}\frac{\dot{q}^A\dot{q}^B}{N^2} - V(q) \right) , \tag{8.6}$$

where q is a shorthand for $q^1 \equiv a$ and $q^2 \equiv \phi$. The minisuperspace DeWitt metric reads explicitly

$$G_{AB} = \begin{pmatrix} -a & 0 \\ 0 & a^3 \end{pmatrix} ,$$

with $\sqrt{-G} = a^2$, where G denotes its determinant here. The indefinite nature of the kinetic term is evident.

Following the general procedure of Section 4.2, one starts with the definition of the canonical momenta. This includes the primary constraint

$$p_N = \frac{\partial L}{\partial \dot{N}}$$

and the dynamical momenta

$$p_a = \frac{\partial L}{\partial \dot{a}} = -\frac{a\dot{a}}{N} , \quad p_\phi = \frac{\partial L}{\partial \dot{\phi}} = \frac{a^3\dot{\phi}}{N} . \tag{8.7}$$

The Hamiltonian is then given by

$$H = p_N \dot{N} + p_a \dot{a} + p_\phi \dot{\phi} - L$$

$$= \frac{N}{2} \left(-\frac{p_a^2}{a} + \frac{p_\phi^2}{a^3} - a + \frac{\Lambda a^3}{3} + m^2 a^3 \phi^2 \right)$$

$$\equiv N \left(1/2 \, G^{AB} p_A p_B + V(q) \right) , \qquad (8.8)$$

where G^{AB} is the inverse DeWitt metric. The explicit form of the potential is

$$V(q) \equiv V(a, \phi) = \frac{1}{2} \left(-a + \frac{\Lambda a^3}{3} + m^2 a^3 \phi^2 \right) \equiv \frac{1}{2} \left(-a + a^3 \mathcal{V}(\phi) \right) .$$

For a general (minimally coupled) field one has to insert the corresponding potential term into $\mathcal{V}(\phi)$. One recognizes that the Λ-term appears on the same footing, so one might also have an effective Λ-term coming from a matter potential. This is the typical case for inflationary scenarios of the early universe.

As in the general formalism, the preservation of the primary constraint, $\{p_N, H\} \approx 0$, leads to the Hamiltonian constraint $H \approx 0$. Due to the ansatz (8.1), no diffeomorphism constraints appear. On the Lagrangian level, $\{p_N, H\} \approx 0$ corresponds to the Friedmann equation

$$\dot{a}^2 = -1 + a^2 \left(\dot{\phi}^2 + \frac{\Lambda}{3} + m^2 \phi^2 \right) . \qquad (8.9)$$

Variation of (8.6) with respect to ϕ yields

$$\ddot{\phi} + \frac{3\dot{a}}{a} \dot{\phi} + m^2 \phi = 0 . \qquad (8.10)$$

The classical equations can only be solved analytically for $m = 0$. In that case one has $p_\phi = a^3 \dot{\phi} = \text{constant} \equiv \mathcal{K}$, leading for $\Lambda = 0$ to

$$\phi(a) = \pm \frac{1}{2} \text{arcosh} \frac{\mathcal{K}}{a^2} . \qquad (8.11)$$

In the case of $m \neq 0$, a typical solution in configuration space behaves as follows: starting away from $\phi = 0$, the trajectory approaches the a-axis and starts to oscillate around it. This model is often used in the context of 'chaotic inflation', cf. Linde (1990), because the part of the trajectory approaching $\phi = 0$ corresponds to an inflationary expansion with respect to the coordinate time t. For a closed Friedmann universe, the trajectory reaches a maximum and recollapses.

Quantization proceeds through implementation of $H \approx 0$ as a condition on the wave function. As in the general theory there is the freedom to choose the factor ordering. The suggestion put forward here is a choice that leads to a kinetic term which is invariant under transformations in configuration space. In fact, this is already the situation in the ordinary Schrödinger equation. It corresponds to the substitution

$$G^{AB} p_A p_B \longrightarrow -\hbar^2 \nabla^2_{\text{LB}} \equiv -\frac{\hbar^2}{\sqrt{-G}} \partial_A (\sqrt{-G} G^{AB} \partial_B)$$

$$= \frac{\hbar^2}{a^2} \frac{\partial}{\partial a} \left(a \frac{\partial}{\partial a} \right) - \frac{\hbar^2}{a^3} \frac{\partial^2}{\partial \phi^2} , \tag{8.12}$$

where the 'Lapace–Beltrami operator' ∇_{LB} is the covariant generalization of the Lapace operator. With this factor ordering, the Wheeler–DeWitt equation, $\hat{H}\psi(a, \phi) = 0$, reads

$$\frac{1}{2} \left(\frac{\hbar^2}{a^2} \frac{\partial}{\partial a} \left(a \frac{\partial}{\partial a} \right) - \frac{\hbar^2}{a^3} \frac{\partial^2}{\partial \phi^2} - a + \frac{\Lambda a^3}{3} + m^2 a^3 \phi^2 \right) \psi(a, \phi) = 0 . \tag{8.13}$$

This equation assumes a particular simple form if the variable $\alpha \equiv \ln a$ is used instead of a,

$$\frac{\mathrm{e}^{-3\alpha}}{2} \left(\hbar^2 \frac{\partial^2}{\partial \alpha^2} - \hbar^2 \frac{\partial^2}{\partial \phi^2} - \mathrm{e}^{4\alpha} + \mathrm{e}^{6\alpha} \left[m^2 \phi^2 + \frac{\Lambda}{3} \right] \right) \psi(\alpha, \phi) = 0 . \tag{8.14}$$

The variable α ranges from $-\infty$ to ∞, and there are thus no problems in connection with a restricted range of the configuration variable.

Equation (8.14) has the form of a Klein–Gordon equation with 'time-and space-dependent' mass term, that is, with non-trivial potential term given by

$$V(\alpha, \phi) = -\mathrm{e}^{4\alpha} + \mathrm{e}^{6\alpha} \left[m^2 \phi^2 + \frac{\Lambda}{3} \right] . \tag{8.15}$$

Since this potential can also become negative, the system can develop 'tachyonic' behaviour with respect to the minisuperspace lightcone defined by the kinetic term in (8.14). This does not lead to any inconsistency since one is dealing with a configuration space here, not with space–time.

The minisuperspace configuration space of this model is conformally flat. This is an artefact of two dimensions. In general, the configuration space is curved. Therefore, the presence of an additional factor-ordering term of the form $\eta \hbar \mathcal{R}$ is possible, where η is a number and $\mathcal{R}$ is the Ricci scalar of configuration space. It has been argued (Misner 1972; Halliwell 1988) that the choice

$$\eta = \frac{d - 2}{4(d - 1)}$$

is preferred, where $d \neq 1$ is the dimension of minisuperspace. This follows from the demand for the invariance of the Wheeler–DeWitt equation under conformal transformations of the DeWitt metric.

One recognizes from (8.14) that α plays the role of an 'intrinsic time'— the variable that comes with the opposite sign in the kinetic term.[1] Since the potential (8.15) obeys $V(\alpha, \phi) \neq V(-\alpha, \phi)$, there is no invariance with respect to

---

[1] In two-dimensional configuration spaces, only the relative sign seems to play a role. However, in higher dimensional minisuperspaces, it becomes clear that the variable connected with the *volume* of the universe is the time-like variable; see Section 5.2.2.

reversal of *intrinsic* time. This is of crucial importance to understand the origin of irreversibility; cf. Section 10.2. Moreover, writing (8.14) in the form

$$-\hbar^2 \frac{\partial^2}{\partial \alpha^2} \psi \equiv h_\alpha^2 \psi \ , \tag{8.16}$$

the 'reduced Hamiltonian' h_α is not self-adjoint, so there is no unitary evolution with respect to α. This is, however, not a problem since α is no external time. Unitarity with respect to an intrinsic time is not expected to hold.

Models such as the one above can thus serve to illustrate the difficulties which arise in the approaches of 'reduced quantization' discussed in Section 5.2. Choosing classically $a = t$ and solving $H \approx 0$ with respect to p_a leads to $p_a + h_a \approx 0$ and, therefore, to a reduced Hamiltonian h_a equivalent to the one in (8.16),

$$h_a = \pm \sqrt{\frac{p_\phi^2}{t} - t^2 + m^2 \phi^2 t^4} \ . \tag{8.17}$$

One can recognize all the problems that are connected with such a formulation— explicit t-dependence of the Hamiltonian, no self-adjointness, complicated expression. One can make alternative choices for t—either $p_a = t$ ('extrinsic time'), $\phi = t$ ('matter time'), or a mixture of all these. This non-uniqueness is an expression of the 'multiple-choice problem' mentioned in Section 5.2. Due to these problems, we shall restrict our attention to the discussion of the Wheeler–DeWitt equation and do not follow the reduced approach any further.

A simple special case of (8.14) is obtained if we set $\Lambda = 0$ and $m = 0$. Setting also again $\hbar = 1$, one gets

$$\left(\frac{\partial^2}{\partial \alpha^2} - \frac{\partial^2}{\partial \phi^2} - e^{4\alpha} \right) \psi(\alpha, \phi) = 0 \ . \tag{8.18}$$

A separation ansatz leads immediately to the special solution

$$\psi_k(\alpha, \phi) = e^{-ik\phi} K_{ik/2} \left(\frac{e^{2\alpha}}{2} \right) \ , \tag{8.19}$$

where $K_{ik/2}$ denotes a Bessel function. A general solution can be found by performing a superposition with a suitable amplitude $A(k)$,

$$\psi(\alpha, \phi) = \int_{-\infty}^{\infty} dk \ A(k) \psi_k(\alpha, \phi) \ . \tag{8.20}$$

Taking, for example, a Gaussian centred at $k = \bar{k}$ with width b,

$$A(k) = \frac{1}{\sqrt{\sqrt{\pi} b}} e^{-\frac{1}{2b^2}(k-\bar{k})^2} \ , \tag{8.21}$$

one obtains the following wave-packet solution for the real part of the wave function (cf. Kiefer 1988),

$$\text{Re} \ \psi(\alpha, \phi) \approx c_{\bar{k}} \cos f_{\bar{k}}(\alpha, \phi) e^{-(b^2/2)(\phi + 1/2 \ \text{arcosh}(\bar{k}/e^{2\alpha}))^2}$$

$$+ c_{\bar{k}} \cos g_{\bar{k}}(\alpha, \phi) e^{-(b^2/2)\left(\phi - 1/2 \ \mathrm{arcosh}(\bar{k}/e^{2\alpha})\right)^2} . \qquad (8.22)$$

The functions f and g are explicitly given in Kiefer (1988) but are not needed here. The wave function (8.22) is a sum of two (modulated) Gaussians of width b^{-1}, which are symmetric with respect to $\phi = 0$ and which follow the classical path given by (8.11).

This example demonstrates an important feature of the quantum theory (see Fig. 8.1). In the classical theory, the 'recollapsing' part of the trajectory in configuration space can be considered as the deterministic successor of the 'expanding' part. In the quantum theory, on the other side, the 'returning' part of the wave packet has to be present 'initially' with respect to intrinsic time (the scale factor a of the universe) in order to yield a wave tube following the classical trajectory. This ensures destructive interference near the classical turning point in order to avoid exponentially growing pieces of the wave function for large a.

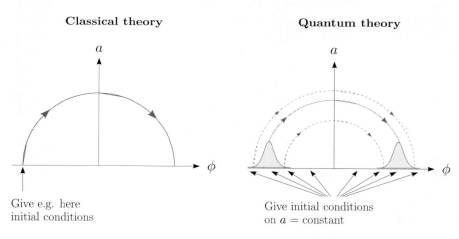

FIG. 8.1. (a) Recollapsing part is deterministic successor of expanding part.

(b) 'Recollapsing' wave packet must be present 'initially'.

Whereas in the above example it is possible to construct wave packets following the classical trajectory without dispersion, this is actually not possible in the general case. Already for $m \neq 0$, one finds that the demand for $\psi \to 0$ as $a \to \infty$ is in conflict with the existence of narrow wave packets all along the classical trajectory (Kiefer 1988). Consider the Wheeler–DeWitt equation (8.14) for $\Lambda = 0$,

$$\left(\frac{\partial^2}{\partial \alpha^2} - \frac{\partial^2}{\partial \phi^2} - e^{4\alpha} + e^{6\alpha} m^2 \phi^2 \right) \psi(\alpha, \phi) = 0 . \qquad (8.23)$$

Within a Born–Oppenheimer type of approximation, one can make the following ansatz for the wave packet,

$$\psi(a, \phi) = \sum_n c_n(\alpha)\varphi_n\left(\sqrt{me^{3\alpha}}\phi\right) , \qquad (8.24)$$

where the φ_n denote the usual eigenfunctions of the harmonic oscillator. From (8.23), one gets the following effective potentials for the $c_n(\alpha)$,

$$V_n(\alpha) = 1/2\left(-e^{4\alpha} + (2n+1)me^{3\alpha}\right) . \qquad (8.25)$$

These potentials become negative for large enough α. In the classical theory, this means that trajectories are drawn into the region with negative V_n and are reflected. In the quantum theory, it means that the wave function is a combination of exponentially increasing or decreasing solutions. In order to fulfil the 'final condition' $\psi \to 0$ for $a \to \infty$, the exponentially *decreasing* solution has to be chosen. The n-dependent reflection expressed by (8.25) leads to an unavoidable *spreading* of the wave packet (Kiefer 1988). This means that the semiclassical approximation does not hold all along the expanding and recollapsing part of the classical trajectory. How, then, does classical behaviour emerge? The answer is provided by adding other degrees of freedom (Section 8.2). They can act as a kind of environment for the minisuperspace variables a and ϕ and force them to behave classically. This process of *decoherence* is discussed in Section 10.1. It has also to be emphasized that wave packets are here always understood as corresponding to *branches* of the full wave function (representing quasiclassical histories), but not to the full wave function itself.

Another example is a closed Friedmann universe with a *non-minimally* coupled scalar field. In the general case, the Wheeler–DeWitt equation can become elliptic instead of hyperbolic for a certain range of field values; see Kiefer (1989). This would modify the 'initial-value problem' in quantum cosmology. No elliptic region occurs, for example, in the simplest case—the case of a conformally coupled field ϕ. Choosing units such that $8\pi G = 1$ and performing a field redefinition

$$\phi \to \chi = \frac{\sqrt{2}\pi a\phi}{6} ,$$

one obtains the following Wheeler–DeWitt equation (Zeh 1988; Kiefer 1990),

$$\hat{H}\psi(a,\chi) \equiv (-H_a + H_\chi)\psi \equiv \left(\frac{\partial^2}{\partial a^2} - \frac{\partial^2}{\partial\chi^2} - a^2 + \chi^2\right)\psi = 0 . \qquad (8.26)$$

This has the form of an 'indefinite oscillator'—two harmonic oscillators distinguished by a relative sign in the Hamiltonian. The model defined by (8.26) is the simplest non-trivial model in quantum cosmology. (It also arises from scalar–tensor theories of gravity; cf. Lidsey 1995.) Its classical solutions are represented by Lissajou ellipses confined to a rectangle in configuration space. The corresponding wave packets are obtained if a normalization condition is imposed with respect to a and χ. In this way, one obtains an ordinary quantum-mechanical

Hilbert space. The wave packets follow here the classical Lissajous ellipses without dispersion (Kiefer 1990; Gousheh and Sepangi 2000). From (8.26), one finds

$$H_a \varphi_n(a) \varphi_n(\chi) = H_\chi \varphi_n(a) \varphi_n(\chi) \ ,$$

where φ_n denote again the usual harmonic-oscillator eigenfunctions. A wave-packet solution can then be constructed according to

$$\psi(a, \chi) = \sum_n A_n \varphi_n(a) \varphi_n(\chi) = \sum_n A_n \frac{H_n(a) H_n(\chi)}{2^n n!} e^{a^2/2 - \chi^2/2} \ , \qquad (8.27)$$

with suitable coefficients A_n. From the properties of the Hermite polynomials, it is evident that the wave packet has to satisfy the 'initial condition' $\psi(0, \chi) = \psi(0, -\chi)$. The requirement of normalizability thus gives a restriction on possible initial conditions.

In a more general oscillator model one would expect to find two different frequencies for a and χ, that is, instead of the potential $-a^2 + \chi^2$, one would have $-\omega_a^2 a^2 + \omega_\chi^2 \chi^2$. The demand for normalizability would then entail the commensurability condition

$$\frac{\omega_\chi}{\omega_a} = \frac{2n_a + 1}{2n_\chi + 1} \ , \qquad (8.28)$$

where n_a and n_χ are integer numbers. Thus, one gets from normalizability a restriction on the 'coupling constants' of this model. It is imaginable that such conditions may also hold in the full theory, for example, for the cosmological constant.[2]

As Page (1991) has demonstrated, one can map various minisuperspace models into each other. He also presents a plenty of classical and quantum solutions for these models. For example, one can find a map between the cases of minimally and non-minimally coupled scalar fields. The fields have to be rescaled and, what is most important, they differ in the range of allowed values. As mentioned above, this has consequences for the initial value problem in quantum gravity (Kiefer 1989). A particular example is the map between the models defined by (8.26) and (8.13) for $m = 0$ and $\Lambda = 0$, respectively. The respective wave functions have, however, different domains for their argument. Another map relates (8.26) to a massive Klein–Gordon equation in the (1+1)-dimensional 'Rindler wedge'.

Up to now we have addressed quantum cosmology in the framework of geometrodynamic variables with (mostly) minimally coupled fields and without supersymmetry (SUSY). It is possible to extend the above discussion to the presence of SUSY, to effective actions coming from string theory ('quantum string cosmology'), and to quantum connection and loop dynamics. We shall add in the following some remarks on each of these frameworks.

---

[2]Quantization conditions for the cosmological constant may also arise in string theory, cf. Bousso and Polchinski (2000) and Feng *et al.* (2001).

Canonical supergravity (SUGRA) was discussed in Section 5.3.6. Again, one can restrict the corresponding action to spatially homogeneous models, in the simplest case to a Friedmann universe. After this specialization, only the spinor indices remain. In this way one can say that full SUGRA with $N = 1$ leads to an effective minisuperspace model with $N = 4$ SUSY—one just has four quantum mechanical generators of SUSY, S_A ($A = 1, 2$) and $\bar{S}_{A'}$ ($A' = 1', 2'$); cf. (5.99). As has been discussed in Section 5.3.6, it is fully sufficient to solve the Lorentz and the SUSY constraints; the Hamiltonian and diffeomorphism constraints follow through the Dirac-bracket relations. This holds, of course, also in minisuperspace. The Lorentz constraints can be implemented by making the following decomposition of the wave function Ψ into the fermionic variables, allowing the coefficients to be functions of the bosonic variables,

$$\Psi = \mathcal{A}(a, \phi, \bar{\phi}) + \mathcal{B}(a, \phi, \bar{\phi})\psi^A\psi_A + i\mathcal{C}(a, \phi, \bar{\phi})\chi^A\psi_A$$
$$+\mathcal{D}(a, \phi, \bar{\phi})\chi^A\chi_A + \mathcal{E}(a, \phi, \bar{\phi})\psi^A\psi_A\chi^A\chi_A . \tag{8.29}$$

Here, a is the scale factor of the Friedmann universe, ϕ is a complex scalar field, χ_A its fermionic superpartner, and ψ_A is the homogeneous degree of freedom coming from the gravitino field. The configuration space of Ψ is thus given by $(\chi_A, \psi_A, a, \phi, \bar{\phi})$. The SUSY constraints then yield a system of coupled first-order differential equations for the coefficient functions $\mathcal{A}$–$\mathcal{E}$. They may be solved in various situations, and particular boundary conditions (Section 8.3) may be imposed. For a detailed discussion, we refer to Moniz (2003) and the references therein.

The method of canonical quantization can also be applied to actions different from the Einstein–Hilbert action for gravity. Popular examples are given by effective actions from string theory; cf. Section 9.2. They contain besides the metric additional fields in the gravitational sector, notably the dilaton field and axion fields. They can act as the starting point for both classical and quantum cosmology; see Gasperini and Veneziano (2003) for an extensive review. String theory possesses an invariance with respect to 'duality transformations' (Chapter 9). This is reflected in the context of Friedmann cosmology by the duality transformation $a \to a^{-1}$ and $t \to -t$. Thereby one can relate different solutions to each other. Classical solutions of particular interest are the 'pre-big bang solutions' for $t < 0$ being duality-related to 'post-big bang solutions' for $t > 0$. The former can describe an accelerated (superinflationary) phase being driven by the kinetic energy of the dilaton. The latter can describe an expanding decelerated phase which is interpreted as the standard evolution of the radiation-dominated universe. The hope is that one can connect both phases and thereby avoid the usual fine-tuning problems of inflationary cosmology. In addition, one can hope to generate primordial gravitational waves with higher amplitude, which could possibly be detected by the space interferometer LISA. However, the transition between the pre- and post-big bang solution proceeds through a regime of a classical singularity (strong coupling and high curvature). This is known as the 'graceful exit problem' (exit from inflation into radiation dominance), which is still an open issue.

The quantum version of string cosmology was first investigated by Bento and Bertolami (1995). The role of boundary conditions with respect to the graceful exit problem was discussed by Dąbrowski and Kiefer (1997). Starting from the tree-level effective action with only metric and dilaton, one first makes the following redefinition in minisuperspace,

$$\beta \equiv \sqrt{3} \ln a \ ,$$

$$\bar{\phi} \equiv \phi - 3 \ln a - \ln \int \frac{\mathrm{d}^3 x}{\lambda_s^3} \ ,$$

where ϕ is the original dilaton and $\lambda_s \equiv \sqrt{\alpha'}$ the string length (Section 3.2). One then finds the following Wheeler–DeWitt equation,

$$\left(-\frac{\partial^2}{\partial \bar{\phi}^2} + \frac{\partial^2}{\partial \beta^2} - \lambda_s^2 V(\beta, \bar{\phi}) \mathrm{e}^{-2\bar{\phi}} \right) \psi(\beta, \bar{\phi}) = 0 \ , \tag{8.30}$$

where $V(\beta, \bar{\phi})$ denotes here the dilaton potential. Since no external time t exists in quantum cosmology, it does not make sense to talk of a transition between the pre- and post-big bang regime. Boundary conditions have to be imposed intrinsically, that is, with respect to the configuration-space variables β and $\bar{\phi}$. Constructing wave packet-solutions to (8.30), one recognizes that the pre- and post-big bang branches just correspond to different solutions (Dąbrowski and Kiefer 1997). An intrinsic distinction between 'expanding' and 'contracting' solution is not possible since the reference phase $\mathrm{e}^{-i\omega t}$ is lacking (Zeh 1988). This can only be achieved if additional degrees of freedom are introduced and a boundary condition of low entropy is imposed; see Section 10.2.

Quantum-cosmological models can, and have been, discussed in more general situations. Cavaglià and Moniz (2001), for example, have investigated the Wheeler–DeWitt equation for effective actions inspired by 'M-theory' (Section 9.1). Lidsey (1995) has discussed general scalar-tensor theories in which it turns out that the duality symmetry of the classical action corresponds to hidden $N = 2$ SUSY. Quantum cosmology for relativity in more than three space dimensions also possesses some interesting features (see e.g. Zhuk 1992).

Quantum cosmology can also be discussed using methods of connection or loop dynamics (Chapter 6). Paternoga and Graham (1998), for example, investigated Bianchi IX models with $\Lambda \neq 0$ in the connection representation. They started from the Chern–Simons state (6.7), which is a solution of the Euclidean quantum constraints for Barbero–Immirzi parameter $\beta = 1$. Through a generalized Fourier transform to the metric transformation, they were able—using inequivalent contours in the transformation formula—to find various solutions to the Wheeler–DeWitt equation.

One can address quantum cosmology directly in the loop representation, see Bojowald (2003) and the references therein. Up to now we have discussed the minisuperspace approximation as it is obtained from restricting the classical theory by imposing symmetries and then going to the quantum theory. In loop quantum cosmology, the reduction is not carried out at the classical level but at the level of the auxiliary Hilbert space $\mathcal{F}$ (see Section 5.1). Thereby results such as volume quantization and the form of the Hamiltonian constraint (discussed in Sections 6.2 and 6.3) survive the reduction. The discrete structure of space thus leaves its imprints in minisuperspace. The volume has discrete values V_n ($n \in \mathbb{Z}$, with n and $-n$ giving the same values for the volume); the first non-vanishing value is $V_2 = l_{\mathrm{P}}^3/6$. For the eigenvalues of the inverse scale factor Bojowald gets

$$(a^{-1})_n = 16 l_{\mathrm{P}}^{-4} \left(\sqrt{V_{n+1}} - \sqrt{V_{n-1}} \right)^2 . \tag{8.31}$$

Its maximum is obtained for $n = 2$ and reads

$$(a^{-1})_{\mathrm{max}} = \frac{32(2 - \sqrt{2})}{3 l_{\mathrm{P}}} , \tag{8.32}$$

so the minimum of the scale factor is of the order of the Planck length, as expected. This is a direct consequence of the discrete nature of space and can be interpreted as the avoidance of the singularity in the quantum theory. In fact, the Wheeler–DeWitt equation in minisuperspace becomes a *difference* equation in n, from which the usual differential form is recovered in the limit $n \gg 1$.

8.1.3 *(2+1)-dimensional quantum gravity*

General relativity in 2+1 dimensions is 'trivial' in the sense that there are no local dynamical degrees of freedom. The Riemann tensor depends linearly on the Ricci tensor and thus the vacuum solutions of Einstein's equations are either flat (for $\Lambda = 0$) or have constant curvature (for $\Lambda \neq 0$). A typical feature is the appearance of *conical* structures which gives rise to a non-trivial global geometry. There may be a finite number of degrees of freedom connected with the topology of space. The quantum theory is, therefore, quantum mechanics, and it is for this reason that we have included 2+1 quantum gravity into the chapter on quantum cosmology, although this framework is not necessarily restricted to cosmological applications. General references on (2+1)-dimensional gravity include Carlip (1998), Brown (1988), and Matschull (1995).

The Planck mass in 2+1 dimensions is given by

$$m_{\mathrm{P}}^{(3)} = \frac{c^2}{G} \tag{8.33}$$

and is, therefore, independent of $\hbar$. Classical GR in 2+1 dimensions thus contains a distinguished mass scale. Planck's constant enters, however, the Planck length,

$$l_{\mathrm{P}}^{(3)} = \frac{\hbar G}{c^3} . \tag{8.34}$$

The classical canonical formalism here employs a foliation of three-dimensional space–time into two-dimensional spaces Σ. An important theorem states that any metric on a compact two-space Σ is conformal to a metric of constant curvature.[3] The curvature is positive for the two-sphere S^2 (having genus $g = 0$), zero for the two-torus T^2 ($g = 1$), and negative for $g > 1$. The two-metric can thus be written as

$$h_{ab}(\mathbf{x}) = \mathrm{e}^{2\xi(\mathbf{x})}\tilde{h}_{ab}(\mathbf{x}) , \qquad (8.35)$$

with $\tilde{h}_{ab}(\mathbf{x})$ denoting a metric of constant curvature. The role of the configuration space is here played by the *moduli space* of Σ—the space of metrics with constant curvature modulo diffeomorphisms. The moduli space has a finite dimension: zero for S^2, 2 for T^2, and $6g - 6$ for $g > 1$. The theory, therefore, describes a finite-dimensional system described by the moduli parameters. It is interesting that a reduced phase space description is possible if 'York's time', cf. (5.16), is being used (see Carlip 1998). A Schrödinger equation can be formulated in the reduced space, which is similar to the equation occurring in spherically symmetric systems (Section 7.2).

What can be said about the Wheeler–DeWitt quantization (Carlip 1998)? The situation is simplest in a first-order formulation, where the connection ω_i^a and the zweibein e_i^a are treated as independent variables. From the Wheeler–DeWitt equation, one finds that the wave functional must be a functional of flat connections. This is related to the fact that (2+1)-dimensional GR can be formulated as a Chern–Simons theory for a vector potential with gauge group $\mathrm{ISO}(2,1)$ (Achúcarro and Townsend 1986; Witten 1988). In fact, the first-order action of GR *is* the Chern–Simons action (6.8).

The second-order formalism is much more complicated. There appear functional derivatives with respect to the scale factor ξ and non-local terms from the solution of the diffeomorphism constraints. Most likely, this approach is inequivalent to reduced quantization, and the Wheeler–DeWitt equation cannot be solved, except perhaps in perturbation theory.

Quantum gravity in 2+1 dimensions provides an example for a theory of the first kind in the sense of Section 1.3: it is a consistent quantum theory of the gravitational field itself. This holds irrespective of the fact that it is non-renormalizable by formal power counting arguments.

Of special interest both classically and quantum mechanically is the existence of a black-hole solution for $\Lambda < 0$, that is, for asymptotic anti-de Sitter space. The solution is called 'BTZ black hole' after the work by Bañados *et al.* (1992) and is characterized by mass and angular momentum. The BTZ hole provides a toy model in which one can study the issues of Hawking radiation and entropy; cf. Chapter 7. One can, in particular, give a microscopic interpretation of black-hole entropy by counting the degrees of freedom on the horizon in a Chern–Simons approach (Carlip 1998).

Interesting features appear if point-like particles (whose existence is allowed in 2+1 gravity) are coupled to gravity; cf. Louko and Matschull (2001), who consider the presence of two such particles. For one massive particle, space–time

---

[3]For open Σ, one has to impose appropriate boundary conditions.

is a product of a conical space with $\mathbb{R}$, the deficit angle of the cone being given by $8\pi G m$, where m is the mass. For two particles this essentially doubles, but care has to be taken to implement the condition of asymptotic flatness. The particles do not interact directly (there is no Newton force in 2+1 gravity) but indirectly through the cone-like structure of space–time. From Wheeler–DeWitt quantization, one finds features that are expected from a quantum theory of gravity: the two particles cannot get closer than a certain minimal distance given by a multiple of the Planck length. This exhibits a discrete structure of space–time. Even if the particles are far apart, it is impossible to localize any single particle below a certain length. All these effects vanish both for $\hbar \rightarrow 0$ and for $G \rightarrow 0$. A generalization to many particles has been done by Matschull (2001). He finds in particular hints for a non-commutative structure of space–time.

8.2 Introduction of inhomogeneities

The minisuperspace models discussed in the last section are easy to deal with, but are not sufficient for a realistic description of the universe. This can only be achieved if inhomogeneous degrees of freedom are introduced. Otherwise, one would not have the chance to understand the emergence of structure in quantum cosmology. In the following, we consider a multipole expansion of the three-metric and a scalar field. We take the universe to be closed, so the expansion is with respect to spherical harmonics on the three-sphere S^3. In order to render the formal treatment manageable, the expansion for the 'higher multipoles' is only performed up to quadratic order in the action. They are thus considered to be small perturbations of the homogeneous background described by a and the homogeneous field ϕ. Since one knows from measurements of the microwave background radiation that the fluctuations were small in the early universe, this approximation may be appropriate for that phase. The multipole expansion is also needed for the description of decoherence (Section 10.1.2). Cosmological perturbations were first studied by Lifshits (1946).

Following Halliwell and Hawking (1985), we make for the three-metric the ansatz

$$h_{ab} = a^2(\Omega_{ab} + \epsilon_{ab}) , \qquad (8.36)$$

where Ω_{ab} denotes the metric on S^3, and the 'perturbation' $\epsilon_{ab}(\mathbf{x}, t)$ is expanded into spherical harmonics,

$$\epsilon_{ab}(\mathbf{x}, t) = \sum_{\{n\}} \left(\sqrt{\frac{2}{3}} a_n(t) \Omega_{ab} Q^n + \sqrt{6} b_n(t) P_{ij}^n + \sqrt{2} c_n(t) S_{ij}^n + 2 d_n(t) G_{ij}^n \right) .$$

$$(8.37)$$

Here $\{n\}$ stands for the three quantum numbers $\{n, l, m\}$, where $n = 1, 2, 3, \ldots$, $l = 0, \ldots, n - 1$, and $m = -l, \ldots, l$. The scalar field is expanded as

$$\Phi(\mathbf{x}, t) = \frac{1}{\sqrt{2\pi}} \phi(t) + \epsilon(\mathbf{x}, t) ,$$

$$\epsilon(\mathbf{x}, t) = \sum_{\{n\}} f_n(t) Q^n \; . \tag{8.38}$$

The scalar harmonic functions $Q^n \equiv Q^n_{lm}$ on S^3 are the eigenfunctions of the Laplace operator on S^3,

$$Q^n_{lm|k}{}^{|k} = -(n^2 - 1) Q^n_{lm} \; , \tag{8.39}$$

where $|k$ denotes the covariant derivative with respect to Ω_{ab}. The harmonics can be expressed as

$$Q^n_{lm}(\chi, \theta, \phi) = \Pi^n_l(\chi) Y_{lm}(\theta, \phi) \; , \tag{8.40}$$

where $\Pi^n_l(\chi)$ are the 'Fock harmonics', and $Y_{lm}(\theta, \phi)$ are the standard spherical harmonics on S^2. They are orthonormalized according to

$$\int_{S^3} \mathrm{d}\mu \; Q^n_{lm} Q^{n'}_{l'm'} = \delta^{nn'} \delta_{ll'} \delta_{mm'} \; , \tag{8.41}$$

where $\mathrm{d}\mu = \sin^2 \chi \sin \theta \mathrm{d}\chi \mathrm{d}\theta \mathrm{d}\varphi$. The scalar harmonics are thus a complete orthonormal basis with respect to which each scalar field on S^3 can be expanded. The remaining harmonics appearing in (8.37) are called tensorial harmonics of scalar type (P^n_{ab}), vector type (S^n_{ab}), and tensor type (G^n_{ab}); see Halliwell and Hawking (1985) and the references therein. At the present order of approximation for the higher multipoles (up to quadratic order in the action), the vector harmonics are pure gauge. The tensor harmonics G^n_{ab} describe gravitational waves and are gauge independent. It is possible to work exclusively with gauge-independent variables (Bardeen 1980), but this is not needed for the following discussion.

Introducing the shorthand notation $\{x_n\}$ for the collection of multipoles a_n, b_b, c_n, d_n, the wave function is defined on an infinite-dimensional configuration space spanned by a (or α) and ϕ (the 'minisuperspace background') and the variables $\{x_n\}$. The Wheeler–DeWitt equation can be decomposed into two parts referring to first and second derivatives in the $\{x_n\}$, respectively. The part with the second derivatives reads (Halliwell and Hawking 1985)

$$\left(H_0 + 2e^{3\alpha} \sum_n H_n(a, \phi, x_n) \right) \Psi(\alpha, \phi, \{x_n\}) = 0 \; , \tag{8.42}$$

where H_0 denotes the minisuperspace part,

$$H_0 \equiv \left(\frac{\partial^2}{\partial \alpha^2} - \frac{\partial^2}{\partial \phi^2} + e^{6\alpha} m^2 \phi^2 - e^{4\alpha} \right) \equiv \left(\frac{\partial^2}{\partial \alpha^2} - \frac{\partial^2}{\partial \phi^2} + V(\alpha, \phi) \right) \; , \tag{8.43}$$

and H_n is a sum of Hamiltonians referring to the scalar, vector, and tensor part of the modes, respectively,

$$H_n = H_n^{(S)} + H_n^{(V)} + H_n^{(T)} \; .$$

We now make the ansatz

$$\Psi(\alpha, \phi, \{x_n\}) = \psi_0(\alpha, \phi) \prod_{n>0} \psi_n(\alpha, \phi; x_n) \qquad (8.44)$$

and insert this into (8.42). Following Kiefer (1987), we get

$$-\frac{\nabla^2 \psi_0}{\psi_0} - 2\frac{\nabla \psi_0}{\psi_0} \sum_n \frac{\nabla \psi_n}{\psi_n} - \sum_n \frac{\nabla^2 \psi_n}{\psi_n}$$

$$-\sum_{n \neq m} \frac{\nabla \psi_n \nabla \psi_m}{\psi_n \psi_m} + V(\alpha, \phi) + 2\mathrm{e}^{3\alpha} \sum_n \frac{H_n \Psi}{\Psi} = 0 \; ,$$

where

$$\nabla \equiv \left(\frac{\partial}{\partial \alpha}, \frac{\partial}{\partial \phi} \right)$$

denotes the 'minisuperspace gradient'. One then gets by separation of variables the two equations

$$-\frac{\nabla^2 \psi_0}{\psi_0} + V(\alpha, \phi) = -2f(\alpha, \phi) \; , \qquad (8.45)$$

$$-2\frac{\nabla \psi_0}{\psi_0} \sum_n \frac{\nabla \psi_n}{\psi_n} - \sum_n \frac{\nabla^2 \psi_n}{\psi_n} - \sum_{n \neq m} \frac{\nabla \psi_n \nabla \psi_m}{\psi_n \psi_m}$$

$$+2\mathrm{e}^{3\alpha} \sum_n \frac{H_n \Psi}{\Psi} = 2f(\alpha, \phi) \; , \qquad (8.46)$$

where $f(\alpha, \phi)$ is an arbitrary function. The second equation can be separated further if one imposes additional assumptions. First, we assume that the dependence of the ψ_n on the minisuperspace variables is weak in the sense that

$$\left| \frac{\nabla \psi_0}{\psi_0} \right| \gg \left| \frac{\nabla \psi_n}{\psi_n} \right| \; , \; n \gg 1 \; .$$

Second, we assume that the terms $\nabla \psi_n / \psi_n$ add up incoherently ('random-phase approximation') so that the third term on the left-hand side of (8.46) can be neglected compared to the first term. One then gets from (8.46)

$$-\frac{\nabla \psi_0}{\psi_0} \nabla \psi_n - \frac{1}{2}\nabla^2 \psi_n + \mathrm{e}^{3\alpha} \left(\frac{H_n \Psi}{\Psi} \right) \psi_n = \varphi_n(\alpha, \phi)\psi \; ,$$

where

$$\sum_n \varphi_n(\alpha, \phi) = f(\alpha, \phi) \; .$$

Since the ψ_n are assumed to vary much less with α and ϕ than ψ_0 does, one would expect the term $\nabla^2 \psi_n$ to be negligible. Finally, assuming that in H_n the x_n-derivatives dominate over the ∇-derivatives, one can substitute

$$\frac{H_n \Psi}{\Psi} \psi_n \approx H_n \psi_n \ .$$

One then arrives at

$$-\frac{\nabla \psi_0}{\psi_0} \nabla \psi_n + e^{3\alpha} H_n \psi_n = \varphi_n \psi_n \ . \tag{8.47}$$

The choice $f = 0$ in (8.46) would entail that ψ_0 is a solution of the minisuperspace Wheeler–DeWitt equation. If one chooses in addition $\varphi_n = 0$, (8.47) reads

$$e^{-3\alpha} \frac{\nabla \psi_0}{\psi_0} \nabla \psi_n = H_n \psi_n \ . \tag{8.48}$$

If ψ_0 were of WKB form, $\psi_0 \approx C \exp(iS_0)$ (with a slowly varying prefactor C), one would get

$$i\frac{\partial \psi_n}{\partial t} = H_n \psi_n \ , \tag{8.49}$$

with

$$\frac{\partial}{\partial t} \equiv e^{-3\alpha} \nabla S_0 \cdot \nabla \ . \tag{8.50}$$

Equation (8.49) is a Schrödinger equation for the multipoles, its time parameter t being defined by the minisuperspace variables α and ϕ. This 'WKB time' controls the dynamics in this approximation. The above derivation reflects the recovery of the Schrödinger equation from the Wheeler–DeWitt equation as has been discussed in Section 5.4.

One could also choose the φ_n in such a way as to minimize the variation of the ψ_n along WKB time,

$$\varphi_n = e^{3\alpha} \langle H_n \rangle \ , \tag{8.51}$$

where

$$\langle H_n \rangle \equiv \frac{\int dx_n \ \psi_n^* H_n \psi_n}{\int dx_n \ \psi_n^* \psi_n} \ .$$

Instead of (8.49) one then obtains

$$i\frac{\partial \psi_n}{\partial t} = (H_n - \langle H_n \rangle) \psi_n \ . \tag{8.52}$$

The expectation value can be absorbed into the ψ_n if they are redefined by an appropriate phase factor. The minisuperspace equation (8.45) then reads

$$-\frac{\nabla^2 \psi_0}{\psi_0} + V(\alpha, \phi) = -2e^{3\alpha} \sum_n \langle H_n \rangle \ . \tag{8.53}$$

The term on the right-hand side corresponds to the back reaction discussed in Section 5.4. From the point of view of the full wave function Ψ, it is just a matter of the splitting between ψ_0 and the ψ_n.

Consider as a particular example the case of the tensor multipoles $\{d_n\}$, which describe gravitational waves. After an appropriate redefinition, the wave functions $\psi_n(\alpha, \phi, d_n)$ obey the Schrödinger equations (Halliwell and Hawking 1985)

$$i\frac{\partial \psi_n}{\partial t} = \frac{1}{2}e^{-3\alpha}\left(-\frac{\partial^2}{\partial d_n^2} + (n^2 - 1)e^{4\alpha}d_n^2\right)\psi_n .$$

(8.54)

This has the form of a Schrödinger equation with 'time-dependent' frequency given by

$$\nu \equiv \frac{\sqrt{n^2 - 1}}{e^\alpha} \overset{n \gg 1}{\approx} ne^{-\alpha} .$$

(8.55)

The (adiabatic) ground-state solution, for example, of this equation would read

$$\psi_n \propto \exp\left(-\frac{n^3 d_n^2}{2\nu^2}\right)\exp\left(-\frac{i}{2}\int^t \mathrm{d}s\, \nu(s)\right) ,$$

(8.56)

see Halliwell and Hawking (1985) and Kiefer (1987). It plays a role in the discussion of primordial fluctuations in inflationary cosmology.

8.3 Boundary conditions

In this section, we shall address the following question which has been neglected so far: what are the appropriate boundary conditions for the Wheeler–DeWitt equation in quantum cosmology?

Since Newton it has become customary to separate the description of Nature into dynamical laws and initial conditions. The latter are usually considered as artificial and can be fixed by the experimentalist, at least in principle. The situation in cosmology is different. The Universe is unique and its boundary conditions are certainly not at our disposal. It has, therefore, been argued that boundary conditions play a key role in the more fundamental framework of quantum cosmology. It has even been claimed that quantum cosmology *is* the theory of initial conditions (see e.g. Hartle 1997; Barvinsky 2001). In the following, we shall briefly review various proposals for boundary conditions.

8.3.1 *DeWitt's boundary condition*

DeWitt (1967*a*) suggested to impose the boundary condition

$$\Psi\left[^{(3)}\mathcal{G}\right] = 0$$

(8.57)

for all three-geometries $^{(3)}\mathcal{G}$ related with 'barriers', for example, singular three-geometries. This could automatically alleviate or avoid the singularities of the classical theory. Ideally, one would expect that a unique solution to the Wheeler–DeWitt equation is obtained after this boundary condition is imposed. Whether this is true remains unsettled. In a sense, the demand for the wave function to go to zero for large scale factors—as has been discussed in Section 8.1 in connection with wave packets—can be interpreted as an implementation of (8.57) in minisuperspace. In the example of the collapsing dust shell in Section 7.4, the wave function is also zero for $r \to 0$, that is, at the region of the classical singularity. In that case this is, however, not the consequence of a boundary condition, but of the dynamics—it is a consequence of unitary time evolution.

8.3.2 No-boundary condition

This proposal goes back to Hawking (1982) and Hartle and Hawking (1983). It is, therefore, also called 'Hartle–Hawking proposal'. A central role in its formulation is played by Euclidean path integrals. In fact, the description of black-hole thermodynamics by such path integrals was one of Hawking's original motivations to introduce this proposal, cf. Hawking (1979). The 'no-boundary condition' states that the wave function Ψ is for a *compact* three-dimensional space Σ given by the sum over all compact Euclidean four-geometries of all topologies that have Σ as their *only* boundary. This means that there does not exist a second, 'initial', boundary on which one would have to specify boundary data. Formally one would write

$$\Psi[h_{ab}, \Phi, \Sigma] = \sum_{\mathcal{M}} \nu(\mathcal{M}) \int_{\mathcal{M}} \mathcal{D}g\mathcal{D}\Phi \; e^{-S_{\mathrm{E}}[g_{\mu\nu}, \Phi]} \; . \qquad (8.58)$$

The sum over $\mathcal{M}$ expresses the sum over all four-manifolds with measure $\nu(\mathcal{M})$. Since it is known that four-manifolds are not classifiable, this cannnot be put into a precise mathematical form. The integral is the quantum gravitational path integral discussed in Section 2.2.1., where S_{E} denotes the Euclidean Einstein–Hilbert action (2.72).

Except in simple minisuperspace models, the path integral in (8.58) cannot be evaluated exactly. It is therefore usually being calculated in a semiclassical ('saddle point') approximation. Since there exist in general several saddle points, one must address the issue of which contour of integration has to be chosen. Depending on the contour only part of the saddle points may contribute to the path integral.

The integral over the four-metric in (8.58) splits into integrals over three-metric, lapse function, and shift vector; cf. Section 5.3.4. In a Friedmann model, only the integral over the lapse function remains and turns out to be an ordinary integral; cf. Halliwell (1988),

$$\psi(a, \phi) = \int \mathrm{d}N \int \mathcal{D}\phi \; e^{-I[a(\tau), \phi(\tau), N]} \; , \qquad (8.59)$$

where we have denoted the Euclidean minisuperspace action by I instead of S_{E}. For notational simplicity, we have denoted the arguments of the wave function by the same letters than the corresponding functions which are integrated over in the path integral. For the Friedmann model containing a scalar field, the Euclidean action reads

$$I = \frac{1}{2} \int_{\tau_1}^{\tau_2} d\tau \ N \left[-\frac{a}{N^2} \left(\frac{da}{d\tau} \right)^2 + \frac{a^3}{N^2} \left(\frac{d\phi}{d\tau} \right)^2 - a + a^3 V(\phi) \right] , \qquad (8.60)$$

where $V(\phi)$ may denote just a mass term, $V(\phi) \propto m^2 \phi^2$, or include a self-interaction such as $V \propto \phi^4$.

How is the no-boundary proposal being implemented in minisuperspace? The imprint of the restriction in the class of contours in (8.58) is to integrate over Euclidean paths $a(\tau)$ with the boundary condition $a(0) = 0$, cf. the discussion in Halliwell (1991). This is supposed to implement the idea of integration over regular four-geometries with no 'boundary' at $a = 0$. (The point $a = 0$ has to be viewed like the pole of a sphere, which is completely regular.) One is often interested in discussing quantum-cosmological models in the context of inflationary cosmology. Therefore, in evaluating (8.59), one might restrict oneself to the region where the scalar field ϕ is slowly varying ('slow-roll approximation' of inflation). One can then neglect the kinetic term of ϕ and integrate over Euclidean paths with $\phi(\tau) \approx$ constant. For $a^2 V < 1$, one gets the following two saddle point actions (Hawking 1984; Halliwell 1991),

$$I_\pm = -\frac{1}{3V(\phi)} \left[1 \pm (1 - a^2 V(\phi))^{3/2} \right] . \qquad (8.61)$$

The action I_- is obtained for a three-sphere being closed off by less than half the four-sphere, while in evaluating I_+, the three-sphere is closed off by more than half the four-sphere.

There exist various arguments in favour of which of the two extremal actions are distinguished by the no-boundary proposal. This can in general only be decided by a careful discussion of integration contours in the complex N-plane; see for example, Halliwell and Louko (1991) or Kiefer (1991). For the present purpose, it is sufficient to assume that I_- gives the dominant contribution (Hartle and Hawking 1983). The wave functions $\psi \propto \exp(-I_\pm)$ are WKB solutions to the minisuperspace Wheeler–DeWitt equation (8.23) in the classically forbidden ('Euclidean') region. Taking into account the standard WKB prefactor, one thus has for the no-boundary wave function (choosing I_-) for $a^2 V < 1$ the expression

$$\psi_{\mathrm{NB}} \propto \left(1 - a^2 V(\phi) \right)^{-1/4} \exp \left(\frac{1}{3V(\phi)} \left[1 - (1 - a^2 V(\phi))^{3/2} \right] \right) . \qquad (8.62)$$

Note that the sign in the exponent has been fixed by the proposal—the WKB approximation would also allow a solution of the form $\propto \exp(- \ldots)$. The continuation into the classically allowed region $a^2 V > 1$ is obtained through the standard WKB connection formulae to read

$$\psi_{\mathrm{NB}} \propto \left(a^2 V(\phi) - 1 \right)^{-1/4} \exp \left(\frac{1}{3V(\phi)} \right) \cos \left(\frac{(a^2 V(\phi) - 1)^{3/2}}{3V(\phi)} - \frac{\pi}{4} \right) . \qquad (8.63)$$

The no-boundary proposal thus picks out a particular WKB solution in the classically allowed region. It is a real solution and can, therefore, be interpreted as a superposition of two complex WKB solutions of the form $\exp(iS)$ and $\exp(-iS)$.

The above wave functions have been obtained for regions of slowly varying ϕ. This is the context of inflationary cosmology in which one has an effective (ϕ-dependent) Hubble parameter of the form

$$H^2(\phi) \approx \frac{4\pi V(\phi)}{3m_{\rm P}^2} \, , \tag{8.64}$$

where in the simplest case one has $V(\phi) = m^2\phi^2$. In the units used here ($3m_{\rm P}^2 = 4\pi$), this reads $H^2(\phi) = V(\phi)$. The radius of the four-sphere is $a = H^{-1} = V^{-1/2}$. The geometric picture underlying the no-boundary proposal is here to imagine the dominant geometry to the path integral as consisting of two parts: half of a four-sphere to which half of de Sitter space is attached. The matching must be made at exactly half the four-sphere because only there is the extrinsic curvature equal to zero. Only for vanishing extrinsic curvature, $K_{ab} = 0$, is continuity guaranteed; cf. Gibbons and Hartle (1990). From (8.61) it is clear that the action corresponding to half of the four-sphere is $I = -1/3V(\phi)$. The solutions of the classical field equations are $a(\tau) = H^{-1}\sin(H\tau)$ in the Euclidean regime ($0 \leq \tau \leq \pi/2H$) and $a(t) = H^{-1}\cosh(Ht)$ in the Lorentzian regime ($t > 0$).

The picture of de Sitter space attached to half a four-sphere (also called 'de Sitter instanton') is often referred to as 'quantum creation from nothing' or 'nucleation' from the Euclidean regime into de Sitter space. However, this is somewhat misleading since it is not a process in time, but corresponds to the emergence *of* time (cf. Butterfield and Isham 1999). Moreover, it is far from clear that this is really the dominant contribution from the path integral: discussions in 2+1 dimensions, where more explicit calculations can be made, indicate that the path integral is dominated by an infinite number of complicated topologies (Carlip 1998). This would cast some doubt on the validity of the above saddle-point approximation in the first place.

An interesting consequence of the no-boundary proposal occurs if the inhomogeneous modes of Section 8.2 are taken into account. In fact, this proposal selects a distinguished vacuum for de Sitter space—the so-called 'Euclidean' or 'Bunch–Davies' vacuum (Laflamme 1987). What is the Euclidean vacuum? In Minkowski space, there exists a distinguished class of equivalent vacua (simply called the 'Minkowski vacuum'), which is invariant under the Poincaré group and therefore the same for all inertial observers. De Sitter space is, like Minkowski space, maximally symmetric: instead of the Poincaré group it possesses $SO(4,1)$ (the 'de Sitter group') as its isometry group, which also has 10 parameters. It turns out that there exists, contrary to Minkowski space, a one-parameter family of inequivalent vacua which are invariant under the de Sitter group; see for example, Birrell and Davies (1982). One of these vacua is distinguished in many respects: it corresponds to the Minkowski vacuum for constant a, and its mode functions are regular on the Euclidean section $t \mapsto \tau = it + \pi/2H$. This second property gives it the name 'Euclidean vacuum'. One expands as in (8.38) the scalar field into its harmonics,

$$\Phi(\mathbf{x}, \tau) - \frac{1}{\sqrt{2\pi}} \phi(\tau) = \sum_{\{n\}} f_n(\tau) Q^n ,$$

but now with respect to Euclidean time τ. One then calculates the Euclidean action for the modes $\{f_n(\tau)\}$ and imposes the following regularity conditions from the no-boundary proposal:

$$f_n(0) = 0 , \ n = 2, 3, \ldots , \quad \frac{\mathrm{d}f_n}{\mathrm{d}\tau}(0) = 0 , \ n = 1 .$$

This then yields for the wave functions ψ_n satisfying the Schrödinger equations (8.49) a solution that just corresponds to the Euclidean vacuum. The reason is that essentially the same regularity conditions are required for the no-boundary proposal and the Euclidean vacuum. In the Lorentzian section, the Euclidean vacuum for modes with small wavelength (satisfying $\lambda \ll H^{-1}$) is just given by the state (8.56); see also Section 10.1. According to the no-boundary proposal, the multipoles thus enter the Lorentzian regime in their ground state. Because of its high symmetry, the de Sitter-invariant vacuum was assumed even before the advent of the no-boundary condition to be a natural initial quantum state (Starobinsky 1979).

Hawking (1984) has put forward the point of view that the Euclidean path integral is the true fundamental concept. The fact that a Euclidean metric usually does not have a Lorentzian section, therefore, does not matter. Only the result—the wave function—counts. If the wave function turns out to be exponentially increasing or decreasing, it describes a classically forbidden region. If it is of oscillatory form, it describes a classically allowed region—this corresponds to the world we live in. Since one has to use in general complex integration contours anyway, it is clear that only the result can have interpretational value, with the formal manipulations playing only the role of a heuristic device.

8.3.3   *Tunnelling condition*

The no-boundary wave function calculated in the last subsection turned out to be real. This is a consequence of the Euclidean path integral; even if complex metrics contribute they should do so in complex-conjugate pairs. The wave function (8.63) can be written as a sum of semiclassical components of the form $\exp(\mathrm{i}S)$, each of which gives rise to a semiclassical world in the sense of Section 5.4 (recovery of the Schrödinger equation). These components become independent of each other only after decoherence is taken into account; see Section 10.1. Alternative boundary conditions may directly give a complex wave function, being of the form $\exp(\mathrm{i}S)$ in the semiclassical approximation. This is achieved by the 'tunnelling proposal' put forward by Vilenkin; see for example, Vilenkin (1988, 2003).[4]

---

[4]Like the no-boundary proposal, this usually refers to closed three-space Σ. A treatment of 'tunnelling' into a universe with open Σ is presented in Zel'dovich and Starobinsky (1984).

The tunnelling proposal is easiest being formulated in minisuperspace. In analogy with, for example, the process of α-decay in quantum mechanics, it is proposed that the wave function consists solely of *outgoing* modes. More generally, it states that it consists solely of outgoing modes at singular boundaries of superspace (except the boundaries corresponding to vanishing three-geometry). In the minisuperspace example above, this is the region of infinite a or ϕ. What does 'outgoing' mean? The answer is clear in quantum mechanics, since there one has a reference phase $\propto \exp(-i\omega t)$. An outgoing plane wave would then have a wave function $\propto \exp(ikx)$. But since there is no external time t in quantum cosmology, one can call a wave function 'outgoing' only by definition (Zeh 1988). In fact, the whole concept of tunnelling loses its meaning if an external time is lacking (Conradi 1998).

We have seen in (5.22) that the Wheeler–DeWitt equation possesses a conserved 'Klein–Gordon current', which here reads

$$j = \frac{i}{2}(\psi^*\nabla\psi - \psi\nabla\psi^*) , \quad \nabla j = 0 \tag{8.65}$$

(∇ denotes again the derivatives in minisuperspace). A WKB solution of the form $\psi \approx C\exp(iS)$ leads to

$$j \approx -|C|^2\nabla S . \tag{8.66}$$

The tunnelling proposal states that this current should point outwards at large a and ϕ (provided, of course, that ψ is of WKB form there). If ψ were real (as is the case in the no-boundary proposal), the current would vanish.

In the above minisuperspace model, we have seen that the eikonal $S(a, \phi)$, which is a solution of the Hamilton–Jacobi equation, is given by the expression, cf. (8.63),

$$S(a, \phi) = \frac{(a^2V(\phi) - 1)^{3/2}}{3V(\phi)} . \tag{8.67}$$

We would thus have to take the solution $\propto \exp(-iS)$ since then j would, according to (8.66), become positive and point outwards for large a and ϕ. For $a^2V > 1$, the tunnelling wave function then reads

$$\psi_T \propto (a^2V(\phi) - 1)^{-1/4}\exp\left(-\frac{1}{3V(\phi)}\right)\exp\left(-\frac{i}{3V(\phi)}(a^2V(\phi) - 1)^{3/2}\right) , \tag{8.68}$$

while for $a^2V < 1$ (the classically forbidden region), one has

$$\psi_T \propto (1 - a^2V(\phi))^{-1/4}\exp\left(-\frac{1}{3V(\phi)}\left(1 - (1 - a^2V(\phi))^{3/2}\right)\right) . \tag{8.69}$$

As for the inhomogeneous modes, the tunnelling proposal also picks out the Euclidean vacuum.

8.3.4 *Comparison of no-boundary and tunnelling wave function*

The important difference between ψ_T and ψ_{NB} in the above example is, except for the real versus complex wave function, the fact that ψ_T contains a factor $\exp(-1/3V)$, whereas ψ_{NB} has a factor $\exp(1/3V)$. Assuming that our branch of the wave function is in some sense dominant, these results for the wave function have been used to calculate the probability for the occurrence of an inflationary phase. More precisely, one has investigated whether the wave function favours large values of ϕ (as would be needed for inflation) or small values. It is clear from the above results that ψ_T seems to favour large ϕ over small ϕ and therefore seems to predict inflation, whereas ψ_{NB} seems to prefer small ϕ and therefore seems to predict no inflation. However, the assumption of the slow-roll approximation (needed for inflation) is in contradiction to sharp probability peaks; see Barvinsky (2001) and the references therein. The reason is that this approximation demands the ϕ-derivatives to be small. A possible way out of this dilemma is to take into account inhomogeneities (the higher multipoles of Section 8.2) and to proceed to the one-loop approximation of the wave function. This ensures the normalizability of the wave function provided that certain restrictions on the particle content of the theory are fulfilled.

The wave function in the Euclidean one-loop approximation is given by (Barvinsky and Kamenshchik 1990)

$$\Psi_{T,NB} = \exp(\pm I - W), \tag{8.70}$$

where the T and NB refer to 'tunnelling' and 'no-boundary', respectively, I is the classical Euclidean action,[5] and W is the one-loop correction to the effective action,

$$W = \frac{1}{2}\text{tr}\ln\frac{F}{\mu^2} \ . \tag{8.71}$$

Here, F represents the second-order differential operator which is obtained by taking the second variation of the action with respect to the fields, while μ is a renormalization mass parameter. It can be shown that the wave function assumes after the analytic extension into the Lorentz regime the form

$$\Psi_{T,NB} = \left(\frac{1}{|\det u|^{1/2}}\right)^R \times \exp\left(\pm I + iS + \frac{1}{2}if^T(Dv)v^{-1}f\right) \ . \tag{8.72}$$

Here, S is the minisuperspace part of the Lorentzian classical action, f denotes the amplitudes of the inhomogeneities geometry and matter (Section 8.2), u denotes the solutions of the linearized equations for all the modes, v those solutions referring to the inhomogeneous modes, D is a first-order differential operator (the Wronskian related to the operator F), and the superscript R denotes the renormalization of the infinite product of basis functions.

---

[5]In the above example, $I = -(3V)^{-1}[1 - (1 - a^2V)^{3/2}]$, and $\exp(+I)$ results from the tunnelling condition, while $\exp(-I)$ results from the no-boundary condition.

Information concerning the normalizability of the wave function of the universe (8.72) can be obtained from the diagonal elements of the density matrix

$$\hat{\rho} = \mathrm{tr}_f |\Psi\rangle\langle\Psi| . \tag{8.73}$$

It was shown in Barvinsky and Kamenshchik (1990) that the diagonal elements (denoted by $\rho(\phi)$) can be expressed as

$$\rho(\phi) \sim \exp(\pm I - \Gamma_{1-\text{loop}}) , \tag{8.74}$$

where $\Gamma_{1-\text{loop}}$ is the one-loop correction to the effective action calculated on the closed compact 'de Sitter instanton'. This quantity is conveniently calculated by the zeta-regularization technique (Birrell and Davies 1982), which allows $\Gamma_{1-\text{loop}}$ to be represented as

$$\Gamma_{1-\text{loop}} = 1/2\ \zeta'(0) - 1/2\ \zeta(0)\ln(\mu^2 a^2) , \tag{8.75}$$

where $\zeta(s)$ is the generalized Riemann zeta function, and a is the radius of the instanton. In the limit $\phi \to \infty$ or, equivalently, $a \to 0$, the expression (8.74) reduces to

$$\rho(\phi) \sim \exp(\pm I)\phi^{-Z-2} , \tag{8.76}$$

where Z is the anomalous scaling of the theory, expressed in terms of $\zeta(0)$ for all the fields included in the model. The requirement of normalizability imposes the following restriction on Z,

$$Z > -1 , \tag{8.77}$$

which is obtained from the requirement that the integral

$$\int^{\infty} \mathrm{d}\phi\ \rho(\phi)$$

converge at $\phi \to \infty$. In view of this condition, it turns out that SUSY models seem to be preferred (Kamenshchik 1990).

Having obtained the one-loop order, one can investigate whether the wave function is peaked at values for the scalar field preferrable for inflation (Barvinsky and Kamenshchik 1998; Barvinsky 2001). It turns out that this works only if the field ϕ is coupled non-minimally: one must have a coupling $-\xi R\phi^2/2$ in the action with $\xi < 0$ and $|\xi| \gg 1$. One can then get a probability peak at a value of ϕ corresponding to an energy scale needed for inflation (essentially the GUT scale). It seems that, again, only the tunnelling wave function can fulfil this condition, although the last word on this has not been spoken. One obtains from this consideration also a restriction on the particle content of the theory in order to obtain $\Delta\phi/\phi \sim \Delta T/T$ in accordance with the observational constraint $\Delta T/T \sim 10^{-5}$ (temperature anisotropy in the cosmic microwave background).

8.3.5 *Symmetric initial condition*

This condition has been proposed by Conradi and Zeh (1991); see also Conradi (1992). For the wave-packet solutions of Section 8.1, we had to demand that ψ goes to zero for $\alpha \to \infty$. Otherwise, the packet would not reflect the behaviour of a classically recollapsing universe. But what about the behaviour at $\alpha \to -\infty$ ($a \to 0$)? Consider again the model of a massive scalar field in a Friedmann universe given by the Wheeler–DeWitt equation (8.23). The potential term vanishes in the limit $\alpha \to -\infty$, so the solutions which are exponentially decreasing for large α become constant in this limit. With regard to finding normalizable solutions, it would be ideal if there was a reflecting potential also at $\alpha \to -\infty$. One can add for this purpose in an *ad hoc* manner, a repulsive (negative) potential that would be of relevance only in the Planck regime. It is, however, expected anyway that the Wheeler–DeWitt equation has to be modified for small length, due to the unification of interactions. One can, for example, choose the 'Planck potential'

$$V_\mathrm{P}(\alpha) = -C^2 \mathrm{e}^{-2\alpha} \, , \tag{8.78}$$

where C is a real constant. Neglecting as in the previous subsections the ϕ-derivatives (corresponding to the slow-roll approximation), one can thereby *select* a solution to the Wheeler–DeWitt equation (Eqn (8.23) supplemented by V_P) that decreases exponentially towards $\alpha \to -\infty$. This implements also DeWitt's boundary condition that $\psi \to 0$ for $\alpha \to -\infty$ (Section 8.3.1).

The 'symmetric initial condition' (SIC) now states that the full wave function depends for $\alpha \to -\infty$ only on α; cf. Conradi and Zeh (1991). In other words, it is a particular superposition (not an ensemble) of all excited states of Φ and the three-metric, that is, these degrees of freedom are completely absent in the wave function. This is analogous to the symmetric vacuum state in field theory before the symmetry breaking into the 'false' vacuum (Zeh 2001). In both cases, the actual symmetry breaking will occur through decoherence (Section 10.1). The resulting wave function coincides approximately with the no-boundary wave function. Like there, the higher multipoles enter the semiclassical Friedmann regime in their ground state. The SIC is also well suited for a discussion of the arrow of time and the dynamical origin of irreversibility; cf. Section 10.2.

9

STRING THEORY

9.1 General introduction

The approaches discussed so far start from the assumption that the gravitational field can be quantized separately. That this can really be done is, however, not clear. One could imagine that the problem of quantum gravity can only be solved within a unified quantum framework of all interactions. The only serious candidate up to now to achieve this goal is superstring theory. Our interest here is mainly to exhibit the role of quantum gravitational aspects. It is not our aim to give an introduction into the many physical and mathematical aspects of string theory. This is done in a series of excellent textbooks, see in particular Green *et al.* (1987), Lüst and Theisen (1989), Polchinski (1998*a, b*), and Kaku (1999). Mohaupt (2003) gives a concise overview with particular emphasis on gravitational aspects. For more details we refer the reader to these references.

String theory started as an attempt to explain the spectrum of hadrons. After the discovery of quantum chromodynamics and its successful predictions, it was abandoned as such. It was, however, realized that string theory could in principle implement a theory of quantum gravity (Scherk and Schwarz 1974; Yoneya 1974). The main reason is the appearance of a massless spin-2 particle in the spectrum of the string. As we have learned in Chapter 2, such a particle necessarily leads to GR in the low-energy limit.

String theory transcends the level of local field theory because its fundamental objects are one-dimensional entities ('strings') instead of fields defined at space–time points. More recently it has turned out that higher dimensional objects ('branes') appear within string theory in a natural way and on an equal footing with strings (see below). We shall nevertheless continue to talk about 'string theory'.

What are the main features of string theory?

1. String theory necessarily contains gravity. The graviton appears as an excitation of closed strings. Open strings do not contain the graviton by themselves, but since they contain closed strings as virtual contributions, the appearance of the graviton is unavoidable there, too.

2. String theory necessarily leads to gauge theories since the corresponding gauge bosons are found in the string spectrum.

3. String theory seems to need supersymmetry (SUSY) for a consistent formulation. Fermions are therefore an essential ingredient.

4. All 'particles' arise from string excitations. Therefore, they are no longer fundamental and their masses should in principle be fixed with respect to the string mass scale.

5. Higher space–time dimensions appear in a natural way, thus implementing the old idea by Kaluza and Klein.

6. As emphasized above, string theory entails a unified quantum description of all interactions.

7. Since one can get chiral gauge couplings from string theory, the hope is raised that one can derive the Standard Model of elementary particles from it (although one is still very far from having achieved this goal).

The bosonic string has already been introduced in Section 3.2. Its fundamental dimensionful parameter is α' or the string length $l_{\mathrm{s}} = \sqrt{2\alpha'\hbar}$ derived from it. In view of the unification idea, one would expect that l_{s} is roughly of the order of the Planck length l_{P}. The starting point is the Polyakov action S_{P} (3.51). Making use of the three local symmetries (two diffeomorphisms and one Weyl transformation), we have put it into the 'gauge-fixed form' (3.58). In this form, the action still possesses an invariance with respect to conformal transformation, which form an infinite-dimensional group in two dimensions. It, therefore, gives rise to an infinite number of generators—the generators L_n of the Virasoro algebra (3.61).

Consider an *open string* with ends at $\sigma = 0$ and $\sigma = \pi$. Variation of the action (3.58) yields after a partial integration the expression

$$
\delta S_{\mathrm{P}} = \frac{1}{2\pi\alpha'} \int_{\mathcal{M}} \mathrm{d}^2\sigma \; \left(\eta^{\alpha\beta}\partial_\alpha\partial_\beta X_\mu \right) \delta X^\mu
$$
$$
+ \int_{\partial\mathcal{M}} \mathrm{d}\tau \; \left(X'_\mu \delta X^\mu \big]_{\sigma=\pi} - X'_\mu \delta X^\mu \big]_{\sigma=0} \right) . \tag{9.1}
$$

The classical theory demands that $\delta S_{\mathrm{P}} = 0$. For the surface term to vanish, one has the following options. One is to demand that $X'_\mu = 0$ for $\sigma = 0$ and $\sigma = \pi$ (*Neumann* condition). This would guarantee that no momentum exits from the ends of the string. Alternatively one can demand the *Dirichlet* condition: $X_\mu = $ constant for $\sigma = 0$ and $\sigma = \pi$. This condition comes automatically into play if the duality properties of the string are taken into account; see Section 9.2.3. The vanishing of the worldsheet integral in the variation of S_{P} leads to the wave equation

$$
\left(\frac{\partial^2}{\partial\tau^2} - \frac{\partial^2}{\partial\sigma^2} \right) X^\mu(\sigma, \tau) = 0 . \tag{9.2}
$$

The solution of this equation for Neumann boundary conditions reads

$$
X^\mu(\sigma, \tau) = x^\mu + 2\alpha'p^\mu\tau + \mathrm{i}\sqrt{2\alpha'} \sum_{n \neq 0} \frac{\alpha_n^\mu}{n} \mathrm{e}^{-\mathrm{i}n\tau} \cos n\sigma . \tag{9.3}
$$

Here, x^μ and p^μ denote position and momentum of the centre of mass, respectively. These would be the only degrees of freedom for a point particle.

The quantities α_n^μ are the Fourier components (oscillator coordinates) and obey $\alpha_{-n}^\mu = (\alpha_n^\mu)^\dagger$ due to the reality of the X^μ. The solution (9.3) describes a standing wave.

In view of the path-integral formulation, it is often convenient to continue the worldsheet formally into the Euclidean regime, that is, to introduce worldsheet coordinates $\sigma^1 \equiv \sigma$ and $\sigma^2 \equiv i\tau$. One can then use the complex coordinate

$$z = e^{\sigma^2 - i\sigma^1} , \tag{9.4}$$

with respect to which the solution (9.3) reads

$$X^\mu(z, \bar{z}) = x^\mu - i\alpha' p^\mu \ln(z\bar{z}) + i\sqrt{\frac{\alpha'}{2}} \sum_{n \neq 0} \frac{\alpha_n^\mu}{n} \left(z^{-n} + \bar{z}^{-n} \right) . \tag{9.5}$$

For a *closed string* the boundary condition $X^\mu(\sigma) = X^\mu(\sigma + 2\pi)$ is sufficient. The solution of (9.2) can then be written as

$$X^\mu(\sigma, \tau) = X_R^\mu(\sigma^-) + X_L^\mu(\sigma^+) , \tag{9.6}$$

where we have introduced the lightcone coordinates $\sigma^+ \equiv \tau + \sigma$ and $\sigma^- \equiv \tau - \sigma$. The index R (L) corresponds to modes which would appear 'rightmoving' ('leftmoving') in a two-dimensional space–time diagram. Explicitly one has

$$X_R^\mu(\sigma^-) = \frac{x^\mu}{2} + \frac{\alpha'}{2} p^\mu \sigma^- + i\sqrt{\frac{\alpha'}{2}} \sum_{n \neq 0} \frac{\alpha_n^\mu}{n} e^{-in\sigma^-} \tag{9.7}$$

and

$$X_L^\mu(\sigma^+) = \frac{x^\mu}{2} + \frac{\alpha'}{2} p^\mu \sigma^+ + i\sqrt{\frac{\alpha'}{2}} \sum_{n \neq 0} \frac{\tilde{\alpha}_n^\mu}{n} e^{-in\sigma^+} , \tag{9.8}$$

where $\tilde{\alpha}_n^\mu$ denotes the Fourier components of the leftmoving modes. One can also give a formulation with respect to z and $\bar{z}$, but this will be omitted here. It is convenient to define

$$\tilde{\alpha}_0^\mu = \alpha_0^\mu = \sqrt{\frac{\alpha'}{2}} p^\mu$$

for the closed string, and

$$\alpha_0^\mu = \sqrt{2\alpha'} p^\mu$$

for the open string.

In Section 3.2, we have introduced the string Hamiltonian; see (3.59). Inserting into this expression, the classical solution for X^μ one obtains

$$H = \frac{1}{2} \sum_{-\infty}^{\infty} \alpha_{-n} \alpha_n \tag{9.9}$$

for the open string, and

$$H = \frac{1}{2} \sum_{-\infty}^{\infty} (\alpha_{-n}\alpha_n + \tilde{\alpha}_{-n}\tilde{\alpha}_n) \tag{9.10}$$

for the closed string, and $\alpha_{-n}\alpha_n$ is a shorthand for $\eta_{\mu\nu}\alpha_{-n}^{\mu}\alpha_n^{\nu}$, etc. In (3.60), we have introduced the quantities L_m for the open string; one has in particular $L_0 = H$. One can analogously define for the closed string

$$L_m = \frac{1}{\pi\alpha'} \int_0^{2\pi} \mathrm{d}\sigma \, \mathrm{e}^{-im\sigma} T_{--} = \frac{1}{2} \sum_{-\infty}^{\infty} \alpha_{m-n}\alpha_n \ , \tag{9.11}$$

$$\tilde{L}_m = \frac{1}{\pi\alpha'} \int_0^{2\pi} \mathrm{d}\sigma \, \mathrm{e}^{im\sigma} T_{++} = \frac{1}{2} \sum_{-\infty}^{\infty} \tilde{\alpha}_{m-n}\tilde{\alpha}_n \ , \tag{9.12}$$

from which one obtains $H = L_0 + \tilde{L}_0$. As we have seen in Section 3.2, the L_m (and $\tilde{L}_m$) vanish as constraints.

Recalling that $\alpha_0^{\mu} = \sqrt{2\alpha'}p^{\mu}$ for the open string and, therefore,

$$\alpha_0^2 = 2\alpha' p^{\mu} p_{\mu} \equiv -2\alpha' M^2 \ ,$$

one obtains from $L_0 = H$ for the mass M of the open string in dependence of the oscillatory string modes, the expression

$$M^2 = \frac{1}{\alpha'} \sum_{n=1}^{\infty} \alpha_{-n}\alpha_n \ , \tag{9.13}$$

and from $H = L_0 + \tilde{L}_0$ for the mass of the closed string,

$$M^2 = \frac{2}{\alpha'} \sum_{n=1}^{\infty} (\alpha_{-n}\alpha_n + \tilde{\alpha}_{-n}\tilde{\alpha}_n) \ . \tag{9.14}$$

The variables x^{μ}, p^{μ}, α_n^{μ}, and $\tilde{\alpha}_n^{\mu}$ obey Poisson-bracket relations which follow from the fundamental Poisson brackets between the X^{μ} and their canonical momenta P^{μ} (Section 3.2). Upon quantization, one obtains

$$[x^{\mu}, p^{\nu}] = i\eta^{\mu\nu} \ , \tag{9.15}$$

$$[\alpha_m^{\mu}, \alpha_n^{\nu}] = m\delta_{m,-n}\eta^{\mu\nu} \ , \tag{9.16}$$

$$[\tilde{\alpha}_m^{\mu}, \tilde{\alpha}_n^{\nu}] = m\delta_{m,-n}\eta^{\mu\nu} \ , \tag{9.17}$$

$$[\alpha_m^{\mu}, \tilde{\alpha}_n^{\nu}] = 0 \ . \tag{9.18}$$

The Minkowski metric $\eta^{\mu\nu}$ appears because of Lorentz invariance. It can cause negative probabilities which must be carefully avoided in the quantum theory.

The task is then to construct a Fock space out of the vacuum state $|0, p^{\mu}\rangle$, which is the ground state of a single string with momentum p^{μ}, *not* the no-string state. The above algebra of the oscillatory modes can be written after rescaling as the usual oscillator algebra of annihilation and creation operators, a_m^μ and $a_m^{\mu\dagger}$,

$$\alpha_m^\mu = \sqrt{m}\, a_m^\mu\ , \quad \alpha_{-m}^\mu = \sqrt{m}\, a_m^{\mu\dagger}\ , m > 0\ .$$

One, therefore, has

$$\alpha_m^\mu |0, p^\mu\rangle = \tilde{\alpha}_m^\mu |0, p^\mu\rangle = 0\ , \quad m > 0\ , \tag{9.19}$$

and

$$\alpha_0^\mu |0, p^\mu\rangle = \tilde{\alpha}_0^\mu |0, p^\mu\rangle = p^\mu |0, p^\mu\rangle\ . \tag{9.20}$$

The rest of the spectrum is generated by the creation operators α_m^μ and $\tilde{\alpha}_m^\mu$ for $m < 0$. In order to implement the conformal generators L_n in the quantum theory, one must address the issue of operator ordering. It was already mentioned in Section 3.2 that this leads to the presence of a central term in the quantum algebra; see (3.62). Consequently, one cannot impose equations of the form $L_n |\psi\rangle = 0$ for all n. This is different from the spirit of the Wheeler–DeWitt equation where all constraints are implemented in this form. Instead, one can achieve this here only for $n > 0$. The demand for the absence of a Weyl anomaly on the worldsheet (see the next section) fixes the number D of the embedding space–time to $D = 26$. This is also called the 'critical dimension'. For the mass spectrum, one then gets in the critical dimension, the following expressions, which differ from their classical counterparts (9.13) and (9.14) by constants (see e.g. Polchinski 1998a): for the open string one has

$$M^2 = \frac{1}{\alpha'}\left(\sum_{n=1}^\infty \alpha_{-n}\alpha_n - 1\right) \equiv \frac{1}{\alpha'}\,(N - 1)\ , \tag{9.21}$$

where N denotes the level of excitation, while for the closed string one has

$$M^2 = \frac{4}{\alpha'}\left(\sum_{n=1}^\infty \alpha_{-n}\alpha_n - 1\right) = \frac{4}{\alpha'}\left(\sum_{n=1}^\infty \tilde{\alpha}_{-n}\tilde{\alpha}_n - 1\right)\ . \tag{9.22}$$

The ground state ($N = 0$) for the open string thus has

$$M^2 = -\frac{1}{\alpha'} < 0\ . \tag{9.23}$$

(In D dimensions, one would have $M^2 = (2 - D)/24\alpha'$.) The corresponding particle describes a *tachyon*—a particle with negative mass-squared—which signals the presence of an unstable vacuum. There may be a different vacuum which is stable, and there is indeed some evidence that this is the case for the bosonic string; cf. Berkovits *et al.* (2000). What is clear is that the presence of SUSY eliminates the tachyon. This is one of the main motivations to introduce the *superstring* (Section 9.2.4). For the first excited state, one finds

$$|e, p\rangle = e_\mu \alpha_{-1}^\mu |0, p\rangle \tag{9.24}$$

with a polarization vector e_μ that turns out to be transversal to the string propagation, $e_\mu p^\mu = 0$, and therefore corresponds in the critical dimension to

$D - 2 = 24$ degrees of freedom. Since for $n = 1$ one has $M^2 = 0$, this state describes a massless vector boson (a 'photon'). In fact, it turns out that had we chosen $D \neq 26$, we would have encountered a breakdown of Lorentz invariance. Excited states for $n > 1$ correspond to massive particles. They are usually neglected because their masses are assumed to be of the order of the Planck mass—this is the mass scale of unification where string theory is of relevance (since we expect *a priori* that $l_s \sim l_P$).

We emphasize that here we are dealing with higher dimensional representations of the Poincaré group, which do not necessarily have analogues in $D = 4$. Therefore, the usual terminology of speaking about photons, etc., should not be taken literally.

For the closed string one has the additional restriction $L_0 = \tilde{L}_0$, leading to

$$\sum_{n=1}^{\infty} \alpha_{-n}\alpha_n = \sum_{n=1}^{\infty} \tilde{\alpha}_{-n}\tilde{\alpha}_n .$$

The ground state is again a tachyon, with mass squared $M^2 = -4/\alpha'$. The first excited state is massless, $M^2 = 0$, and described by

$$|e, p\rangle = e_{\mu\nu}\alpha^{\mu}_{-1}\tilde{\alpha}^{\nu}_{-1}|0, p\rangle , \qquad (9.25)$$

where $e_{\mu\nu}$ is a transversal polarization tensor, $p^{\mu}e_{\mu\nu} = 0$. The state (9.25) can be decomposed into its irreducible parts. One thereby obtains a symmetric traceless tensor, a scalar, and an antisymmetric tensor. The symmetric tensor describes a spin-2 particle in $D = 4$ and can therefore—in view of the uniqueness features discussed in Chapter 2—be identified with the *graviton*. It is at this stage that string theory makes its first contact with quantum gravity. The perturbation theory discussed in Chapter 2 will thus be implemented in string theory. But as we shall see in the next section, string theory can go beyond it.

The scalar is usually referred to as the *dilaton*, Φ. In $D = 4$, the antisymmetric tensor has also spin zero and is in this case called the *axion*. The fact that massless fields appear in the open- and the closed-string spectrum is very interesting. Both the massless vector boson as well as the graviton couple to conserved currents and thereby introduce the principle of gauge invariance into string theory. Higher excited states lead also for closed strings to massive ('heavy') particles.

Up to now we have discussed oriented strings, that is, strings whose quantum states have no invariance under $\sigma \rightarrow -\sigma$. We note that one can also have non-oriented strings by demanding this invariance to hold. For closed strings this invariance would correspond to an exchange between right- and leftmoving modes. It turns out that the graviton and the dilaton are also present for non-oriented strings, but not the axion.

9.2 Quantum gravitational aspects

9.2.1 *The Polyakov path integral*

We have seen in the last section that the graviton appears in a natural way in the spectrum of closed strings. Linearized quantum gravity is, therefore, automatically contained in string theory. Here we discuss other aspects which are relevant in the context of quantum gravity.

One can generalize the Polyakov action (3.51) to the situation of a string moving in a general D-dimensional curved space–time. It makes sense to take into account besides gravity other massless fields that arise in string excitations— the dilaton and the 'axion'. One, therefore, formulates the generalized Polyakov action as

$$S_{\mathrm{P}} \equiv S_\sigma + S_\phi + S_B$$
$$= -\frac{1}{4\pi\alpha'} \int \mathrm{d}^2\sigma \ \left(\sqrt{h}h^{\alpha\beta}\partial_\alpha X^\mu \partial_\beta X^\nu g_{\mu\nu}(X)\right.$$
$$\left. -\alpha'\sqrt{h}\, {}^{(2)}R\Phi(X) + \epsilon^{\alpha\beta}\partial_\alpha X^\mu \partial_\beta X^\nu B_{\mu\nu}(X)\right) . \qquad (9.26)$$

The fields $g_{\mu\nu}$ (D-dimensional metric of the embedding space), Φ (dilaton), and $B_{\mu\nu}$ (antisymmetric tensor field) are *background fields*, that is, they will not be integrated over in the path integral. The fields X^μ define again the embedding of the worldsheet into the D-dimensional space which is also called 'target space'. An action in which the coefficients of the kinetic term depend on the fields themselves (here, $g_{\mu\nu}$ depends on X) is for historic reasons called a *non-linear sigma model*. This is why the first part on the right-hand side of (9.26) is abbreviated as S_σ. The second part S_ϕ is, in fact, independent of the string parameter α', since in natural units (where $\hbar = 1$), the dilaton is dimensionless. We emphasize that (9.26) describes a quantum field theory on the worldsheet, not the target space. For the latter, one uses an effective action (see below).

In string theory, it has been proven fruitful to employ a path-integral approach (Section 2.2). In the Euclidean formulation, where $\sigma^1 = \sigma$ and $\sigma^2 = i\tau$, the starting point would be

$$Z = \int \mathcal{D}X\mathcal{D}h \ \mathrm{e}^{-S_{\mathrm{P}}} , \qquad (9.27)$$

where X and h are a shorthand for the embedding variables and the worldsheet metric, respectively. Only these variables are to be integrated over. In order to get a sensible expression, one must employ the gauge-fixing procedure outlined in Section 2.2.3. The invariances on the worldsheet involve two local diffeomorphisms and one Weyl transformation. Since $h_{ab}(\sigma^1, \sigma^2)$ has three independent parameters, one can *fix* it to a given 'fiducial' form $\tilde{h}_{ab}$, for example, $\tilde{h}_{ab} = \delta_{ab}$ ('flat gauge') or $\tilde{h}_{ab} = \exp[2\omega(\sigma^1, \sigma^2)]\delta_{ab}$ ('conformal gauge'). As discussed in Section 2.2.3, the Faddeev–Popov determinant can be written as a path integral over (anticommuting) ghost fields. The action in (9.27) has then to be replaced by the full action $S_{\mathrm{P}} + S_{\mathrm{ghost}} + S_{\mathrm{gf}}$, that is, augmented by ghost and gauge-fixing action.

The full action is invariant under BRST transformations, which were already briefly mentioned in Section 2.2.3. This is an important concept, since it encodes the information about gauge invariance at the gauge-fixed level. For this reason, we shall give a brief introduction here (see e.g. Weinberg 1996 for more details). BRST transformations mix commuting and anticommuting fields (ghosts) and are generated by the 'BRST charge' Q_B. Be ϕ_a a general set of first-class constraints, see Section 3.1.2,

$$\{\phi_a, \phi_b\} = f_{ab}^c \phi_c . \tag{9.28}$$

The BRST charge then reads

$$Q_B = \eta^a \phi_a - 1/2\, P_c f_{ab}^c \eta^b \eta^a , \tag{9.29}$$

where η^a denotes the Faddeev–Popov ghosts and P_a their canonically conjugate momenta ('anti-ghosts') obeying $[\eta^a, P_b]_+ = \delta_b^a$. We have assumed here that the physical fields are bosonic; for a fermion there would be a plus sign in (9.29). One can show that Q_B is nilpotent,

$$Q_B^2 = 0 . \tag{9.30}$$

This follows from (9.28) and the Jacobi identities for the structure constants.

BRST invariance of the path integral leads in the quantum theory to the demand that physical states should be BRST-invariant, that is,

$$\hat{Q}_B |\Psi\rangle = 0 . \tag{9.31}$$

This condition is less stringent than the Dirac condition which states that physical states be annihilated by all constraints. Equation (9.31) can be fulfilled for the quantized bosonic string, which is not the case for the Dirac conditions. The quantum version of (9.30) reads

$$\left[\hat{Q}_B, \hat{Q}_B\right]_+ = 0 . \tag{9.32}$$

For this to be fulfilled, the total central charge of the X^μ-fields and the Faddeev–Popov ghosts must vanish,

$$c_{\text{tot}} = c + c_{\text{ghost}} = D - 26 = 0 , \tag{9.33}$$

since it turns out that the ghosts have central charge -26. The string must therefore move in 26 dimensions. In the case of the superstring (see Section 9.2.4), the corresponding condition leads to $D = 10$. The condition (9.32) thus carries information about quantum anomalies (here, the Weyl anomaly) and their possible cancellation by ghosts.[1] One can prove the 'no-ghost theorem' (see e.g. Polchinski 1998a): the Hilbert space arising from BRST quantization has a positive inner product and is isomorphic to the Hilbert space of transverse string excitations.

---

[1] One can also discuss non-critical strings living in $D \neq 26$ dimensions. They have a Weyl anomaly, which means that different gauge choices are inequivalent.

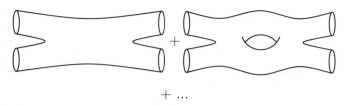

$+ \ldots$

FIG. 9.1. The first two contributions to the scattering of two closed strings.

How can the path integral (9.27) be evaluated? The sum over all 'paths' contains, in particular, a sum over all worldsheets, that is, a sum over all Riemann surfaces. In this sum, all topologies have to be taken into account. Figure 9.1 shows as an example the first two topologies which arise in the scattering of two closed strings. It is in this way that string *interactions* arise—as amplitudes in the path integral. Unlike the situation in four dimensions, the classification of these surfaces in two dimensions is well known. Consider as an example the dilaton part of the action (9.26),

$$S_\phi = \frac{1}{4\pi} \int \mathrm{d}^2\sigma \sqrt{h}\,^{(2)}R\,\Phi(X) \ . \tag{9.34}$$

If ϕ were constant, $\phi(X) = \lambda$, this would yield

$$S_\phi = \chi\lambda = \lambda(2 - 2g) \ , \tag{9.35}$$

where χ is the Euler number and g the genus of the surface. (We assume for simplicity here that only handles are present and no holes or cross-caps). This then gives the contribution

$$\mathrm{e}^{-2\lambda(1-g)} \equiv \alpha^{g-1}$$

to the path integral, and we have introduced

$$\alpha = \mathrm{e}^{2\lambda} \equiv g_\mathrm{c}^2 \ , \tag{9.36}$$

which plays the role of the 'fine-structure constant' for the loop expansion; g_c denotes the string-coupling constant for closed strings. Adding a handle corresponds to emission and re-absorption of a closed string.[2] The parameter g (meaning g_c or g_o, depending on the situation) is the expansion parameter for string loops. It must be emphasized that one has only *one* diagram at each order of the perturbation theory, in contrast to Feynman diagrams in quantum field theory. The reason is that point-like interactions are avoided. Such a 'smearing' can be done consistently in string theory, but no other such theory is known. In this way the usual divergences of quantum field theory seem to be avoided,

[2]For the open string one finds $g_\mathrm{o}^2 \propto \mathrm{e}^\lambda$.

although no proof is yet known demonstrating finiteness at all loop orders. The sum as a whole does not converge and is not even Borel summable (i.e. the terms in the sum increase with $n!$, where n is the number of handles), see Gross and Periwal (1988). One therefore expects that this is an asymptotic series like in QED and thus must be an approximation to some non-perturbative theory.

In discussing scattering amplitudes, one has also to specify the ingoing and outgoing string states, which have to be given at infinity in the spirit of an 'S-matrix'. This is done with the help of 'vertex operators': in the example of Fig. 9.1, such an operator would correspond to four point-like insertions in the worldsheet. Vertex operators do not describe interactions, but instead the creation or annihilation of a string state at a position on the worldsheet. Vertex operators have to be included into the expression for the path integral.

It must also be emphasized that the gauge choice fixes the worldsheet metric only locally (e.g. to $\tilde{h}_{ab} = \delta_{ab}$). There may be, however, additional global degrees of freedom described by a finite number of parameters. These parameters are called *moduli*. In the case of the torus ($g = 1$), for example, this is the Teichmüller parameter or modulus $\tau \in \mathbb{C}$. These parameters have to be summed over in the path integral.

For a string propagating in flat space–time, the demand for the absence of the Weyl anomaly leads to the restriction $D = 26$. What about the string in a curved space–time as described by (9.26)? The requirement that no Weyl anomaly be present on the worldsheet leads at the tree level to the following set of consistency equations, which follow from the vanishing of renormalization-group beta functions,

$$0 = R_{\mu\nu} - \frac{1}{4}H_{\mu}{}^{\lambda\rho}H_{\nu\lambda\rho} + 2\nabla_{\mu}\nabla_{\nu}\Phi + \mathcal{O}(\alpha') \ , \tag{9.37}$$

$$0 = \nabla_{\lambda}H^{\lambda}{}_{\mu\nu} - 2\nabla^{\lambda}\Phi H_{\lambda\mu\nu} + \mathcal{O}(\alpha') \ , \tag{9.38}$$

$$0 = \frac{D - 26}{6\alpha'} + \nabla_{\mu}\nabla^{\mu}\Phi - \frac{1}{2}\nabla_{\mu}\Phi\nabla^{\mu}\Phi - \frac{1}{24}H_{\mu\nu\rho}H^{\mu\nu\rho} + \mathcal{O}(\alpha') \ . \tag{9.39}$$

Here we have introduced the field strength $H_{\mu\nu\rho}$ connected with the antisymmetric tensor field,

$$H_{\mu\nu\rho} = 3! \ \partial_{[\mu}B_{\nu\rho]} \ .$$

The above set of equations builds the bridge to the concept of effective action in string theory. As in Section 2.2.3, effective actions are useful to connect the full theory with phenomenology.

9.2.2 *Effective actions*

The consistency equations (9.37)–(9.39) follow as field equations from the following *effective action* in D space–time dimensions,

$$S_{\text{eff}} = \frac{1}{2\kappa_0^2}\int \mathrm{d}^Dx \ \sqrt{-g} \ \mathrm{e}^{-2\Phi}\left(R - \frac{2(D - 26)}{3\alpha'} - \frac{1}{12}H_{\mu\nu\rho}H^{\mu\nu\rho}\right)$$

$$+4\nabla_\mu\Phi\nabla^\mu\Phi + \mathcal{O}(\alpha')\Big) \ . \tag{9.40}$$

Higher orders exhibit powers of α' and are thus genuine string corrections. They contain, for example, higher curvature terms, which means that one has to replace in (9.37)

$$R_{\mu\nu} \to R_{\mu\nu} + \frac{\alpha'}{2}R_{\mu\kappa\lambda\tau}R_\nu{}^{\kappa\lambda\tau} + \dots \ .$$

The approximation of a classical space–time metric is well defined only if the curvature scale r_c associated with it obeys

$$r_c \gg l_s \ .$$

It has been speculated that a 'non-metric phase' will appear if this condition is violated (see e.g. Horowitz (1990), Greene (1997), and the references therein). The expansion of the effective action into powers of α' is thus a low-energy expansion.

One can make a comparison of the effective action with a 'Jordan–Brans–Dicke' type of action, which contains an additional scalar field in the gravitational sector. There one would have a kinetic term of the form $-4\omega\nabla_\mu\Phi\nabla^\mu\Phi$, where ω denotes the Brans–Dicke parameter (GR is recovered for $\omega \to \infty$). Comparison with (9.40) exhibits that string theory would correspond to $\omega = -1$. If the field Φ were really massless—as suggested by (9.40)—this would be in conflict with observations because the additional interaction of matter fields with Φ *in addition* to the metric would violate the equivalence principle (cf. Lämmerzahl 2003). The latter has been tested with great accuracy. One would, however, expect that the dilaton gets a mass term from the higher order terms in α' so that no conflict with observation would arise. Because of the natural occurrence of the dilaton from string theory, theories with a scalar field in addition to the metric ('scalar-tensor theories') are widely studied, cf. Fuji and Maeda (2003).

We emphasize that (9.37) are the Einstein equations describing the coupling of the dilaton and the axion to the metric. It is interesting that these space–time equations follow from the Weyl invariance on the worldsheet. This gives one of the most important connections between the string and gravity.

One can also perform a Weyl transformation in space–time in order to put S_{eff} into a form in which the first term is just the space–time Ricci scalar without the dilaton. This is sometimes called the 'Einstein frame' in contrast to the 'string' or 'Jordan frame' of (9.40). While this new form is convenient for various situations, the physical form is given by (9.40), as can be seen from the behaviour of test particles—the physical metric is $g_{\mu\nu}$, not the Weyl-transformed metric. The Jordan frame is also distinguished by the fact that the standard fields are coupled minimally to the metric.

We have already emphasized above that space–time metric, dilaton, and axion play only the role of background fields. The simplest solution for them is

$$g_{\mu\nu} = \eta_{\mu\nu} \ , \quad B_{\mu\nu} = 0 \ , \quad \Phi = \text{const.} = \lambda \ .$$

It is usually claimed that quite generally the stationary points of S_{eff} correspond to possible ground states ('vacua') of the theory. String theory may, in fact, predict a huge number of such vacua, cf. Douglas (2003). It is clear from (9.39) that $D = 26$ is a necessary condition for the solution with constant background fields. Thus, we have recovered the old consistency condition for the string in flat space–time. There are now, however, solutions of (9.39) with $D \neq 26$ and $\Phi \neq$ constant, which would correspond to a solution with a large cosmological constant $\propto (D - 26)/6\alpha'$, in conflict with observation.

The parameter κ_0 in (9.40) does not have a physical significance by itself since it can be changed by a shift in the dilaton. The physical gravitational constant (in D dimensions) reads

$$8\pi G_D = 2\kappa_0^2 e^{2\lambda} . \tag{9.41}$$

Apart from α'-corrections, one can also consider loop corrections to (9.40). Since g_c is determined by the value of the dilaton, see (9.36), the tree-level action (9.40) is of order g_c^{-2}. The one-loop approximation is obtained at order g_c^0, the two-loop approximation at order g_c^2, and so on.[3]

In Section 9.1, we have seen that the graviton appears as an excitation mode for closed strings. What is the connection to the appearance of gravity in the effective action (9.40)? Such a connection is established through the ansatz

$$g_{\mu\nu} = \bar{g}_{\mu\nu} + \sqrt{32\pi G} f_{\mu\nu} ,$$

cf. (2.76), and making a perturbation expansion in the effective action with respect to $f_{\mu\nu}$. It then turns out that the term of order $f_{\mu\nu}$ just yields the vertex operator for the string graviton state (see e.g. Mohaupt 2003). Moreover, it is claimed that exponentiating this graviton vertex operator leads to a 'coherent state' of gravitons. The connection between the graviton as a string mode and gravity in the effective action thus proceeds via a comparison of scattering amplitudes. For example, the amplitude for graviton–graviton scattering from the scattering of strings at tree level coincides with the field-theoretic amplitude of the corresponding process at tree level as being derived from S_{eff}. The reason for this coincidence is the vanishing of the Weyl anomaly for the worldsheet. The coincidence continues to hold at higher loop order and at higher orders in $\alpha' \sim l_s^2$. Since the string amplitude contains the parameter α' and the effective action contains the gravitational constant G_D, the comparison of the amplitudes yields a connection between both; see for example, Veneziano (1993),

$$G_D \sim g^2 l_s^{D-2} , \tag{9.42}$$

where g is again the string-coupling constant (here we do not distinguish between open and closed strings and write for simplicity just g for the string coupling). An analogous relation holds between gauge couplings for grand unified theories and the string length.

---

[3]For open strings, also odd orders of the coupling (g_o) appear.

Since we do not live in 26 dimensions, a connection must be made to the four-dimensional world. This is usually done through compactification of the additional dimensions. In this way, one obtains a relation between the four-dimensional gravitational constant and the string length,

$$G \sim g^2 l_{\mathrm{s}}^2 \; , \tag{9.43}$$

in which the details of the compactification enter (geometric factors). Ideally, one would like to recover in this way other parameters such as particle masses or the number of families from the details of compactification. One is, however, still far away from this goal.

The finiteness of the string length l_{s} leads to an automatic cutoff at high momenta. It thus seems impossible to resolve arbitrarily small distances in an operational sense. In fact, one can derive from gedanken experiments of scattering situations, a generalized uncertainty relation of the form

$$\Delta x > \frac{\hbar}{\Delta p} + \frac{l_{\mathrm{s}}^2}{\hbar} \Delta p \; , \tag{9.44}$$

cf. Veneziano (1993) and the references therein. This seems to match the idea of a minimal length which we have also encountered in the canonical approach (Section 6.2), although for example, D-branes (Section 9.2.3) can probe smaller scales.

How many fundamental constants appear in string theory? This has been a matter of some debate; see Duff *et al.* (2002). We adopt here the standpoint already taken in Chapter 1 that three dimensionful constants are needed, which can be taken to be c, $\hbar$, and l_{s}.

9.2.3   *T-duality and branes*

In this subsection, we shall introduce the concept of T-duality, from which one is led in a natural way to the concept of D-branes. The 'T' arises from the fact that one assumes here that the higher dimensions are compactified on tori.

The classical solutions for the closed string are given in Section 9.1. In order to introduce the concept of duality, we take the left- and rightmoving modes of the closed string as independent, that is, we write

$$X_{\mathrm{R}}^{\mu}(\sigma^-) = \frac{x^{\mu}}{2} + \sqrt{\frac{\alpha'}{2}} \alpha_0^{\mu}(\tau - \sigma) + \cdots \; , \tag{9.45}$$

$$X_{\mathrm{L}}^{\mu}(\sigma^-) = \frac{\tilde{x}^{\mu}}{2} + \sqrt{\frac{\alpha'}{2}} \tilde{\alpha}_0^{\mu}(\tau + \sigma) + \cdots \; , \tag{9.46}$$

where $\cdots$ stands for 'oscillators' (this part does not play a role in the following discussion). The sum of both thus yields

$$X^{\mu} = \frac{1}{2}(x^{\mu} + \tilde{x}^{\mu}) + \sqrt{\frac{\alpha'}{2}}(\alpha_0^{\mu} + \tilde{\alpha}_0^{\mu})\tau + \sqrt{\frac{\alpha'}{2}}(\tilde{\alpha}_0^{\mu} - \alpha_0^{\mu})\sigma + \cdots \; . \tag{9.47}$$

The oscillators are invariant under $\sigma \to \sigma + 2\pi$, but the X^μ transform as

$$X^\mu \to X^\mu + 2\pi \sqrt{\frac{\alpha'}{2}} (\tilde{\alpha}_0^\mu - \alpha_0^\mu) \ . \tag{9.48}$$

We now distinguish between a non-compact direction of space and a compact direction; see in particular Polchinski (1998a) for the following discussion. In the non-compact directions, the X^μ must be unique. One then obtains for them from (9.48)

$$\tilde{\alpha}_0^\mu = \alpha_0^\mu = \sqrt{\frac{\alpha'}{2}} p^\mu \ . \tag{9.49}$$

This is the situation encountered before and one is back at Eqns (9.7) and (9.8). For a compact direction the situation is different. Assume that there is one compact direction with radius R in the direction $\mu = 25$. The coordinate $X^{25} \equiv X$ thus has period $2\pi R$. Under $\sigma \to \sigma + 2\pi$, X can now change by $2\pi w R$, $w \in \mathbb{Z}$, where w is called the 'winding number'. These modes are called 'winding modes' because they can wind around the compact dimension. Since $\exp(2\pi i R p^{25})$ generates a translation around the compact dimension which must lead to the same state, the momentum $p^{25} \equiv p$ must be discretized,

$$p = \frac{n}{R} \ , \quad n \in \mathbb{Z} \ . \tag{9.50}$$

From

$$p = \frac{1}{\sqrt{2\alpha'}} (\tilde{\alpha}_0 + \alpha_0)$$

(α_0 is a shorthand for α_0^{25}, etc.) one gets for these 'momentum modes', the relation

$$\tilde{\alpha}_0 + \alpha_0 = \frac{2n}{R} \sqrt{\frac{\alpha'}{2}} \ . \tag{9.51}$$

For the winding modes, one has from (9.48)

$$2\pi \sqrt{\frac{\alpha'}{2}} (\tilde{\alpha}_0 - \alpha_0) = 2\pi w R$$

and therefore

$$\tilde{\alpha}_0 - \alpha_0 = w R \sqrt{\frac{2}{\alpha'}} \ . \tag{9.52}$$

This then yields

$$\alpha_0 = \left(\frac{n}{R} - \frac{wR}{\alpha'} \right) \sqrt{\frac{\alpha'}{2}} \equiv p_R \sqrt{\frac{\alpha'}{2}} \ , \tag{9.53}$$

$$\tilde{\alpha}_0 = \left(\frac{n}{R} + \frac{wR}{\alpha'}\right)\sqrt{\frac{\alpha'}{2}} \equiv p_{\mathrm{L}}\sqrt{\frac{\alpha'}{2}} \ . \tag{9.54}$$

For the mass spectrum, one obtains

$$M^2 = -\sum_{\mu \neq 25} p^\mu p_\mu + p^2 = \frac{4}{\alpha'}\left(\sum_{n=1}^\infty \alpha_{-n}\alpha_n - 1\right) + \frac{2(\alpha_0)^2}{\alpha'}$$

$$= \frac{4}{\alpha'}\left(\sum_{n=1}^\infty \tilde{\alpha}_{-n}\tilde{\alpha}_n - 1\right) + \frac{2(\alpha_0)^2}{\alpha'} \ . \tag{9.55}$$

The expressions in parentheses correspond to the excitation level of the right- and leftmoving modes, respectively, in the non-compact dimensions (the sum there runs over $\mu \neq 25$).

Of particular interest are the limiting cases $R \to \infty$ and $R \to 0$. For $R \to \infty$ one gets

$$\alpha_0 \to -\frac{wR}{\sqrt{2\alpha'}}\ , \quad \tilde{\alpha}_0 \to \frac{wR}{\sqrt{2\alpha'}}\ . \tag{9.56}$$

Since $M^2 \to \infty$ for $w \neq 0$, all states become infinitely massive in this case. For $w = 0$, on the other hand, one gets a *continuum* of states for all n. For $R \to 0$, one gets

$$\alpha_0 \to \frac{n}{R}\sqrt{\frac{\alpha'}{2}}\ , \quad \tilde{\alpha}_0 \to \frac{n}{R}\sqrt{\frac{\alpha'}{2}}\ . \tag{9.57}$$

For $n \neq 0$, all states get infinitely massive, while for $n = 0$, one has a continuum for all w. These are the winding states that can wind around the extra dimension without cost of energy (note that $H_0 = (\alpha_0^2 + \tilde{\alpha}_0^2) \to 0$ for $n = 0$). This is a typical feature of string theory related to the presence of one-dimensional objects and has no counterpart in field theory. We recognize from (9.53) and (9.54) that there is a symmetry between R and the 'dual radius' R_{D},

$$R \leftrightarrow R_{\mathrm{D}} \equiv \frac{\alpha'}{R} = \frac{l_{\mathrm{s}}^2}{2R}\ , \tag{9.58}$$

which corresponds to

$$n \leftrightarrow w\ , \quad \alpha_0 \leftrightarrow -\alpha_0\ , \quad \tilde{\alpha}_0 \leftrightarrow \tilde{\alpha}_0\ . \tag{9.59}$$

The mass spectrum in both cases is identical. The above duality between R and R_{D} is called 'T-duality'. It is an exact symmetry of perturbation theory (and beyond) for closed strings. The critical value is, of course, obtained for $R \sim R_{\mathrm{D}} \sim l_{\mathrm{s}}$, which is just the length of the string, as expected.

It is useful for the discussion below to go again to Euclidean space,

$$z = \mathrm{e}^{\sigma^2 - \mathrm{i}\sigma^1}\ , \quad \sigma^1 = \sigma\ , \quad \sigma^2 = \mathrm{i}\tau\ .$$

Equation (9.47) then reads

$$X^\mu(z, \bar{z}) \ = \ X_{\mathrm{R}}^\mu(z) + X_{\mathrm{L}}^\mu(\bar{z})$$

$$= \frac{x^\mu + \tilde{x}^\mu}{2} - \mathrm{i}\sqrt{\frac{\alpha'}{2}}(\tilde{\alpha}_0^\mu + \alpha_0^\mu)\sigma^2 + \sqrt{\frac{\alpha'}{2}}(\tilde{\alpha}_0^\mu - \alpha_0^\mu)\sigma^1 + \dots$$

$$\xrightarrow{\mathrm{T}} -X_\mathrm{R}^\mu(z) + X_\mathrm{L}^\mu(\bar{z}) . \tag{9.60}$$

What happens for the open string? In the limit $R \to 0$, it has no possibility to wind around the compactified dimension and therefore seems to live only in $D - 1$ dimensions. There are, however, closed strings present in the theory of open strings, and the open string can in particular have vibrations into the 25th dimension. Only the *endpoints* of the open string are constrained to lie on a $(D-1)$-dimensional hypersurface. For the vibrational part in the 25th dimension, one can use the expression (9.60) for the closed string. Choosing $X^{25} \equiv X$ in this equation yields

$$X = \frac{x + \tilde{x}}{2} + \sqrt{\frac{\alpha'}{2}}(\tilde{\alpha}_0 - \alpha_0)\sigma^1 - \mathrm{i}\sqrt{\frac{\alpha'}{2}}(\tilde{\alpha}_0 + \alpha_0)\sigma^2 + \dots , \tag{9.61}$$

so that one gets for its dual

$$X_\mathrm{D} = \frac{x + \tilde{x}}{2} + \sqrt{\frac{\alpha'}{2}}(\tilde{\alpha}_0 + \alpha_0)\sigma^1 - \mathrm{i}\sqrt{\frac{\alpha'}{2}}(\tilde{\alpha}_0 - \alpha_0)\sigma^2 + \dots , \tag{9.62}$$

that is, $X \leftrightarrow X_\mathrm{D}$ corresponds to $-\mathrm{i}\sigma^2 \leftrightarrow \sigma^1$, leading in particular to

$$X_\mathrm{D}' \equiv \frac{\partial X_\mathrm{D}}{\partial \sigma^1} \leftrightarrow \mathrm{i}\frac{\partial X}{\partial \sigma^2} \equiv \mathrm{i}\dot{X} .$$

Integration yields

$$X_\mathrm{D}(\pi) - X_\mathrm{D}(0) = \int_0^\pi \mathrm{d}\sigma^1 \, X_\mathrm{D}' = \mathrm{i}\int_0^\pi \mathrm{d}\sigma^1 \, \dot{X} .$$

Inserting $\dot{X}$ following from (9.3) yields

$$X_\mathrm{D}(\pi) - X_\mathrm{D}(0) = 2\pi\alpha' p = \frac{2\pi\alpha' n}{R} = 2\pi n R_\mathrm{D} , \; n \in \mathbb{Z} ,$$

where only vibrational modes were considered. The dual coordinates thus obey a Dirichlet-type condition (an exact Dirichlet condition would arise for $n = 0$, cf. the remarks before (9.2)). Since, therefore, $X_\mathrm{D}(\pi)$ and $X_\mathrm{D}(0)$ differ by a multiple of the internal circumference $2\pi R_\mathrm{D}$ of the dual space, the endpoints must lie on a $(D - 1)$-dimensional hypersurface. Consideration of several open strings reveals that it is actually the same hypersurface. This hypersurface is called a D-*brane* where 'D' refers to the Dirichlet condition holding normal to the brane. Sometimes one refers to it more precisely as the 'Dp-brane', where p is the number of space dimensions.

A D-brane is a *dynamical* object since momentum can leak out of the string and is absorbed by the brane (this cannot happen with a Neumann boundary condition which still holds in the directions tangential to the plane). A D-brane is a soliton of string theory and can be described by an action that resembles an action proposed long ago by Born and Infeld to describe non-linear electrodynamics (it was then meant as a candidate for a modification of linear electrodynamics at short distances). A D-brane can carry generalized electric and magnetic charges.

The above discussion can be extended to the presence of gauge fields. The reason behind this is that open strings allow additional degrees of freedom called 'Chan–Paton factors'. These are 'charges' i and j ($i, j = 1, \ldots, n$) which reside at the endpoints of the string (historically one was thinking about quark–antiquark pairs). One can introduce a U(n) symmetry acting on (and only on) these charges. One can get from this the concept of U(n) gauge bosons living *on* the branes (one can have n branes at different positions).

Interestingly, n coinciding D-branes give rise to '$n \times n$'-matrices for the embedding variables X^μ and the gauge fields A_a. One thus arrives at space–times coordinates that do not commute, giving rise to the notion of non-commutative space–time. It has been argued that the D-brane action corresponds to a Yang–Mills action on a non-commutative worldvolume. Details are reviewed, for example, in Douglas and Nekrasov (2002).

The concept of D-branes is especially interesting concerning gravitational aspects. First, it plays a crucial role for the derivation of the black-hole entropy from counting microscopic degrees of freedom (Section 9.2.5). The second point has to do with the fact that these branes allow one to localize gauge and matter fields on the branes, whereas the gravitational field can propagate through the full space–time. This gives rise to a number of interesting features discussed in the context of 'brane worlds'; cf. Section 9.2.6.

9.2.4 *Superstrings*

So far, we have not yet included fermions, which are necessary for a realistic description of the world. Fermions are implemented by the introduction of SUSY, which we have already discussed in Sections 2.3 and 5.3.6. In contrast to the discussion there, we shall here introduce SUSY on the worldsheet, not on space–time. This will help us to get rid of problems of the bosonic string, such as the presence of tachyons. A string with SUSY is called 'superstring'. Worldsheet SUSY will be only indirectly related to space–time SUSY. We shall be brief in our discussion and refer the reader to, for example, Polchinski (1998b) for more details.

We start by introducing on the worldsheet the superpartners for the X^μ— they are called ψ^μ_A, where μ is a space–time vector index and A is a worldsheet spinor index ($A = 1, 2$). The ψ^μ_A are taken to be Majorana spinors and have, thus, two real components,

$$\psi^\mu = \begin{pmatrix} \psi_- \\ \psi_+ \end{pmatrix} . \qquad (9.63)$$

They are not to be confused with the gravitinos, which are trivial in two dimensions.

The 'superversion' of the (flat) Polyakov action (3.58) is the 'RNS action' (named after Ramond, Neveu, and Schwarz). It reads

$$S_{\text{RNS}} = -\frac{1}{4\pi\alpha'} \int_{\mathcal{M}} \mathrm{d}^2\sigma \; \left(\partial_\alpha X^\mu \partial^\alpha X_\mu + i\bar{\psi}^\mu \rho^\alpha \partial_\alpha \psi_\mu\right) \;, \qquad (9.64)$$

where the ρ^α denote two-dimensional Dirac matrices (suppressing worldsheet spinor indices). Thus, they obey

$$\left[\rho^\alpha, \rho^\beta\right]_+ = 2\eta^{\alpha\beta} \;. \qquad (9.65)$$

The RNS action is invariant under global SUSY transformations on the worldsheet. The classical equations of motion are (9.2) and

$$\rho^\alpha \partial_\alpha \psi^\mu = 0 \;. \qquad (9.66)$$

To make SUSY on the worldsheet explicit, one can formulate the theory in a two-dimensional 'superspace' described by worldsheet coordinates (σ, τ) and two additional Grassmann coordinates θ^A. This must not be confused with the superspace of canonical gravity discussed in Chapters 4 and 5! We shall not elaborate on this formalism here.

Compared to the bosonic string, new features arise in the formulation of the *boundary conditions*. Considering first the open string, the demand for a vanishing surface term in the variation of the action allows two possible boundary conditions for the fermions:

1. *Ramond* (R) boundary conditions: ψ is periodic (on a formally doubled worldsheet), and the sum in the mode expansion is over $n \in \mathbb{Z}$.
2. *Neveu–Schwarz* (NS) boundary conditions: ψ is antiperiodic, and the sum is over $n \in \mathbb{Z} + 1/2$ (this is possible because a relative sign is allowed for spinors).

It turns out that for consistency one must really have both types of boundary conditions in the theory, the R-sector and the NS-sector of Hilbert space.

For the closed string one can demand that ψ_+ and ψ_- are either periodic (R) or antiperiodic (NS). Since they are independent, one gets four types of boundary conditions: R–R, NS–R, R–NS, and NS–NS. Again, they must all be taken into account for consistency.

What about quantization? From the X^μ-part, one gets as before the commutation relations (9.16)–(9.18) for the α_n^μ and the $\tilde{\alpha}_n^\mu$. For the superstring, one has in addition anticommutators for the fermionic modes. They are of the type

$$\left[b_m^\mu, b_n^\nu\right]_+ = \eta^{\mu\nu} \delta_{m,-n} \;, \qquad (9.67)$$

etc. One finds a SUSY extension of the Virasoro algebra, which again has a central charge. The demand for the vanishing of the Weyl anomaly leads this time to $D = 10$ dimensions for the superstring.

Consider first the case of closed strings.[4] The NS–NS sector yields, like in the bosonic case, a tachyonic ground state. At $M = 0$, one finds again a graviton, a dilaton, and an antisymmetric tensor field ('axion' in four dimensions). The R–R sector yields antisymmetric tensor gauge fields, while the NS–R sector gives space–time fermions. The R–NS sector contains an exchange of left- and rightmoving fermions compared to the NS–R sector. Among these fermions are the massless gravitinos. In this sense, string theory contains space–time supergravity (SUGRA); see Section 2.3. An important notion (for both open and closed strings) is the 'GSO projection' (named after Gliozzi, Scherk, and Olive). It removes the tachyon and makes the spectrum supersymmetric. Moreover, it must necessarily be implemented in the quantum theory. In the R–R sector, the GSO projection applied on ground states can yield states of the opposite or of the same chirality. In the first case, one talks about type IIA superstring (which is non-chiral), in the latter case, about type IIB superstring (which is chiral). Types IIA and IIB are oriented closed superstrings with $N = 2$ SUSY. After the GSO projection, there is no longer a tachyon in the NS–NS sector, but one still has the graviton, the dilaton, and the axion as massless states (for both types IIA and IIB). In the NS–R and the R–NS sectors, one is left with two gravitinos and two dilatinos (the SUSY partners of the dilaton), which have opposite chiralities for type IIA and the same chirality for type IIB.

In the case of open strings, one gets in the NS sector a tachyonic ground state and a massless gauge boson. In the R-sector, all states are space–time spinors. Again, one gets rid of the tachyon by applying the GSO projection. This leads to the type I superstring—the only consistent theory with open (and closed) strings (the strings here are non-oriented). It must have the gauge group SO(32) and has $N = 1$ SUSY. In the closed-string sector of type I theory, one must project type IIB onto states which are invariant under worldsheet parity in order to get non-oriented strings. There remain the graviton, the dilaton, a two-form field, one gravitino, and one dilatino. From the open-string sector, one gets massless vector and spinor fields.

In addition to types I, IIA, and IIB, there exists a consistent hybrid construction for closed strings combining the bosonic string with type II superstrings. This is referred to as 'heterotic string'; the rightmoving part is taken from type II and the leftmoving part is from the bosonic string. It possesses $N = 1$ SUSY. There exist two different versions referring to gauge groups SO(32) and $E_8 \times E_8$, respectively. Anomaly-free chiral models for particle physics can, thus, be constructed from string theory for these gauge groups. This has raised the hope that the Standard Model of strong and electroweak interactions can be derived from string theory—a hope, however, which up to now has not been realized.

To summarize, one has found (the weak-field limit of) *five* consistent string theories in $D = 10$ dimensions. To find the theory in four dimensions, one has to invoke a compactification procedure. Since no principle has yet been found to

---

[4]We neglect all massive states in our discussion.

fix this, there are a plenty of consistent string theories in four dimensions and it is not clear which one to choose.

This is, however, not yet the end of the story. Type IIA theory also contains D0-branes ('particles'); cf. Witten (1995). If one has n such D0-branes, their mass M is given by

$$M = \frac{n}{g\sqrt{\alpha'}} \ . \qquad (9.68)$$

In the perturbative regime $g \ll 1$, this state is very heavy, while in the strong-coupling regime $g \to \infty$, it becomes lighter than any perturbative excitation. The mass spectrum (9.68) resembles a Kaluza–Klein spectrum; cf. the beginning of Section 9.2.6. It thus signals the presence of an *11th dimension* with radius

$$R_{11} = g\sqrt{\alpha'} \ . \qquad (9.69)$$

The 11th dimension cannot be seen in string perturbation theory, which is a perturbation theory for small g. Since $D = 11$ is the maximal dimension in which SUSY can exist, this suggests a connection with 11-dimensional SUGRA. It is generally believed that the five string theories are the perturbative limits of one fundamental theory of which 11-dimensional SUGRA is a low-energy limit. This fundamental theory, about which little is known, is called *M-theory*. A particular proposal of this theory is 'matrix theory' which employs only a finite number of degrees of freedom connected with a system of D0-branes (Banks *et al.* 1997). Its fundamental scale is the 11-dimensional Planck length. The understanding of M-theory is indeed very limited. It is, for example, not yet possible to give a full non-perturbative calculation of graviton–graviton scattering, one of the important processes in quantum gravity (see Chapter 2).

In Section 9.2.3, we have discussed the notion of T-duality, which connects descriptions of small and large radii. There is a second important notion of duality called 'S-duality', which relates the five consistent superstring theories to each other. Thereby the weak-coupling sector ($g \ll 1$) of one theory can be connected to the strong-coupling sector ($g \gg 1$) of another (or the same) theory.

9.2.5   *Black-hole entropy*

In Section 7.3, we have reviewed attempts to calculate the Bekenstein–Hawking entropy (7.17) by counting microscopic degrees of freedom in canonical quantum gravity; see also the remarks in Section 8.1.3 on the situation in (2+1)-dimensional gravity. What can string theory say about this issue? It turns out that one can give a microscopic foundation for *extremal* black holes and black holes that are close to extremality. 'Extremal' is here meant with respect to the generalized electric and magnetic charges that can be present in the spectrum of string theory. It is analogous to the situation for an extremal Reissner–Nordström black hole (Section 7.1). We shall be brief in the following and refer the reader to, for example, Horowitz (1998) and Peet (1998) for reviews.

The key idea to the string calculation of (7.17) is the notion of S-duality discussed at the end of the last subsection. A central role is played by so-called

'BPS states' (named after Bogomolnyi, Prasad, and Sommerfield), which have the important property that they are invariant under a non-trivial subalgebra of the full SUSY algebra. As a consequence, their mass is fixed in terms of their charges and their spectrum is preserved while going from a weak-coupling limit of string theory to a strong-coupling limit. In the weak-coupling limit, a BPS state can describe a bound state of D-branes whose entropy S_s can be easily calculated. In the strong-coupling limit, the state can describe an extremal black hole whose entropy can be calculated by (7.17). Interestingly, both calculations lead to the same result. This was first shown by Strominger and Vafa (1996) for an extremal hole in five dimensions. It has to be emphasized that all these calculations are being done in the semiclassical regime in which the black hole is not too small. Its final evaporation has therefore not yet been addressed.

We give here only some heuristic arguments why this result can hold and refer to the above references for details. The level density d_N of a highly excited string state with level of excitation N is (for open strings) in the limit $N \to \infty$ given by

$$d_N \sim e^{4\pi\sqrt{N}} \approx e^{M/M_0} , \qquad (9.70)$$

where (9.21) has been used, and

$$M_0 \equiv \frac{1}{4\pi\sqrt{\alpha'}} . \qquad (9.71)$$

The temperature $T_0 \equiv M_0/k_B$ connected with M_0 is called 'Hagedorn temperature' or 'temperature of the hell' because the free energy diverges when T_0 is approached (signalling a phase transition). The expression for d_N can be understood as follows. Dividing a string with energy M into two parts with energies M_1 and M_2, respectively, one would expect that $M = M_1 + M_2$. The number of states would then obey

$$d_N(M) = d_N(M_1)d_N(M_2) = d_N(M_1 + M_2) ,$$

from which a relation of the form (9.70) follows. Using from statistical physics the formula $d_N = \exp(S_s)$, one finds for the 'string entropy'

$$S_s \propto M \propto \sqrt{N} . \qquad (9.72)$$

Since the gravitational constant depends on the string coupling, see (9.43), the effective Schwarzschild radius $R_S = 2GM$ increases if g is increased and a black hole can form if g becomes large enough. On the other hand, starting from a black hole and decreasing g, one finds that once R_S is smaller than l_s, a highly excited string state is formed; cf. Horowitz and Polchinski (1997). It seems, however, that (9.72) is in contradiction with the Bekenstein–Hawking entropy, which is (for the Schwarzschild case) proportional to M^2, not M. That this is not a problem follows from the g-dependence of the gravitational constant. Following

Horowitz (1998), we compare the mass $M \sim \sqrt{N}/l_s$ of a string with the mass $M_{BH} \sim R_S/G$ of the black hole when $R_S \approx l_s$,

$$M_{BH} \sim \frac{R_S}{G} \approx \frac{l_s}{G} \sim M \sim \frac{\sqrt{N}}{l_s} \ ,$$

leading to $l_s^2/G \sim \sqrt{N}$. The entropy of the black hole is then given by

$$S_{BH} \sim \frac{R_S^2}{G} \approx \frac{l_s^2}{G} \sim \sqrt{N}$$

and thus comparable to the string entropy (9.72). Strings thus possess enough states to yields the Bekenstein–Hawking entropy. It is most remarkable that the exact calculation yields (for BPS states) an exact coincidence between both entropies. The fact that $R_S \approx l_s$ in the above estimate does not mean that the black hole is small: eliminating in the above expressions l_s in favour of the Planck length $l_P \sim \sqrt{G}$, one finds

$$R_S \sim N^{1/4} l_P \ .$$

Since $N \gg 1$, the Schwarzschild radius R_S is much bigger than the Planck length—again a consequence of the fact that G varies with g, while l_s is fixed.

The exact calculations mentioned above refer to extremal black holes, for which the Hawking temperature is zero (see Section 7.1). It was, however, possible to generalize the result to near-BPS states. Hawking radiation is then non-vanishing and corresponds to the emission of a closed string from a D-brane. If the D-brane state is traced out, the radiation is described by a thermal state. This is in accordance with Section 7.2 where it has been argued that unitarity is preserved for the full system and the mixed appearance of Hawking radiation arises from quantum entanglement with an 'environment', leading to decoherence (cf. Section 10.1). In the present case, the environment would be a system of D-branes. One would thus not expect any information-loss paradox to be present. In fact, the calculations in string theory preserve—in their range of validity—unitarity.

It is interesting that the calculations can recover the exact cross-sections for the black hole including the grey-body factor $\Gamma_{\omega l}$ in (7.12). Unfortunately, it was not yet possible to extend these exact results to generic black holes. An exact treatment of the Schwarzschild black hole, for example, remains elusive.

Sub-leading corrections to both S_{BH} and S_s have been calculated and shown to be connected with higher curvature terms; cf. Mohaupt (2001) and the references therein.

The Bekenstein–Hawking entropy is calculated from the surface of the horizon, not from the volume inside. The idea that for a gravitating system the information is located on the *boundary* of some spatial region is called the 'holographic principle' (see Bousso (2002) for a review). More generally, the principle states that the number of degrees of freedom in a volume of a spatial region is equal to that of a system residing on the boundary of that region. Apparently this gives rise to non-local features.

A realization of the holographic principle in string theory seems to be the 'AdS/CFT-correspondence'; see Aharony *et al.* (2000) for a review. This is the second approach besides matrix theory to learn something about M-theory. The AdS/CFT-correspondence states that non-perturbative string theory in a background space–time which is asymptotically anti-de Sitter (AdS) is dual to a conformal field theory (CFT) defined in a flat space–time of one less dimension. As a concrete example may serve type IIB string theory on an asymptotically $AdS_5 \times S^5$ space–time (called the 'bulk'). This is supposed to be dual to a CFT which is (3+1)-dimensional SUSY Yang–Mills theory with gauge group $U(n)$. Since the conformal boundary of AdS_5 is $\mathbb{R} \times S^3$, whose dimension agrees with that of the CFT, one claims that the CFT is defined on the boundary of AdS space. This cannot be meant literally, since a boundary cannot in general be separated from the enclosed volume because of quantum entanglement between both. It should also be mentioned that in a more recent version of AdS/CFT, one can compare genuine string calculations with calculations in gauge theories; cf. Berenstein *et al.* (2002).

The AdS/CFT-correspondence (also called 'AdS/CFT conjecture') associates fields in string theory with operators in CFT and compares expectation values and symmetries in both theories. An equivalence on the level of the quantum states in both theories has not been shown and is, moreover, unlikely to hold.

9.2.6 *Brane worlds*

String theory employs higher dimensions for its formulation. The idea of using higher dimensions for unified theories goes back to the pioneering work of Kaluza and Klein in the 1920s; see for example, Lee (1984) for a collection of reviews and an English translation of the original papers. In the simplest version, there is one additional space dimension, which is compactified to a circle with circumference $2\pi R$. We label the usual four dimensions by coordinates x^μ and the fifth dimension by y. One can get easily in such a scenario particle masses in four dimensions from a massless five-dimensional field. Assuming for simplicity that the metric is flat, the dynamical equation for a massless scalar field Φ in five dimensions is given by the wave equation

$$\Box_5 \Phi(x^\mu, y) = 0 \ , \tag{9.73}$$

where $\Box_5$ is the five-dimensional d'Alembert operator. Making a Fourier expansion with respect to the fifth dimension,

$$\Phi(x^\mu, y) = \sum_n \varphi_n(x^\mu) \mathrm{e}^{iny/R} \ , \ n \in \mathbb{Z} \ , \tag{9.74}$$

one obtains for the $\varphi_n(x^\mu)$ an effective equation of the form

$$\left(\Box_4 - \frac{n^2}{R^2} \right) \varphi_n(x^\mu) = 0 \ , \tag{9.75}$$

where $\Box_4$ is the four-dimensional d'Alembert operator. Equation (9.75) is nothing but the four-dimensional Klein–Gordon equation for a massive scalar field $\varphi_n(x^\mu)$ with mass

$$m_n = \frac{|n|}{R} \, . \tag{9.76}$$

From the four-dimensional point of view one thus has a whole 'Kaluza–Klein tower' of particles with increasing masses. For low energies $E \lesssim 1/R$, the massive Kaluza–Klein modes remain unexcited and only the massless mode for $n = 0$ remains. The higher dimensions only show up for energies beyond $1/R$. Since no evidence has been seen yet at accelerators for the massive modes, the size of the fifth dimension must be very small, definitely smaller than about 10^{-17} cm.

The Kaluza–Klein scenario has been generalized in various directions. In Section 9.2.3, we have seen that the notion of T-duality in string theory gives rise to the concept of D-branes. Gauge and matter fields are localized on the brane, whereas gravity can propagate freely through the higher dimensions (the bulk). One can, therefore, assume that our observed four-dimensional world is, in fact, such a brane being embedded in higher dimensions. This gives rise to various 'brane-world scenarios' which often are very loosely related to string theory itself, taking from there only the idea of a brane without giving necessarily a dynamical justification from string theory. A general review is Rubakov (2001).

In one scenario, the so-called 'ADD' approach, the brane tension (energy per unit three-volume of the brane) and therefore its gravitational field is neglected; see Arkani-Hamed *et al.* (1998) and Antoniadis *et al.* (1998). A key ingredient in this approach is to take the extra dimensions compact (as in the standard Kaluza–Klein approach) but *not* microscopically small. The reason for this possibility is the fact that only gravity can probe the extra dimensions, and the gravitational attraction has only been tested down to distances of about 0.2 mm. Any value for R with $R \lesssim 0.1$ mm would thus be allowed. This could give a clue for understanding the 'hierarchy problem' in particles physics—the problem why the electroweak scale (at about 1 TeV) is so much smaller than the Planck scale $m_\mathrm{P} \approx 10^{19}$ GeV. This works as follows.[5] One starts with the Einstein–Hilbert action in $D = 4 + d$ dimensions,

$$S_{\mathrm{EH}} = \frac{1}{16\pi G_D} \int \mathrm{d}^4 x \mathrm{d}^d y \, \sqrt{-g_D} \, {}^{(D)}R \, , \tag{9.77}$$

where the index D refers to the corresponding quantities in D dimensions. We denote by m_* the D-dimensional Planck mass, that is,

$$G_D = \frac{1}{m_*^{D-2}} = \frac{1}{m_*^{d+2}} \, ,$$

---

[5]It was also suggested that this huge discrepancy in scales may have a cosmological origin; cf. Hogan (2000).

where $d = D - 4$ is the number of extra dimensions. Assuming that the D-dimensional metric is (approximately) independent of the extra dimensions labelled by y, one gets from (9.77) an effectively four-dimensional action,

$$S_{\text{EH}} = \frac{V_d}{16\pi G_D} \int \mathrm{d}^4 x \sqrt{-g_4}\,^{(4)}R \ , \tag{9.78}$$

where $V_d \sim R^d$ denotes the volume of the extra dimensions. Comparison with the four-dimensional Einstein–Hilbert action (1.1) gives the connection between m_* and the four-dimensional Planck mass,

$$m_{\text{P}} \sim m_*(m_*R)^{d/2} \ . \tag{9.79}$$

The four-dimensional Planck mass is thus big (compared to the weak scale) because the size of the extra dimensions is big. Thereby the hierarchy problem is transferred to a different problem: why is R so big? This reformulation has two advantages. First, a unified theory such as string theory might give an explanation for the size of R. Second, it opens the possibility to observe the extra dimensions, either through scattering experiments at colliders or through sub-mm tests of Newton's law; see Rubakov (2001) and the references therein. Higher dimensional theories generically predict a violation of the Newtonian $1/r$-potential at some scale. No sign of the extra dimensions, however, has been seen up to now.

Both in the traditional Kaluza–Klein and the ADD scenario, the full metric factorizes[6] into the four-dimensional part describing our macroscopic dimensions and the (compact) part referring to the extra dimensions. Such an assumption is, however, not obligatory. If factorization does not hold, one talks about a 'warped metric'. The extra dimensions can be compact or infinite in size. A warped metric occurs, for example, if the gravitational field produced by the brane is taken into account. We shall briefly describe one particular model with two branes put forward by Randall and Sundrum (1999). There is one extra dimension, and the bulk has an AdS geometry.

The action of this model is given by the following expression in which the index $I = \pm$ enumerates the two branes with tensions $\sigma_\pm$,

$$S[G, g, \phi] = S_5[G] + \sum_I \int_{\Sigma_I} \mathrm{d}^4 x \left(L_{\text{m}}(\phi, \partial\phi, g) - g^{1/2}\sigma_I + \frac{1}{8\pi G_5}[K] \right) \ ,$$

$$S_5[G] = \frac{1}{16\pi G_5} \int_{M^5} \mathrm{d}^5 x\, G^{1/2} \left({}^{(5)}R(G) - 2\Lambda_5 \right) \ . \tag{9.80}$$

Here, $S[G, g, \phi]$ is the action of the five-dimensional gravitational field with the metric $G = G_{AB}(x, y)$, $A = (\mu, 5)$, $\mu = 0, 1, 2, 3$ propagating in the bulk

---

[6]Factorization here means that the D-dimensional metric can be put into block-diagonal form, in which one block is the four-dimensional metric, and where the various blocks do not depend on the coordinates referring to the other blocks.

space–time ($x^A = (x, y)$, $x = x^\mu$, $x^5 = y$), and matter fields ϕ are confined to the branes Σ_I, which are four-dimensional time-like surfaces embedded in the bulk. The branes carry the induced metrics $g = g_{\mu\nu}(x)$ and the matter field Lagrangians $L_m(\phi, \partial\phi, g)$. The bulk part of the action contains the five-dimensional gravitational and cosmological constants, G_5 and Λ_5, while the brane parts have four-dimensional cosmological constants σ_I. The bulk cosmological constant Λ_5 is negative and is thus capable of generating an AdS geometry, while the brane cosmological constants play the role of brane tensions σ_I and, depending on the model, can be of either sign. The Einstein–Hilbert bulk action in (9.80) is accompanied by the brane surface terms, cf. (1.1), containing the jump of the extrinsic curvature trace $[K]$ associated with both sides of each brane.

The fifth dimension has the topology of a circle labelled by the coordinate y, $-d < y \leq d$, with an orbifold $\mathbb{Z}_2$-identification of points y and $-y$.[7] The branes are located at antipodal fixed points of the orbifold, $y = y_\pm$, $y_+ = 0$, $|y_-| = d$. When they are empty, $L_m(\phi, \partial\phi, g_{\mu\nu}) = 0$, and their tensions are opposite in sign and fine-tuned to the values of Λ_5 and G_5,

$$\Lambda_5 = -\frac{6}{l^2} , \qquad \sigma_+ = -\sigma_- = \frac{3}{4\pi G_5 l} , \qquad (9.81)$$

this model admits a solution with an AdS metric in the bulk (l is its curvature radius),

$$ds^2 = dy^2 + e^{-2|y|/l}\eta_{\mu\nu}dx^\mu dx^\nu , \qquad (9.82)$$

$0 = y_+ \leq |y| \leq y_- = d$, and with a flat induced metric $\eta_{\mu\nu}$ on both branes. The metric on the negative tension brane is rescaled by the 'warp factor' $\exp(-2d/l)$. This could provide a solution to the hierarchy problem in this model; see Randall and Sundrum (1999). With the fine tuning (9.81) this solution exists for an arbitrary brane separation d—the two flat branes stay in equilibrium. Their flatness is the result of a compensation between the bulk cosmological constant and the brane tensions.

More generally one considers the Randall–Sundrum model with small matter sources for metric perturbations $h_{AB}(x, y)$ on the background of this solution,

$$ds^2 = dy^2 + e^{-2|y|/l}\eta_{\mu\nu}dx^\mu dx^\nu + h_{AB}(x, y)\,dx^A dx^B , \qquad (9.83)$$

such that this five-dimensional metric induces on the branes two four-dimensional metrics of the form

$$g_{\mu\nu}^\pm(x) = a_\pm^2\,\eta_{\mu\nu} + h_{\mu\nu}^\pm(x) . \qquad (9.84)$$

Here the scale factors $a_\pm = a(y_\pm)$ can be expressed in terms of the interbrane distance,

---

[7] An orbifold is a coset space $M/\mathcal{G}$, where $\mathcal{G}$ is a group of discrete symmetries of the manifold M. Here we deal with the special case of an $S^1/\mathbb{Z}_2$-orbifold.

$$a_+ = 1, \quad a_- = e^{-2d/l} \equiv a , \qquad (9.85)$$

and $h_{\mu\nu}^{\pm}(x)$ are the perturbations by which the brane metrics $g_{\mu\nu}^{\pm}(x)$ differ from the (conformally) flat metric in the Randall–Sundrum solution (9.82).

Instead of using the Kaluza–Klein formalism with its infinite tower of modes, one can employ an alternative formalism which captures more the spirit of the holographic principle and the AdS/CFT correspondence. This results in the calculation of a non-local effective action for the branes; see Barvinsky *et al.* (2003*a*) for details. This action is a functional of the induced metrics on both branes. One can also derive a reduced action which depends only on one brane (the visible brane, which is taken here to be σ_+). These effective actions contain *all* the physical information that is available on the brane(s). Interesting applications are inflationary cosmology and gravitational-wave interferometry. The latter arises because this model has *light* massive graviton modes in addition to the massless graviton. The mixing of these modes can lead to gravitational-wave oscillations analogous to neutrino oscillations. The parameters of these oscillations depend crucially on the size of the extra dimension. Such an effect can in principle be observed with current gravitational-wave interferometers. This mechanism and its phenomenology are discussed in Barvinsky *et al.* (2003*b*) to which we refer the reader for more details.

10

QUANTUM GRAVITY AND THE INTERPRETATION OF QUANTUM THEORY

10.1 Decoherence and the quantum universe

The central concept in quantum theory is the superposition principle. It consists of a kinematical and a dynamical part; cf. Joos *et al.* (2003). The kinematical part declares that for any two physical states Ψ_1 and Ψ_2, the sum $c_1\Psi_1 + c_2\Psi_2$ (with $c_1, c_2 \in \mathbb{C}$) is again a physical state. This expresses the linear structure of Hilbert space and gives rise to the important notion of quantum *entanglement* between systems. The dynamical part refers to the linearity of the Schrödinger equation. If $\Psi_1(t)$ and $\Psi_2(t)$ are solutions, then the sum $c_1\Psi_1(t) + c_2\Psi_2(t)$ is again a solution.

The superposition principle remains untouched in most approaches to quantum gravity. This holds in particular for quantum GR and string theory, which are both discussed in this book. There exist suggestions about a gravity-induced breakdown of the superposition principle; see for example, Penrose (1996) and the discussion in Chapter 8 of Joos *et al.* (2003). However, no such mechanism was developed to a technical level comparable with the approaches discussed here.

If the superposition principle is universally valid, quantum gravity allows the superposition of macroscopically different metrics. This has drastic consequences in particular for quantum cosmology (Chapter 8) where it would be difficult to understand why we observe a classical universe at all. In some interpretations of quantum mechanics, notably the Copenhagen interpretation(s), an external observer is invoked who 'reduces' the wave function from the superposition to the observed component. In quantum cosmology, on the other hand, no such external measuring agency is available, since the universe contains by definition everything. A reduction (or collapse) of the wave function by external observers would then be impossible. How, then, does the classical appearance of our universe emerge? In the last decades, one has reached an understanding of how classical properties can emerge within quantum mechanics. It is an amazing fact that the key role in this process is played by the superposition principle itself, through the process of *decoherence*. Following Kiefer (2003 *b*), we shall give a brief introduction into decoherence in quantum mechanics and then extrapolate decoherence into the realm of quantum cosmology (Section 10.1.2). An exhaustive treatment can be found in Joos *et al.* (2003); see also Zurek (2003).

10.1.1 *Decoherence in quantum mechanics*

If quantum theory is universally valid, every system should be described in quantum terms, and it would be inconsistent to draw an *a priori* border line between a quantum system and a classical apparatus. John von Neumann was the first who analysed in 1932 the measurement process within quantum mechanics; see von Neumann (1932). He considers the coupling of a system (S) to an apparatus (A), see Fig. 10.1.

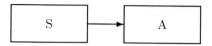

FIG. 10.1. Original form of the von Neumann measurement model.

If the states of the measured system that are discriminated by the apparatus are denoted by $|n\rangle$ (e.g. spin up and spin down), an appropriate interaction Hamiltonian has the form

$$H_{\text{int}} = \sum_n |n\rangle\langle n| \otimes \hat{A}_n \ . \tag{10.1}$$

The operators $\hat{A}_n$ act on the states of the apparatus and are rather arbitrary, but must, of course, depend on the 'quantum number' n. Equation (10.1) describes an 'ideal' interaction during which the apparatus becomes correlated with the system state, without changing the latter. There is thus no disturbance of the system by the apparatus—on the contrary, the apparatus is disturbed by the system (in order to yield a measurement result).

If the measured system is initially in the state $|n\rangle$ and the device in some initial state $|\Phi_0\rangle$, the evolution according to the Schrödinger equation with the Hamiltonian (10.1) reads

$$|n\rangle|\Phi_0\rangle \xrightarrow{\ t\ } \exp\left(-\mathrm{i}H_{\text{int}}t\right)|n\rangle|\Phi_0\rangle = |n\rangle \exp\left(-\mathrm{i}\hat{A}_n t\right)|\Phi_0\rangle$$
$$\equiv |n\rangle|\Phi_n(t)\rangle \ . \tag{10.2}$$

The resulting apparatus states $|\Phi_n(t)\rangle$ are often called 'pointer states'. A process analogous to (10.2) can also be formulated in classical physics. The essential new quantum features now come into play when one considers a *superposition* of different eigenstates (of the measured 'observable') as the initial state. The linearity of time evolution immediately leads to

$$\left(\sum_n c_n|n\rangle\right)|\Phi_0\rangle \xrightarrow{\ t\ } \sum_n c_n|n\rangle|\Phi_n(t)\rangle \ . \tag{10.3}$$

But this state is a superposition of macroscopic measurement results (of which Schrödinger's cat is just one drastic example)! To avoid such a bizarre state, and to avoid the apparent conflict with experience, von Neumann introduced a dynamical collapse of the wave function as a new law. The collapse should then select one component with the probability $|c_n|^2$.

Can von Neumann's conclusion and the introduction of the collapse be avoided? The crucial observation that enforces an extension of von Neumann's measurement theory is the fact that macroscopic objects (such as measurement devices) are so strongly coupled to their natural environment that a unitary treatment as in (10.2) is by no means sufficient and has to be modified to include the environment (Zeh 1970); cf. Fig. 10.2.

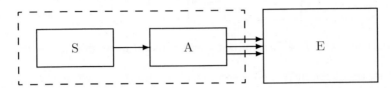

FIG. 10.2. Realistic extension of the von Neumann measurement model including the environment. Classical properties emerge through the unavoidable, irreversible interaction of the apparatus with the environment.

Fortunately, this can be easily done to a good approximation, since the interaction with the environment has in many situations the same form as given by the Hamiltonian (10.1): the measurement device is itself 'measured' (passively recognized) by the environment, according to

$$\left(\sum_n c_n |n\rangle |\Phi_n\rangle \right) |E_0\rangle \quad \xrightarrow{t} \quad \sum_n c_n |n\rangle |\Phi_n\rangle |E_n\rangle. \tag{10.4}$$

This is again a macroscopic superposition, now including the myriads of degrees of freedom pertaining to the environment (gas molecules, photons, etc.). However, most of these environmental degrees of freedom are inaccessible. Therefore, they have to be integrated out from the full state (10.4). This leads to the reduced density matrix for system plus apparatus, which contains all the information that is available there. It reads

$$\rho_{\text{SA}} \approx \sum_n |c_n|^2 |n\rangle\langle n| \otimes |\Phi_n\rangle\langle\Phi_n| \quad \text{if} \quad \langle E_n|E_m\rangle \approx \delta_{nm} , \tag{10.5}$$

since under realistic conditions, different environmental states are orthogonal to each other. Equation (10.5) is identical to the density matrix of an ensemble of measurement results $|n\rangle|\Phi_n\rangle$. System and apparatus thus seem to be in one of the states $|n\rangle$ and $|\Phi_n\rangle$, given by the probability $|c_n|^2$.

Both system and apparatus thus assume classical properties through the unavoidable, irreversible interaction with the environment. This dynamical process, which is fully described by quantum theory, is called decoherence. It is based on the quantum entanglement between apparatus and environment. Under ordinary macroscopic situations, decoherence occurs on an extremely short time-scale, giving the impression of an instantaneous collapse or a 'quantum jump'. Recent experiments were able to demonstrate the continuous emergence of classical properties in mesoscopic systems (Hornberger *et al.* 2003; Joos *et al.* 2003). Therefore, one would never ever be able to observe a weird superposition such as Schrödinger's cat, because the information about this superposition would almost instantaneously be delocalized into unobservable correlations with the environment, resulting in an *apparent* collapse for the cat state.

The interaction with the environment distinguishes the local basis with respect to which classical properties (unobservability of interferences) hold. This 'pointer basis' must obey the condition of robustness, that is, it must keep its classical appearance over the relevant time-scales; cf. Zurek (2003). Classical properties are thus not intrinsic to any object, but only defined by their interaction with other degrees of freedom. In simple (Markovian, i.e. local in time) situations, the pointer states are given by localized Gaussian states (Diósi and Kiefer 2000). They are, in particular, relevant for the localization of macroscopic objects.

The ubiquitous occurrence of decoherence renders the interpretational problem of quantum theory currently a 'matter of taste' (Zeh 1994). Provided one adopts a realistic interpretation without additional variables,[1] the alternatives would be an Everett interpretation or the assumption of a collapse for the total system (including the environment). The latter would have to entail an explicit modification of quantum theory, since one would have to introduce non-linear or stochastic terms into the Schrödinger equation in order to achieve this goal. The Everett interpretation assumes that all components of the full quantum state exist and are real. Decoherence produces robust macroscopic branches, one of which corresponds to the observed world. Interferences with the other branches are suppressed, so decoherence readily explains the observation of an apparent collapse of the wave function, independent of whether there is a real collapse for the total system or not. The question is thus whether one applies Ockham's razor[2] to the equations or the intuition (Zeh 1994): either one has to complicate the formalism in order to have just one macroscopic branch or one retains the linear structure of quantum theory and has to accept the existence of 'many worlds'.

10.1.2 *Decoherence in quantum cosmology*

In this subsection, we investigate the question: how can one understand the classical appearance of global space–time variables such as the radius (scale factor) of the universe? If decoherence is the fundamental process, we have to identify a 'system' and an 'environment'. More precisely, we have to differentiate between

---

[1]Bohm's interpretation would be an example for a realistic interpretation with additional variables.

[2]'Pluralitas non est ponenda sine necessitate'.

relevant and irrelevant variables. All degrees of freedom exist, of course, within the universe. It has been suggested by Zeh (1986) that the irrelevant degrees of freedom are the variables describing density fluctuations and gravitational waves. Their interaction with the scale factor and other homogeneous degrees of freedom (such as an inflaton field) can render the latter classical. In a sense, then, a classical space–time arises from a 'self-measurement' of the universe.

The following discussion will roughly follow Joos $et\ al.$ (2003). Calculations for decoherence in quantum cosmology can be performed in the framework of quantum geometrodynamics (Kiefer 1987), using the formalism of the Wheeler–DeWitt equation presented in Chapters 5 and 8. A prerequisite is the validity of the semiclassical approximation (Section 5.4) for the global variables. This brings an approximate time parameter t into play. The irrelevant degrees of freedom (density fluctuations, gravitational waves) are described by the inhomogeneous variables of Section 8.2. In Kiefer (1987), the relevant system was taken to be the scale factor ('radius') a of the universe together with a homogeneous scalar field ϕ (the 'inflaton'); cf. the model discussed in Section 8.1.2. The inhomogeneous modes of Section 8.2 can then be shown to $decohere$ the global variables a and φ.

An open problem in Kiefer (1987) was the issue of regularization; the number of fluctuations is infinite and would cause divergences, so an $ad\ hoc$ cut-off was suggested to consider only modes with wavelength bigger than the Planck length. The problem was again addressed in Barvinsky $et\ al.$ (1999a) where a physically motivated regularization scheme was introduced. In the following, we shall briefly review this approach.

As a (semi)classical solution for a and ϕ, one may use

$$\phi(t) \approx \phi, \tag{10.6}$$

$$a(t) \approx \frac{1}{H(\phi)} \cosh H(\phi)t\ , \tag{10.7}$$

where $H^2(\phi) = 4\pi V(\phi)/3m_{\rm P}^2$ is the Hubble parameter generated by the inflaton potential $V(\phi)$, cf. Section 8.3.2. It is approximately constant during the inflationary phase in which ϕ slowly 'rolls down' the potential. We take into account fluctuations of a field $f(t, \mathbf{x})$ which can be a field of any spin (not necessarily a scalar field Φ). Space is assumed to be a closed three-sphere, so $f(t, \mathbf{x})$ can be expanded into a discrete series of spatial orthonormal harmonics $Q^n(\mathbf{x})$; cf. Section 8.2.,

$$f(t, \mathbf{x}) = \sum_{\{n\}} f_n(t) Q^n(\mathbf{x})\ . \tag{10.8}$$

One can thus represent the fluctuations by the degrees of freedom f_n (in Section 8.2, f_n were the modes of a scalar field Φ).

The aim is now to solve the Wheeler–DeWitt equation in the semiclassical approximation. This leads to the following solution:

$$\Psi(t|\phi, f) = \frac{1}{\sqrt{v_\phi^*(t)}} \, e^{-I(\phi)/2 + iS_0(t,\phi)} \prod_n \psi_n(t, \phi|f_n) \; . \tag{10.9}$$

The time t that appears here is the semiclassical ('WKB') time and is defined by the background-degrees of freedom a and ϕ through the 'eikonal' S_0, which is a solution of the Hamilton–Jacobi equation; cf. (8.50). Since ϕ is determined by a within a semiclassical branch of the wave function, only one variable (a or ϕ) occurs in the argument of Ψ. The wave functions ψ_n for the fluctuations f_n obey each an approximate Schrödinger equation (8.49) with respect to t, and their Hamiltonians H_n have the form of a ('time-dependent') harmonic-oscillator Hamiltonian. The first exponent in (10.9) contains the Euclidean action $I(\phi)$ from the classically forbidden region (the 'de Sitter instanton') and is independent of t. Its form depends on the boundary conditions imposed, and we shall here choose the no-boundary condition of Section 8.3.2., which amounts to $I(\phi) \approx -3m_{\rm P}^4/8V(\phi)$. The detailed form is, however, not necessary for the discussion below. The function $v_\phi(t)$ in (10.9) is the so-called basis function for ϕ and is a solution of the classical equation of motion.

For the ψ_n, we shall take the de Sitter-invariant vacuum state (the Euclidean vacuum discussed in Section 8.3.2). It reads

$$\psi_n(t, \phi|f_n) = \frac{1}{\sqrt{v_n^*(t)}} \exp\left(-\frac{1}{2}\Omega_n(t)f_n^2\right), \tag{10.10}$$

$$\Omega_n(t) = -ia^3(t)\frac{\dot{v}_n^*(t)}{v_n^*(t)} \; . \tag{10.11}$$

The functions v_n are the basis functions of the de Sitter-invariant vacuum state; they satisfy the classical equation of motion

$$F_n\left(\frac{\rm d}{{\rm d}t}\right) v_n \equiv \left(\frac{\rm d}{{\rm d}t}a^3\frac{\rm d}{{\rm d}t} + a^3m^2 + a(n^2 - 1)\right)v_n = 0 \tag{10.12}$$

with the boundary condition that they should correspond to a standard Minkowski positive-frequency function for constant a. In the simple special case of a spatially flat section of de Sitter space, one would have

$$av_n = \frac{e^{-in\eta}}{\sqrt{2n}}\left(1 - \frac{i}{n\eta}\right), \tag{10.13}$$

where η is the conformal time defined by $a{\rm d}\eta = {\rm d}t$. We note that it is the corresponding *negative*-frequency function that enters the exponent of the Gaussian, see (10.11).

An important property of these vacuum states is that their norm is conserved *along any semiclassical solution* (10.6) and (10.7),

$$\left\langle \psi_n|\psi_n\right\rangle \equiv \int {\rm d}f_n|\psi_n(f_n)|^2 = \sqrt{2\pi}[\Delta_n(\phi)]^{-1/2}, \tag{10.14}$$

$$\Delta_n(\phi) \equiv ia^3(v_n^* \dot{v}_n - \dot{v}_n^* v_n) = \text{constant} . \tag{10.15}$$

Note that $\Delta_n(\phi)$ is just the (constant) Wronskian corresponding to (10.12).[3] We must emphasize that Δ_n is a non-trivial function of the background variable ϕ, since it is defined on the full configuration space and not only along semiclassical trajectories. In a sense, it gives the weights in the 'Everett branches'. It is therefore *not* possible to normalize the ψ_n artificially to one, since this would be inconsistent with respect to the full Wheeler–DeWheeler equation (Barvinsky *et al.* 1999*a*).

The solution (10.9) forms the basis for our discussion of decoherence. Since the $\{f_n\}$ are interpreted as the environmental degrees of freedom, they have to be integrated out to get the reduced density matrix, cf. (10.5), for ϕ or a (a and ϕ can be used interchangeably, since they are connected by t). The reduced density matrix thus reads here

$$\rho(t|\phi, \phi') = \int df \ \Psi(t|\phi, f)\Psi^*(t|\phi', f) , \tag{10.16}$$

where Ψ is given by (10.9), and it is understood that $df = \prod_n df_n$. After the integration, one finds

$$\rho(t|\phi, \phi') = C \frac{1}{\sqrt{v_\phi^*(t)v_\phi'(t)}} \exp\left[-\frac{1}{2}I - \frac{1}{2}I' + i(S_0 - S_0') \right]$$
$$\times \prod_n \left[v_n^* v_n'(\Omega_n + \Omega_n'^*) \right]^{-1/2} , \tag{10.17}$$

where C is a numerical constant. The diagonal elements $\rho(t|\phi, \phi)$ describe the probabilities for certain values of the inflaton field to occur.

It is convenient to rewrite the expression for the density matrix (10.17) in the form

$$\rho(t|\phi, \phi') = C \frac{\Delta_\phi^{1/4} \Delta_\phi'^{1/4}}{\sqrt{v_\phi^*(t)v_\phi'(t)}} \exp\left(-\frac{1}{2}\Gamma - \frac{1}{2}\Gamma' + i(S_0 - S_0') \right)$$
$$\times \boldsymbol{D}(t|\phi, \phi') , \tag{10.18}$$

where

$$\boldsymbol{\Gamma} = I(\phi) + \boldsymbol{\Gamma}_{1-\text{loop}}(\phi) \tag{10.19}$$

is the full Euclidean effective action including the classical part and the one-loop part (cf. Section 2.2.4). The latter comes from the next-order WKB approximation and is important for the normalizability of the wave function with respect to ϕ. The last factor in (10.18) is the *decoherence factor*

---

[3]The corresponding Wronskian for the homogeneous mode ϕ is $\Delta_\phi \equiv ia^3(v_\phi^* \dot{v}_\phi - \dot{v}_\phi^* v_\phi)$.

$$\boldsymbol{D}(t|\phi, \phi') = \prod_n \left(\frac{4\operatorname{Re}\Omega_n \operatorname{Re}\Omega_n'^*}{(\Omega_n + \Omega_n'^*)^2} \right)^{1/4} \left(\frac{v_n \, v_n'^*}{v_n^* \, v_n'} \right)^{1/4} . \tag{10.20}$$

It is equal to one for coinciding arguments. While the decoherence factor is time-dependent, the one-loop contribution to (10.18) does not depend on time and may play a role only at the onset of inflation. In a particular model with non-minimal coupling (Barvinsky *et al.* 1997), the size of the non-diagonal elements is at the onset of inflation approximately equal to those of the diagonal elements. The universe would thus be essentially quantum at this stage, that is, in a non-classical state.

The amplitude of the decoherence factor can be rewritten in the form

$$|\boldsymbol{D}(t|\phi, \phi')| = \exp \frac{1}{4} \sum_n \ln \frac{4\operatorname{Re}\Omega_n \operatorname{Re}\Omega_n'^*}{|\Omega_n + \Omega_n'^*|^2} . \tag{10.21}$$

The convergence of this series is far from being guaranteed. Moreover, the divergences might not be renormalizable by local counterterms in the bare quantized action. We shall now analyse this question in more detail.

We start with a minimally coupled massive scalar field. Equation (10.12) for the basis functions reads

$$\frac{\mathrm{d}}{\mathrm{d}t} \left(a^3 \frac{\mathrm{d}v_n}{\mathrm{d}t} \right) + a^3 \left(\frac{n^2 - 1}{a^2} + m^2 \right) v_n = 0 . \tag{10.22}$$

The appropriate solution to this equation is

$$v_n(t) = (\cosh Ht)^{-1} P_{-\frac{1}{2}+i\sqrt{m^2/H^2-9/4}}^{-n}(i \sinh Ht) , \tag{10.23}$$

where P denotes an associated Legendre function of the first kind. The corresponding expression for (10.11) is for large mass m given by

$$\Omega_n = a^2 \left[\sqrt{n^2 + m^2 a^2} + i \sinh Ht \left(1 + \frac{1}{2} \frac{m^2 a^2}{n^2 + m^2 a^2} \right) \right] + O\left(\frac{1}{m} \right) . \tag{10.24}$$

The leading contribution to the amplitude of the decoherence factor is therefore

$$\ln |\boldsymbol{D}(t|\phi, \phi')| \simeq \frac{1}{4} \sum_{n=0}^{\infty} n^2 \ln \frac{4a^2 a'^2 \sqrt{n^2 + m^2 a^2} \sqrt{n^2 + m^2 a'^2}}{\left(a^2 \sqrt{n^2 + m^2 a^2} + a'^2 \sqrt{n^2 + m^2 a'^2} \right)^2} . \tag{10.25}$$

The first term, n^2, in the sum comes from the degeneracy of the eigenfunctions. This expression has divergences which *cannot* be represented as additive functions of a and a'. This means that no one-argument counterterm to $\boldsymbol{\Gamma}$ and $\boldsymbol{\Gamma}'$ in (10.18) can cancel these divergences of the amplitude (Paz and Sinha 1992). One might try to apply standard regularization schemes from quantum field theory, such as dimensional regularization. The corresponding calculations have been performed in Barvinsky *et al.* (1999*a*) and will not be given here. The important result is that, although they render the sum (10.25) convergent, they lead to a *positive* value of this expression. This means that the decoherence factor would diverge for $(\phi - \phi') \to \infty$ and thus spoil one of the crucial properties of a density matrix—the boundedness of $\operatorname{tr} \hat{\rho}^2$. The dominant term in the decoherence factor would read

$$\ln |\boldsymbol{D}| = \frac{\pi}{24}(ma)^3 + O(m^2), \quad a \gg a' , \tag{10.26}$$

and would thus be unacceptable for a density matrix. Reduced density matrices are usually not considered in quantum field theory, so this problem has not been encountered before. A behaviour such as in (10.26) is even obtained in the case of massless conformally invariant fields, for which one would expect a decoherence factor equal to one, since they decouple from the gravitational background. How, then, does one have to proceed in order to obtain a sensible regularization?

The crucial point is to perform a *redefinition* of environmental fields and to invoke a physical principle to fix this redefinition. The situation is somewhat analogous to the treatment of the S-matrix in quantum field theory: off-shell S-matrix and effective action depend on the parametrization of the quantum fields, in analogy to the non-diagonal elements of the reduced density matrix. In Laflamme and Louko (1991) and Kiefer (1992) it has been proposed within special models to rescale the environmental fields by a power of the scale factor. It was therefore suggested in Barvinsky *et al.* (1999*a*) to redefine the environmental fields by a power of the scale factor that corresponds to the conformal weight of the field (which is defined by the conformal invariance of the wave equation). For a scalar field in four space–time dimensions, this amounts to a multiplication by a,

$$v_n(t) \to \tilde{v}_n(t) = a \, v_n(t) , \tag{10.27}$$

$$\tilde{\Omega}_n = -\mathrm{i}a\frac{\mathrm{d}}{\mathrm{d}t} \ln \tilde{v}_n^* . \tag{10.28}$$

An immediate test of this proposal is to see whether the decoherence factor is equal to one for a massless conformally invariant field. In this case, the basis functions and frequency functions read, respectively,

$$\tilde{v}_n^*(t) = \left(\frac{1 + \mathrm{i}\sinh Ht}{1 - \mathrm{i}\sinh Ht}\right)^{n/2} , \tag{10.29}$$

$$\tilde{\Omega}_n = -\mathrm{i}a\frac{\mathrm{d}}{\mathrm{d}t} \ln \tilde{v}_n^*(t) = n . \tag{10.30}$$

Hence, $\tilde{\boldsymbol{D}}(t|\phi, \phi') \equiv 1$. The same holds for the electromagnetic field (which in four space–time dimensions is conformally invariant). It is interesting to note that the degree of decoherence caused by a certain field depends on the space–time dimension, since its conformal properties are dimension-dependent.

For a massive minimally coupled field, the new frequency function reads

$$\tilde{\Omega}_n = \left[\sqrt{n^2 + m^2 a^2} + i \sinh Ht \left(\frac{1}{2} \frac{m^2 a^2}{n^2 + m^2 a^2} \right) \right] + O(1/m) \, . \qquad (10.31)$$

Note that, in contrast to (10.24), there is no factor of a^2 in front of this expression. Since (10.31) is valid in the large-mass limit, it corresponds to modes that evolve adiabatically on the gravitational background, the imaginary part in (10.31) describing particle creation.

It turns out that the imaginary part of the decoherence factor has at most logarithmic divergences and, therefore, affects only the phase of the density matrix. Moreover, these divergences decompose into an *additive* sum of one-argument functions and can thus be cancelled by adding counterterms to the classical action S_0 (and S_0') in (10.18) (Paz and Sinha 1992). The real part is simply convergent and gives a finite decoherence amplitude. This result is formally similar to the result for the decoherence factor in QED (Kiefer 1992).

For $a \gg a'$ (far off-diagonal terms), one gets the expression

$$|\tilde{\boldsymbol{D}}(t|\phi, \phi')| \simeq \exp\left[-\frac{(ma)^3}{24} \left(\pi - \frac{8}{3} \right) + O(m^2) \right] \, . \qquad (10.32)$$

Compared with the naively regularized (and inconsistent) expression (10.26), π has effectively been replaced by $8/3 - \pi$. In the vicinity of the diagonal, one obtains

$$\ln|\tilde{\boldsymbol{D}}(t|\phi, \phi')| = -\frac{m^3 \pi a(a - a')^2}{64} \, , \qquad (10.33)$$

a behaviour similar to (10.32).

An interesting case is also provided by minimally coupled massless scalar fields and by gravitons. They share the basis- and frequency functions in their respective conformal parametrisations,

$$\tilde{v}_n^*(t) = \left(\frac{1 + i \sinh Ht}{1 - i \sinh Ht} \right)^{n/2} \left(\frac{n - i \sinh Ht}{n + 1} \right) , \qquad (10.34)$$

$$\tilde{\Omega}_n = \frac{n(n^2 - 1)}{n^2 - 1 + H^2 a^2} - i \frac{H^2 a^2 \sqrt{H^2 a^2 - 1}}{n^2 - 1 + H^2 a^2} \, . \qquad (10.35)$$

They differ only by the range of the quantum number n ($2 \leq n$ for inhomogeneous scalar modes and $3 \leq n$ for gravitons) and by the degeneracies of the nth eigenvalue of the Laplacian,

$$\dim(n)_{\text{scal}} = n^2 \, , \qquad (10.36)$$

$$\dim(n)_{\text{grav}} = 2(n^2 - 4). \qquad (10.37)$$

For far off-diagonal elements one obtains the decoherence factor

$$|\tilde{\boldsymbol{D}}(t|\phi, \phi')| \sim e^{-C(Ha)^3}, \quad a \gg a', \quad C > 0 \, , \qquad (10.38)$$

while in the vicinity of the diagonal one finds

$$|\tilde{\boldsymbol{D}}(t|\phi,\phi')| \sim \exp\left(-\frac{\pi^2}{32}(H-H')^2 t^2 e^{4Ht}\right) , \qquad (10.39)$$

$$\sim \exp\left(-\frac{\pi^2 H^4 a^2}{8}(a-a')^2\right) , \quad Ht \gg 1 . \qquad (10.40)$$

These expressions exhibit a rapid disappearance of non-diagonal elements during the inflationary evolution. The universe thus assumes classical properties at the onset of inflation. This justifies the use of classical cosmology since then.

The decohering influence of fermionic degrees of freedom has to be treated separately (Barvinsky et al. 1999b). It turns out that they are less efficient in producing decoherence. In the massless case, for example, their influence is fully absent.

The above analysis of decoherence was based on the state (10.9). One might, however, start with a quantum state which is a superposition of many semiclassical components, that is, many components of the form $\exp(iS_0^k)$, where each S_0^k is a solution of the Hamilton–Jacobi equation for a and ϕ. Decoherence between different such semiclassical branches has also been the subject of intense investigation (Halliwell 1989; Kiefer 1992). The important point is that decoherence between different branches is usually weaker than the above discussed decoherence within one branch. Moreover, it usually follows from the presence of decoherence within one branch. In the special case of a superposition of (10.9) with its complex conjugate, one can immediately recognize that decoherence between the semiclassical components is smaller than within one component: in the expression (10.20) for the decoherence factor, the term $\Omega_n + \Omega_n'^*$ in the denominator is replaced by $\Omega_n + \Omega_n'$. Therefore, the imaginary parts of the frequency functions add up instead of partially cancelling each other and (10.20) becomes smaller. One also finds that the decoherence factor is equal to one for vanishing expansion of the semiclassical universe (Kiefer 1992).

We note that the decoherence between the $\exp(iS_0)$ and $\exp(-iS_0)$ components can be interpreted as a *symmetry breaking* in analogy to the case of sugar molecules (Joos et al. 2003). There, the Hamiltonian is invariant under space reflections, but the state of the sugar molecules exhibits chirality. Here, the Hamiltonian in the Wheeler–DeWitt equation is invariant under complex conjugation, while the 'actual states' (i.e. one decohering WKB component in the total superposition) are of the form $\exp(iS_0)$ and are thus intrinsically complex. It is therefore not surprising that the recovery of the classical world follows only for complex states, in spite of the real nature of the Wheeler–DeWitt equation (see in this context Barbour 1993). Since this is a prerequisite for the derivation of the Schrödinger equation, one might even say that *time* (the WKB time parameter in the Schrödinger equation) arises from symmetry breaking.

The above considerations thus lead to the following picture. The universe was essentially 'quantum' at the onset of inflation. Mainly due to bosonic fields, decoherence set in and led to the emergence of many 'quasi-classical branches' which are dynamically independent of each other. Strictly speaking, the very concept of time makes only sense after decoherence has occurred. In addition to the horizon problem etc., inflation thus also solves the 'classicality problem'. It remains, of course, unclear why inflation happened in the first place (if it really did). Looking back from our universe (our semiclassical branch) to the past, one would notice that at the time of the onset of inflation our component would interfere with other components to form a timeless quantum-gravitational state. The universe would thus cease to be transparent to earlier times (because there was no time). This demonstrates in an impressive way that quantum-gravitational effects are not *a priori* restricted to the Planck scale.

The lesson to be drawn is thus that the universe can appear classically *only* if experienced from within. A hypothetical 'outside view' would only see a static quantum world. The most natural interpretation of quantum cosmology is an Everett-type interpretation, since the 'wave function of the universe' contains by definition all possible branches.[4] As macroscopic observers, however, we have access only to a tiny part of the cosmological wave function—the robust macroscopic branch which we follow. Incidentally, the original motivation for Everett to develop his interpretation was quantum gravity. To quote from Everett (1957):

> The task of quantizing general relativity raises serious questions about the meaning of the present formulation and interpretation of quantum mechanics when applied to so fundamental a structure as the space-time geometry itself. This paper seeks to clarify the foundations of quantum mechanics. It presents a reformulation of quantum theory in a form believed suitable for application to general relativity.

10.1.3 *Decoherence of primordial fluctuations*

We have seen in the last subsection how important global degrees of freedom such as the scale factor of the universe can assume classical properties through interaction with irrelevant degrees of freedom such as density perturbations or gravitational waves. There are, however, situations when part of these 'irrelevant' variables become relevant themselves. According to the inflationary scenario of the early universe, all structure in the universe arises from quantum fluctuations. This was first discussed by Mukhanov and Chibisov (1981), see e.g. Börner (2003) for a review. We would thus owe our existence entirely to the uncertainty relations. In order to serve as the seeds for structure (galaxies, structures of galaxies), these quantum fluctuations have to become classical. Their imprint is seen in the anisotropy spectrum of the cosmic microwave background. The quantum-to-classical transition again relies heavily on the notion of decoherence; see Polarski and Starobinsky (1996), and Kiefer *et al.* (1998), as well as section 4.2.4 of Joos *et al.* (2003) for a review. It happens when the wavelength of the primordial quantum fluctuations becomes much bigger than the horizon size H_I^{-1} during the inflationary regime, where H_I denotes the Hubble parameter of inflation (which is approximately constant). The quantum state becomes highly squeezed during this phase. A prerequisite is the classical nature

---

[4]There exist also attempts to extend the 'Bohm interpretation' of quantum theory to quantum cosmology; cf. Pinto-Neto and Santini (2002), and Błaut and Kowalski-Glikman (1998).

of the background variables discussed in the last subsection, which is why one could talk about a 'hierarchy of classicality'.

Density fluctuations arise from the scalar part of the metric perturbations (plus the corresponding matter part). In addition one has of course the tensor perturbations of the metric. They correspond to gravitons (Chapter 2). Like for the scalar part the tensor part evolves into a highly squeezed state during inflation, and decoherence happens for them, too. The primordial gravitons would manifest themselves in a stochastic background of gravitational waves, which could probably be observed with the space-borne interferometer LISA to be launched in a couple of years. Its observation would constitute a direct test of linearized quantum gravity.

The decoherence time turns out to be of the order

$$t_d \sim \frac{H_I}{g} , \tag{10.41}$$

where g is a dimensionless coupling constant of the interaction with other 'irrelevant' fields causing decoherence. The ensuing coarse-graining brought about by the decohering fields causes an entropy increase for the primordial fluctuations (Kiefer et $al.$ 2000). The entropy production rate turns out to be given by $\dot{S} = H$, where H is the Hubble parameter of a general expansion. During inflation, H is approximately constant and the entropy increases linear with t. In the post-inflationary phases (radiation- and matter-dominated universe), $H \propto t^{-1}$ and the entropy increases only logarithmically in time. The main part of the entropy for the fluctuations is thus created during inflation. Incidentally, this behaviour resembles the behaviour for chaotic systems, although no chaos is involved here. The role of the Lyapunov coefficient is played by the Hubble parameter, and the Kolmogorov entropy corresponds to the entropy production mentioned here.

Decoherence also plays an important role for quantum black holes and in the context of wormholes and string theory; see section 4.2.5 of Joos et $al.$ (2003).

10.2   Arrow of time

One of the most intriguing open problems is the origin of irreversibility in our universe, also called the problem of the arrow of time (Zeh 2001). Since quantum gravity may provide the key for its solution, this topic will be briefly reviewed here, following Kiefer (2004 b) with modifications. More details and references can be found in Zeh (2001).

Although most of the fundamental laws of nature do not distinguish between past and future, there are many classes of phenomena which exhibit an arrow of time. This means that their time-reversed version is, under ordinary conditions, never observed. The most important ones are the following:

- Radiation arrow (advanced versus retarded radiation);
- Second Law of Thermodynamics (increase of entropy);
- Quantum theory (measurement process and emergence of classical properties);

- Gravitational phenomena (expansion of the universe and emergence of structure by gravitational condensation).

The expansion of the universe is distinguished because it does not refer to a class of phenomena; it is a single process. It has, therefore, been suggested that it is the common root for all other arrows of time—the 'master arrow'. We shall see in the course of our discussion that this seems indeed to be the case. But first we shall consider in more detail the various arrows of time.

The *radiation arrow* is distinguished by the fact that fields interacting with local sources are usually described by *retarded* solutions, which in general lead to a damping of the source. Advanced solutions are excluded. They would describe the reversed process, during which the field propagates coherently towards its source, leading to its excitation instead of damping. This holds, in fact, for all wave phenomena. In electrodynamics, a solution of Maxwell's equations can be described by

$$A^\mu = \text{source term plus boundary term}$$
$$= A^\mu_{\text{ret}} + A^\mu_{\text{in}}$$
$$= A^\mu_{\text{adv}} + A^\mu_{\text{out}} \, ,$$

where A^μ is the vector potential. The important question is then why the observed phenomena obey $A^\mu \approx A^\mu_{\text{ret}}$ or, in other words, why

$$A^\mu_{\text{in}} \approx 0 \tag{10.42}$$

holds instead of $A^\mu_{\text{out}} \approx 0$. Equation (10.42) is called a 'Sommerfeld radiation condition'. One believes that the radiation arrow can be traced back to thermodynamics: due to the absorption properties of the material which constitutes the walls of the laboratory in which electrodynamic experiments are being performed, ingoing fields will be absorbed within a very short time and (10.42) will be fulfilled. For the thermal properties of absorbers, the Second Law of Thermodynamics (see below) is responsible.

The condition (10.42) also seems to hold for the universe as a whole ('darkness of the night sky'). The so-called Olbers' paradox can be solved by noting that the universe is, in fact, not static, but has a finite age and is much too young to have enough stars for a bright night sky. This is, of course, not yet sufficient to understand the validity of (10.42) for the universe as a whole. In an early stage, the universe was a hot plasma in thermal equilibrium. Only the expansion of the universe and the ensuing redshift of the radiation are responsible for the fact that radiation has decoupled from matter and cooled to its present value of about three Kelvin—the temperature of the approximately isotropic cosmic background radiation with which the night sky 'glows'. During the expansion, a strong thermal non-equilibrium could develop, which enabled the formation of structure.

The second arrow is described by the *Second Law* of Thermodynamics: for a closed system entropy does not decrease. The total change of entropy is given by

$$\frac{\mathrm{d}S}{\mathrm{d}t} = \underbrace{\left(\frac{\mathrm{d}S}{\mathrm{d}t}\right)_{\text{ext}}}_{\mathrm{d}S_{\text{ext}}=\delta Q/T} + \underbrace{\left(\frac{\mathrm{d}S}{\mathrm{d}t}\right)_{\text{int}}}_{\geq 0} ,$$

so that according to the Second Law, the second term is non-negative. As the increase of entropy is also relevant for physiological processes, the Second Law is responsible for the subjective experience of irreversibility, in particular for the ageing process. If applied to the universe as a whole, it would predict the increase of its total entropy, which would seem to lead to its 'heat death' ('Wärmetod').

The laws of thermodynamics are based on microscopic statistical laws which are time-symmetric. How can the Second Law be derived from such laws? Already in the nineteenth century objections were formulated against a statistical foundation of the Second Law. These were, in particular,

- Loschmidt's reversibility objection ('Umkehreinwand'), and
- Zermelo's recurrence objection ('Wiederkehreinwand').

Loschmidt's objection states that a reversible dynamics must lead to an equal amount of transitions from an improbable to a probable state and from a probable to an improbable state. With overwhelming probability, the system should be in its most probable state, that is, in thermal equilibrium. Zermelo's objection is based on a theorem by Poincaré, according to which every system comes arbitrarily close to its initial state (and therefore to its initial entropy) after a finite amount of time. This objection is irrelevant, since the corresponding 'Poincaré times' are bigger than the age of the universe already for systems with few particles. The reversibility objection can only be avoided if a special boundary condition of low entropy holds for the early universe. Therefore, for the derivation of the Second Law, one needs a special *boundary condition*.

Such a boundary condition must either be postulated or derived from a fundamental theory. The formal description of entropy increase from such a boundary condition is done by master equations; cf. Joos *et al.* (2003). These are equations for the 'relevant' (coarse-grained) part of the system. In an open system, the entropy can of course decrease, provided the entropy capacity of the environment is large enough to at least compensate this entropy decrease. This is crucial for the existence of life, and a particular efficient process in this respect is photosynthesis. The huge entropy capacity of the environment comes in this case from the high temperature gradient between the hot Sun and the cold empty space: few high-energy photons (with small entropy) arrive on Earth, while many low-energy photons (with high entropy) leave it. Therefore, also the thermodynamic arrow of time points towards cosmology: how can gravitationally condensed objects like the Sun come from in the first place?

Another important arrow of time is the quantum-mechanical arrow. The Schrödinger equation is time-reversal invariant, but the measurement process, either through

- a dynamical *collapse* of the wave function, or
- an Everett *branching*

distinguishes a direction; cf. Section 10.1. We have seen that growing entanglement with other degrees of freedom leads to decoherence. The local entropy thereby increases. Again, decoherence only works if a special initial condition—a condition of weak entanglement—holds. But where can this come from?

The last of the main arrows is the gravitational arrow of time. Although the Einstein field equations are time-reversal invariant, gravitational systems in Nature distinguish a certain direction: the universe as a whole *expands*, while local systems such as stars form by *contraction*, for example, from gas clouds. It is by this gravitational contraction that the high temperature gradients between stars such as the Sun and the empty space arise. Because of the negative heat capacity for gravitational systems, homogeneous states possess a low entropy, whereas inhomogeneous states possess a high entropy—just the opposite than for non-gravitational systems.

An extreme case of gravitational collapse is the formation of black holes. We have seen in Section 7.1 that black holes possess an intrinsic entropy, the 'Bekenstein–Hawking entropy' (7.17). This entropy is much bigger than the entropy of the object from which the black hole has formed. If all matter in the observable universe were in a single gigantic black hole, its entropy would be $S_{\mathrm{BH}} \approx 10^{123} k_{\mathrm{B}}$ (Penrose 1981). Black holes thus seem to be the most efficient objects for swallowing information. A 'generic' universe would thus basically consist of black holes. Since this is not the case, our universe must have been started with a very special initial condition. Can this be analysed further? Close to the big bang, the classical theory of general relativity breaks down. A possible answer can thus only come from quantum gravity.

In the following, we shall adopt the point of view that the origin of irreversibility can be traced to the structure of the Wheeler–DeWitt equation. As can be seen, for example, from the minisuperspace case in (8.13), the potential term is highly asymmetric with respect to the scale factor $a \equiv \exp(\alpha)$: in particular, the potential term vanishes near the 'big bang' $\alpha \to -\infty$. This property is robust against the inclusion of (small) perturbations, that is, degrees of freedom describing density fluctuations or gravitational waves (cf. Section 8.2). Denoting these variables ('modes') again by $\{f_n\}$, one has for the total Hamiltonian in the Wheeler–DeWitt equation, an expression of the form (cf. Zeh 2001)

$$\hat{H} = \frac{\partial^2}{\partial \alpha^2} + \sum_n \left(-\frac{\partial^2}{\partial f_n^2} + V_n(\alpha, f_n) \right) + V_{\mathrm{int}}(\alpha, \{f_n\}) , \qquad (10.43)$$

where the last term describes the interaction between the modes (assumed to be small), and the V_n describe the interaction of the mode f_n with the scale factor

α. Both terms have, in fact, the property that they vanish for $\alpha \to -\infty$. It is, therefore, possible to impose in this limit a separating solution of $\hat{H}\Psi = 0$,

$$\Psi \xrightarrow{\alpha \to -\infty} \psi(\alpha) \prod_n \chi_n(f_n) \, , \tag{10.44}$$

that is, a solution of lacking entanglement. If this is taken as an 'initial condition', the Wheeler–DeWitt equation automatically—through the occurrence of the potentials in (10.43)—leads to a wave function, which for increasing α becomes *entangled* between α and all modes. This, then, leads to an increase of local entropy, that is, an increase of the entropy which is connected with the subset of 'relevant' degrees of freedom. Calling the latter $\{y_i\}$, one has

$$S(\alpha, \{y_i\}) = -k_B \mathrm{tr}(\rho \ln \rho) \, , \tag{10.45}$$

where ρ is the reduced density matrix corresponding to α and $\{y_i\}$. It is obtained by tracing out all irrelevant degrees of freedom in the full wave function. Entropy thus increases with increasing scale factor—this would be the gravitational arrow of time. It is also the arrow of time that is connected with decoherence. It is, therefore, the root for both the quantum mechanical and the thermodynamical arrow of time. Quantum gravity could thus really yield the master arrow, the formal reason being the asymmetric appearance of α in the Wheeler–DeWitt equation: the potential goes to zero near the big bang, but becomes highly nontrivial for increasing size of the universe. It is an interesting question whether a boundary condition of the form (10.44) would automatically result from one of the proposals discussed in Section 8.3. The symmetric initial condition (Section 8.3.5) is an example where this can be achieved—in fact, this condition was tailored for this purpose.

In the case of a classically recollapsing universe, the boundary condition (10.44) has interesting consequences: since it is formulated at $\alpha \to \infty$, increasing entropy is always correlated with increasing α, that is, increasing size of the universe; cf. also Fig. 8.1. Consequently, the arrow of time formally reverses near the classical turning point (Kiefer and Zeh 1995). It turns out that this region is fully quantum, so no paradox arises; it just means that there are many quasi-classical components of the wave function, each describing a universe that is experienced from within as expanding. All these components interfere destructively near the classical turning point, which would then constitute the 'end' of evolution. This would be analogous to the quantum region at the onset of inflation discussed in Section 10.1.2. Quantum gravity would thus in principle be able to provide a foundation for the origin of irreversibility—a remarkable achievement.

10.3 Outlook

Where do we stand? We have not yet achieved the final goal of having constructed a consistent quantum theory of gravity checked by experiments. It seems, however, clear what the main problems are. They are both of a conceptual and a mathematical nature. On the conceptual side, the most important task is to get rid of an external background space–time. This is drastically different from ordinary quantum field theory which heavily relies on Minkowski space (and its Poincaré symmetry) or a given curved space–time. An expression of this issue is the 'problem of time' with its connected problems of Hilbert space and the role of the probability interpretation. On the mathematical side, the main task is to construct a non-perturbative, anomaly-free framework from which definite and testable predictions can be made. An example would be the prediction for the final evaporation stage of black holes.

In this book I have presented two main approaches—quantum GR (in both covariant and canonical versions) and string theory. Both rely on the linear structure of quantum theory, that is, the general validity of the superposition principle. In this sense also string theory is a rather conservative approach, in spite of its 'exotic' features such as higher dimensions. This has to be contrasted with the belief of some of the founders of quantum mechanics (especially Heisenberg) that quantum theory has already to be superseded by going from the level of atoms to the level of nuclei.

What are the predictions of quantum gravity? Can it, for example, predict low-energy coupling constants and masses? As is well known, only a fine-tuned combination of the low-energy constants leads to a universe like ours in which human beings can exist. It would thus appear strange if a fundamental theory possessed just the right constants to achieve this. Hogan (2000) has argued that grand unified theories constrain relations among parameters, but leave enough freedom for a selection. In particular, he suggests that one coupling constant and two light fermion masses are *not* fixed by the symmetries of the fundamental theory.[5] One could then determine this remaining free constants only by the (weak form of the) *anthropic principle*: they have values such that a universe like ours is possible. The cosmological constant, for example, must not be much bigger than the presently observed value, because otherwise the universe would expand much too fast to allow the formation of galaxies. The universe is, however, too special to be explainable on purely anthropic grounds. In Section 10.2, we have mentioned that the maximal entropy would be reached if all the matter in the observable universe were collected into a single gigantic black hole. This entropy would be (in units of k_B) about 10^{123}, which is exceedingly more than the observed entropy of about 10^{88}. The 'probability' for our universe would then be about $\exp(10^{88})/\exp(10^{123})$, which is about $\exp(-10^{123})$. From the anthropic principle alone one would not need such a special universe. As for the cosmological constant, for example, one could imagine its calculation from a fundamental theory. Taking the presently observed value for Λ, one can construct a mass according to

---

[5]String theory contains only one fundamental dimensionful parameter, the string length. The connection to low energies may nonetheless be non-unique due to the existence of many different possible 'vacua'.

$$\left(\frac{\hbar^2 \Lambda^{1/2}}{G}\right)^{1/3} \approx 15 \text{ MeV} , \tag{10.46}$$

which in elementary particle physics is not an unusally big or small value. The observed value of Λ could thus emerge together with medium-size particle mass scales.

Since fundamental theories are expected to contain only one dimensionful parameter, low-energy constants emerge from fundamental quantum *fields*. An important example in string theory (Chapter 9) is the dilaton field from which one can calculate the gravitational constant. In order that these fields mimic physical constants, two conditions have to be satisfied. First, decoherence must be effective in order to guarantee a classical behaviour of the field. Second, this 'classical' field must then be approximately constant in large-enough space–time regions, within the limits given by experimental data. The field may still vary over large times or large spatial regions and thus mimic a 'time- or space-varying constant', cf. Uzan (2003).

The last word on any physical theory has to be spoken by experiment (observation). Apart from the possible determination of low-energy constants and their dependence on space and time, what could be the principle tests of quantum gravity?

1. *Black-hole evaporation*: A key test would be the final evaporation phase of a black hole. For this one would need to observe primordial black holes (Carr 2003). These are black holes that are not the end result of stellar collapse, but which can result from strong density perturbations in the early universe. In the context of inflation, their initial mass can be as small as 1 g. Primordial black holes with initial mass of about 5×10^{14} g would evaporate at the present age of the universe. Unfortunately, no such object has yet been observed. Especially promising may be models of inflationary cosmology with a distinguished scale (Bringmann *et al.* 2002).

2. *Cosmology*: Quantum aspects of the gravitational field may be observed in the anisotropy spectrum of the cosmic microwave background. First, future experiments may be able to see the contribution of the gravitons generated in the early universe. This important effect was already emphasized by Starobinsky (1979). The production of gravitons by the cosmological evolution would be an effect of linear quantum gravity. Second, quantum-gravitational correction terms from the Wheeler–DeWitt equation may leave their impact on the anisotropy spectrum (Section 5.4). Third, a discreteness in the inflationary perturbations could manifest itself in the spectrum (Hogan 2002).

3. *Discreteness of space and time*: Both in string theory and quantum general relativity there are hints of a discrete structure of space–time. This could be seen through the observation of effects violating local Lorentz invariance (Amelino-Camelia 2002), for example, in the dispersion relation of the electromagnetic waves coming from gamma-ray bursts. It has even

be suggested that effects of a discrete space–time could be seen in atomic interferometry (Percival 1997).

4. *Signatures of higher dimensions*: An important feature of string theory is the existence of additional space–time dimensions (Section 9.2.6). They could manifest themselves in scattering experiments at the Large Hadron Collider (LHC) at CERN, which will start to operate in around 2006. It is also imaginable that they cause observable deviations from the standard cosmological scenario.

Of course, there may be other possibilities which are not yet known and which could offer great surprises.

Quantum gravity has been studied since the end of the 1920s. No doubt, much progress has been made since then. I hope that this book has given some impressions from this progress. The final goal has not yet been reached. The belief expressed here is that a consistent and experimentally successful theory of quantum gravity will be available in the future. However, it may still take a while before this time is reached.

REFERENCES

Achúcarro, A. and Townsend, P. K. (1986). A Chern-Simons action for three-dimensional anti-de Sitter supergravity theories. *Phys. Lett. B*, **180**, 89–92.

Aharony, O., Gubser, S. S., Maldacena, J., Ooguri, H., and Oz, Y. (2000). Large N field theories, string theory and gravity. *Phys. Rep.*, **323**, 183–386.

Altshuler, B. L. and Barvinsky, A. O. (1996). Quantum cosmology and physics of transitions with a change of the spacetime signature. *Phys. Usp.*, **39**, 429–59.

Amelino-Camelia, G. (2002). Quantum-gravity phenomenology: status and prospects. *Mod. Phys. Lett. A*, **17**, 899–922.

Anselmi, D. (2003). Absence of higher derivatives in the renormalization of propagators in quantum field theories with infinitely many couplings. *Class. Quantum Grav.*, **20**, 2355–78.

Antoniadis, I., Arkani-Hamed, N., Dimopoulos, S., and Dvali, G. (1998). New dimensions at a millimeter to a fermi and superstring at TeV. *Phys. Lett. B*, **436**, 257–63.

Arkani-Hamed, N., Dimopoulos, S., and Dvali, G. (1998). The hierarchy problem and new dimensions at a millimeter. *Phys. Lett. B*, **429**, 263–72.

Arnowitt, R., Deser, S., and Misner, C. W. (1962). The dynamics of general relativity. In *Gravitation: an introduction to current research* (ed. L. Witten), pp. 227–65. Wiley, New York.

Ashtekar, A. (1986). New variables for classical and quantum gravity. *Phys. Rev. Lett.*, **57**, 2244–7.

Ashtekar, A. (1988). *New perspectives in canonical gravity*. Bibliopolis, Napoli.

Ashtekar, A. (1991). *Lectures on non-perturbative canonical gravity*. World Scientific, Singapore.

Ashtekar, A., Baez J., Corichi, A., and Krasnov, K. (1998). Quantum geometry and black hole entropy. *Phys. Rev. Lett.*, **80**, 904–7.

Ashtekar, A., Krasnov, K., and Baez, J. (2000). Quantum geometry of isolated horizons and black hole entropy. *Adv. Theor. Math. Phys.*, **4**, 1–94.

Ashtekar, A. and Lewandowski, J. (1997). Quantum theory of geometry I: area operators. *Class. Quantum Grav.*, **14**, A55–82.

Ashtekar, A. and Lewandowski, J. (1998). Quantum theory of geometry II: volume operators. *Adv. Theor. Math. Phys.*, **1**, 388-429.

Ashtekar, A., Marolf, D., and Mourão, J. (1994). Integration on the space of connections modulo gauge transformations. In *Proc. Cornelius Lanczos Int. Centenary Conf.* (ed. J. D. Brown *et al.*), pp. 143–60. SIAM, Philadelphia.

Ashtekar, A. and Pierri, M. (1996). Probing quantum gravity through exactly soluble midi-superspaces I. *J. Math. Phys.*, **37**, 6250–70.

Audretsch, J., Hehl, F. W., and Lämmerzahl, C. (1992). Matter wave interferometry and why quantum objects are fundamental for establishing a gravitational theory. In *Relativistic gravity research* (ed. J. Ehlers and G. Schäfer), pp. 368–407. Lecture Notes in Physics 410. Springer, Berlin.

Baierlein, R. F., Sharp, D. H., and Wheeler. J. A. (1962). Three-dimensional geometry as carrier of information about time. *Phys. Rev.*, **126**, 1864–65.

Bañados, M., Teitelboim, C., and Zanelli, J. (1992). The black hole in three-dimensional space–time. *Phys. Rev. Lett.*, **69**, 1849–51.

Banks, T. (1985). TCP, quantum gravity, the cosmological constant and all that ... . *Nucl. Phys. B*, **249**, 332–60.

Banks, T., Fischler, W., Shenker, S. H., and Susskind, L. (1997). M theory as a matrix model: a conjecture. *Phys. Rev. D*, **55**, 5112–28.

Barbero, J. F. (1995). Real Ashtekar variables for Lorentzian space–times. *Phys. Rev. D*, **51**, 5507–10.

Barbour, J. B. (1986). Leibnizian time, Machian dynamics, and quantum gravity. In *Quantum concepts in space and time* (ed. R. Penrose and C. J. Isham), pp. 236–46. Oxford University Press, Oxford.

Barbour, J. B. (1989). *Absolute or relative motion? Vol. 1: The discovery of dynamics*. Cambridge University Press, Cambridge.

Barbour, J. B. (1993). Time and complex numbers in canonical quantum gravity. *Phys. Rev. D*, **47**, 5422–9.

Barbour, J. B. (1994). The timelessness of quantum gravity: I. The evidence from the classical theory. *Class. Quantum Grav.*, **11**, 2853–73.

Barbour, J. B. and Bertotti, B. (1982). Mach's principle and the structure of dynamical theories. *Proc. R. Soc. Lond. A*, **382**, 295–306.

Barbour, J. B., Foster, B., and Ó Murchadha, N. (2002). Relativity without relativity. *Class. Quantum Grav.*, **19**, 3217–48.

Bardeen, J. M. (1980). Gauge invariant cosmological perturbations. *Phys. Rev. D*, **22**, 1882–905.

Bartnik, R. and Fodor, G. (1993). On the restricted validity of the thin sandwich conjecture. *Phys. Rev. D*, **48**, 3596–99.

Barvinsky, A. O. (1989). Perturbative quantum cosmology: the probability measure on superspace and semiclassical expansion. *Nucl. Phys. B*, **325**, 705–23.

Barvinsky, A. O. (1990). Effective action method in quantum field theory. In *Gauge theories of fundamental interactions* (ed. M. Pawlowski and R. Raczka), pp. 265–327. World Scientific, Singapore.

Barvinsky, A. O. (1993a). Unitarity approach to quantum cosmology. *Phys. Rep.*, **230**, 237–367.

Barvinsky, A. O. (1993b). Operator ordering in theories subject to constraints of the gravitational type. *Class. Quantum Grav.*, **10**, 1985–99.

Barvinsky, A. O. (1998). Solution of quantum Dirac constraints via path integral. *Nucl. Phys. B*, **520**, 533–60.

Barvinsky, A. O. (2001). Quantum cosmology at the turn of millennium. Presented at 9th Marcel Grossmann Meeting, Rome, Italy, 2-9 Jul 2000. Available on http://arxiv.org/abs/gr-qc/0101046.

Barvinsky, A. O. and Kamenshchik, A. Yu. (1990). One loop quantum cosmology: the normalizability of the Hartle–Hawking wave function and the probability of inflation. *Class. Quantum Grav.*, **7**, L181–6.

Barvinsky, A. O. and Kamenshchik, A. Yu. (1998). Effective equations of motion and initial conditions for inflation in quantum cosmology. *Nucl. Phys. B*, **532**, 339–60.

Barvinsky, A. O. and Kiefer, C. (1998). Wheeler–DeWitt equation and Feynman diagrams. *Nucl. Phys. B*, **526**, 509–39.

Barvinsky, A. O. and Krykhtin, V. (1993). Dirac and BFV quantization methods in the 1-loop approximation: closure of the quantum constraint algebra and the conserved inner product. *Class. Quantum Grav.*, **10**, 1957–84.

Barvinsky, A. O. and Kunstatter, G. (1996). Exact physical black hole states in generic 2-D dilaton gravity. *Phys. Lett. B*, **389**, 231–6.

Barvinsky, A. O. and Nesterov, D. V. (2001). Effective equations in quantum cosmology. *Nucl. Phys. B*, **608**, 333–74.

Barvinsky, A. O. and Vilkovisky, G. A. (1987). The effective action in quantum field theory: two-loop approximation. In *Quantum field theory and quantum statistics*, Vol I (ed. I. Batalin, C. J. Isham, and G. A. Vilkovisky), pp. 245–75. Adam Hilger, Bristol.

Barvinsky, A. O., Kamenshchik, A. Yu., and Karmazin, I. P. (1993). The renormalization group for nonrenormalizable theories: Einstein gravity with a scalar field. *Phys. Rev. D*, **48**, 3677–94.

Barvinsky, A. O., Kamenshchik, A. Yu., and Mishakov, I. V. (1997). Quantum origin of the early inflationary universe. *Nucl. Phys. B*, **491**, 387–426.

Barvinsky, A. O., Kamenshchik, A. Yu., Kiefer, C., and Mishakov, I. V. (1999*a*). Decoherence in quantum cosmology at the onset of inflation. *Nucl. Phys. B*, **551**, 374–96.

Barvinsky, A. O., Kamenshchik, A. Yu., and Kiefer, C. (1999*b*). Effective action and decoherence by fermions in quantum cosmology. *Nucl. Phys. B*, **552**, 420–44.

Barvinsky, A. O., Kamenshchik, A. Yu., Rathke, A., and Kiefer, C. (2003*a*). Nonlocal braneworld action: an alternative to the Kaluza–Klein description. *Phys. Rev. D*, **67**, 023513.

Barvinsky, A. O., Kamenshchik, A. Yu., Rathke, A., and Kiefer, C. (2003*b*). Radion-induced gravitational wave oscillations and their phenomenology. *Ann. Phys. (Leipzig)*, 8th series, **12**, 343–70.

Batalin, I. A. and Fradkin, E. S. (1983). Operator quantization of relativistic dynamical systems subject to first class constraints. *Phys. Lett. B*, **128**, 303–8.

Batalin, I. A. and Vilkovisky, G. A. (1977). Relativistic S matrix of dynamical systems with boson and fermion constraints. *Phys. Lett. B*, **69**, 309–12.

Becker, K. and Becker, M. (1999). Quantum gravity corrections for Schwarzschild black holes. *Phys. Rev. D*, **60**, 026003.

Beig, R. and Ó Murchadha (1987). The Poincaré group as the symmetry group of canonical general relativity. *Ann. Phys. (NY)*, **174**, 463–98.

Bekenstein, J. D. (1973). Black holes and entropy. *Phys. Rev. D*, **7**, 2333–46.

Bekenstein, J. D. (1974). The quantum mass spectrum of the Kerr black hole. *Lett. Nuovo Cim.*, **11**, 467–70.

Bekenstein, J. D. (1999). Quantum black holes as atoms. In *Proc. 8th Marcel Grossmann Meet.* (ed. T. Piran and R. Ruffini), pp. 92–111. World Scientific, Singapore.

Bekenstein, J. D. (2001). The limits of information. *Stud. Hist. Philos. Mod. Phys.*, **32**, 511–24.

Bekenstein, J. D. and Mukhanov, V. F. (1995). Spectroscopy of the quantum black hole. *Phys. Lett. B*, **360**, 7–12.

Belinskii, V. A., Khalatnikov, I. M., and Lifshitz, E. M. (1982). A general solution of the Einstein equations with a time singularity. *Adv. Phys.*, **31**, 639–67.

Bell, J. S. (1987). *Speakable and unspeakable in quantum mechanics*. Cambridge University Press, Cambridge.

Bento, M. C. and Bertolami, O. (1995). Scale factor duality: a quantum cosmological approach. *Class. Quantum Grav.*, **12**, 1919–26.

Berenstein, D., Maldacena, J., and Nastase, H. (2002). Strings in flat space and pp waves from $\mathcal{N} = 4$ super Yang Mills. *JHEP*, 04(2002), 013.

Bergmann, P. (1989). The canonical formulation of general relativistic theories: the early years, 1930–1959. In *Einstein and the history of general relativity* (ed. D. Howard and J. Stachel), pp. 293–99. Birkhäuser, Boston.

Bergmann, P. and Komar, A. (1972). The coordinate group symmetries of general relativity. *Int. J. Theor. Phys.*, **5**, 15–28.

Berkovits, N., Shen, A., and Zwiebach, B. (2000). Tachyon condensation in superstring field theory. *Nucl. Phys. B*, **587**, 147–78.

Bern, Z. (2002). Perturbative quantum gravity and its relation to gauge theory. *Living reviews in relativity*, available on
www.livingreviews.org/Articles/Volume5/2002-5bern.

Berry, M. V. (1984). Quantal phase factors accompanying adiabatic changes. *Proc. Roy. Soc. Lond. A*, **392**, 45–57.

Bertlmann, R. A. (1996). *Anomalies in quantum field theory*. Clarendon Press, Oxford.

Bertoni, C., Finelli, F., and Venturi, G. (1996). The Born–Oppenheimer approach to the matter-gravity system and unitarity. *Class. Quantum Grav.*, **13**, 2375–83.

Birrell, N. D. and Davies, P. C. W. (1982). *Quantum fields in curved space*. Cambridge University Press, Cambridge.

Bjerrum-Bohr, N. E. J., Donoghue, J. F., and Holstein, B. R. (2003*a*). Quantum gravitational corrections to the nonrelativistic scattering potential of two masses. *Phys. Rev. D*, **67**, 084033.

Bjerrum-Bohr, N. E. J., Donoghue, J. F., and Holstein, B. R. (2003*b*). Quantum corrections to the Schwarzschild and Kerr metrics. *Phys. Rev. D*, **68**, 084005.

Blagojević, M. (2002). *Gravitation and gauge symmetries*. Institute of Physics, Bristol.

Błaut, A. and Kowalski-Glikman, J. (1998). The time evolution of quantum universe in the quantum potential picture. *Phys. Lett. A*, **245**, 197–202.

Böhm, M., Denner, A., and Joos, H. (2001). *Gauge theories of the strong and electroweak interaction*. Teubner, Stuttgart.

Bohr, N. (1949). Discussion with Einstein on epistemological problems in atomic physics. In *Albert Einstein: Philosopher-Scientist* (ed. P. A. Schilpp), pp. 200–41. Library of Living Philosophers, Vol. VII, Evanston, Illinois.

Bohr, N. and Rosenfeld, L. (1933). Zur Frage der Messbarkeit der elektromagnetischen Feldgrössen. *Det Kgl. Danske Videnskabernes Selskab. Mathematisk - fysiske Meddelelser*, **XII**, 8, 3–65; English translation in Wheeler, J. A. and Zurek, W. H. (ed.), *Quantum theory and measurement*. Princeton University Press, Princeton (1983).

Bojowald, M. (2003). Initial conditions for a universe. *Gen. Rel. Grav.*, **35**, 1877–83.

Börner, G. (2003). *The early universe—facts and fiction*, 4th edn. Springer, Berlin.

Boulanger, N., Damour, T., Gualtieri, L., and Henneaux, M. (2001). Inconsistency of interacting, multi-graviton theories. *Nucl. Phys. B*, **597**, 127–71.

Boulware, D. G. and Deser, S. (1967). Stress-tensor commutators and Schwinger terms. *J. Math. Phys.*, **8**, 1468–77.

Bousso, R. (2002). The holographic principle. *Rev. Mod. Phys.*, **74**, 825–74.

Bousso, R. and Polchinski, J. (2000). Quantization of four-form fluxes and dynamical neutralization of the cosmological constant. *JHEP*, 06(2000), 006.

Briggs, J. S. and Rost, J. M. (2001). On the derivation of the time-dependent equation of Schrödinger. *Found. Phys.*, **31**, 693–712.

Bringmann, T., Kiefer, C., and Polarski, D. (2002). Primordial black holes from inflationary models with and without broken scale invariance. *Phys. Rev. D*, **65**, 024008.

Bronstein, M. (1936). Quantentheorie schwacher Gravitationsfelder. *Physikalische Zeitschrift der Sowjetunion*, **9**, 140–57.

Brotz, T. and Kiefer, C. (1997). Semiclassical black hole states and entropy. *Phys. Rev. D*, **55**, 2186–91.

Brout, R. and Venturi, G. (1989). Time in semiclassical gravity. *Phys. Rev. D*, **39**, 2436–9.

Brown, J. D. (1988). *Lower dimensional gravity*. World Scientific, Singapore.

Brown, J. D. and Kuchař, K. V. (1995). Dust as a standard of space and time in canonical quantum gravity. *Phys. Rev. D*, **51**, 5600–29.

Brown, J. D. and York, J. W. (1989). Jacobi's action and the recovery of time in general relativity. *Phys. Rev. D*, **40**, 3312–18.

Brügmann, B. (1994). Loop representations. In *Canonical gravity: from classical to quantum* (ed. J. Ehlers and H. Friedrich), pp. 213–53. Lecture Notes in Physics 434. Springer, Berlin.

Buchbinder, I. L., Odintsov, S. D., and Shapiro, I. L. (1992). *Effective action in quantum gravity*. Institute of Physics Publishing, Bristol.

Butterfield, J. and Isham, C. J. (1999). On the emergence of time in quantum gravity. In *The arguments of time* (ed. J. Butterfield), pp. 111–68. Oxford University Press, Oxford.

Callan, C. G., Giddings, S. B., Harvey, S. A., and Strominger, A. (1992). Evanescent black holes. *Phys. Rev. D*, **45**, R1005–9.

Cangemi, D., Jackiw, R., and Zwiebach, B. (1996). Physical states in matter-coupled dilaton gravity. *Ann. Phys. (NY)*, **245**, 408–44.

Carlip, S. (1998). *Quantum gravity in 2+1 dimensions*. Cambridge University Press, Cambridge.

Carlip, S. (2000). Logarithmic corrections to black hole entropy, from the Cardy formula. *Class. Quantum Grav.*, **17**, 4175–86.

Carlip, S. (2001). Quantum gravity: a progress report. *Rep. Prog. Phys.*, **64**, 885–942.

Carlip, S. and Teitelboim, C. (1995). The off-shell black hole. *Class. Quantum Grav.*, **12**, 1699–704.

Caroll, S. M., Friedman, D. Z., Ortiz, M. E., and Page, D. N. (1994). Physical states in canonically quantized supergravity. *Nucl. Phys. B*, **423**, 661–85.

Carr, B. J. (2003). Primordial black holes as a probe of cosmology and high energy physics. In *Quantum gravity: from theory to experimental search* (ed. D. Giulini, C. Kiefer, and C. Lämmerzahl), pp. 301–21. Lecture Notes in Physics 631. Springer, Berlin.

Carrera, M. and Giulini, D. (2001). Classical analysis of the van Dam-Veltman discontinuity. Available on http://arxiv.org/abs/gr-qc/0107058.

Cavaglià, M. and Moniz, P. V. (2001). Canonical and quantum FRW cosmological solutions in M-theory. *Class. Quantum Grav.*, **18**, 95–120.

Chen, P. and Tajima, T. (1999). Testing Unruh radiation with ultraintense lasers. *Phys. Rev. Lett.*, **83**, 256–9.

Choquet-Bruhat, Y. and York, J. (1980). The Cauchy problem. In *General relativity and gravitation*, Vol I (ed. A. Held), pp. 99–172. Plenum Press, New York.

Conradi, H. D. (1992). Initial state in quantum cosmology. *Phys. Rev. D*, **46**, 612–9.

Conradi, H. D. (1998). Tunneling of macroscopic universes. *Int. J. Mod. Phys. D*, **7**, 189–200.

Conradi, H. D. and Zeh, H. D. (1991). Quantum cosmology as an initial value problem. *Phys. Lett. A*, **154**, 321–6.

Csordás, A. and Graham, R. (1995). Exact quantum state for N=1 supergravity. *Phys. Rev. D*, **52**, 6656–9.

Dąbrowski, M. P. and Kiefer, C. (1997). Boundary conditions in quantum string cosmology. *Phys. Lett. B*, **397**, 185–92.

Damour, T. and Taylor, J. H. (1991). On the orbital period change of the binary pulsar PSR 1913+16. *Astrophys. J.*, **366**, 501–11.

Dasgupta, A. and Loll, R. (2001). A proper-time cure for the conformal sickness in quantum gravity. *Nucl. Phys. B*, **606**, 357–79.

Davies, P. C. W. (1975). Scalar particle production in Schwarzschild and Rindler metrics. *J. Phys. A*, **8**, 609–16.

D'Eath, P. D. (1984). Canonical quantization of supergravity. *Phys. Rev. D*, **29**, 2199–219.

D'Eath, P. D. (1996). *Supersymmetric quantum cosmology*. Cambridge University Press, Cambridge.

Demers, J.-G. and Kiefer, C. (1996). Decoherence of black holes by Hawking radiation. *Phys. Rev. D*, **53**, 7050–61.

Deser, S. (1970). Self-interaction and gauge invariance. *Gen. Rel. Grav.* **1**, 9–18.

Deser, S. (1987). Gravity from self-interaction in a curved background. *Class. Quantum Grav.* **4**, L99–105.

Deser, S. (1989). Quantum gravity: whence, whither. In *Trends in theoretical physics* (ed. P. J. Ellis and Y. C. Tang), pp. 175–91. Addison-Wesley, Reading.

Deser, S. (2000). Infinities in quantum gravities. *Ann. Phys. (Leipzig)*, 8th series, **9**, 299–306.

Deser, S., van Nieuwenhuizen, P., and Boulware, D. (1975). Uniqueness and nonrenormalisability of quantum gravitation. In *General relativity and gravitation* (ed. G. Shaviv and J. Rosen), pp. 1–18. Wiley, New York.

DeWitt, B. S. (1962). The quantization of geometry. In *Gravitation: an introduction to current research* (ed. L. Witten), pp. 266–381. Wiley, New York.

DeWitt, B. S. (1964). Gravity: a universal regulator? *Phys. Rev. Lett.*, **13**, 114–18.

DeWitt, B. S. (1965). *Dynamical theory of groups and fields*. Blackie & Son Limited, London.

DeWitt, B. S. (1967a). Quantum theory of gravity. I. The canonical theory. *Phys. Rev.*, **160**, 1113–48.

DeWitt, B. S. (1967b). Quantum theory of gravity. II. The manifestly covariant theory. *Phys. Rev.*, **162**, 1195–239.

DeWitt, B. S. (1967c). Quantum theory of gravity. III. Applications of the covariant theory. *Phys. Rev.*, **162**, 1239–62.

DeWitt, B. S. (1970). Spacetime as a sheaf of geodesics in superspace. In *Relativity* (ed. M. Carmeli, S. I. Fickler, and L. Witten), pp. 359–74.

DeWitt, B. S. (1979). Quantum gravity: the new synthesis. In *General relativity. An Einstein centenary survey* (ed. S. W. Hawking and W. Israel), pp. 680–745. Cambridge University Press, Cambridge.

DeWitt, B. S. (2003). *The global approach to quantum field theory*, Vols. I and II. Clarendon Press, Oxford.

Diósi, L., Gisin, N., and Strunz, W. T. (2000). Royal road to coupling classical and quantum mechanics. *Phys. Rev. A*, **61**, 22108.

Diósi, L. and Kiefer, C. (2000). Robustness and diffusion of pointer states. *Phys. Rev. Lett.*, **85**, 3552–5.

Dirac, P. A. M. (1964). *Lectures on quantum mechanics*. Belfer Graduate School of Science, Yeshiva University, New York.

Dreyer, O. (2003). Quasinormal modes, the area spectrum, and black hole entropy. *Phys. Rev. Lett.*, **90**, 081301.

Donoghue, J. F. (1994). General relativity as an effective field theory. *Phys. Rev. D*, **50**, 3874–88.

Donoghue, J. F. and Torma, T. (1999). Infrared behavior of graviton–graviton scattering. *Phys. Rev. D*, **60**, 024003.

Douglas, M. R. (2003). The statistics of string/M theory vacua. *JHEP*, 05(2003), 046.

Douglas, M. R. and Nekrasov, N. A. (2002). Noncommutative field theory. *Rev. Mod. Phys.*, **73**, 977–1029.

Duff, M. J., Okun, L. B., and Veneziano, G. (2002). Trialogue on the number of fundamental constants. *JHEP*, 03(2002), 023.

Ehlers, J. (1973). The nature and structure of spacetime. In *The physicist's conception of nature* (ed. J. Mehra), pp. 71–91. D. Reidel, Dordrecht.

Ehlers, J. (1995). Machian ideas and general relativity. In *Mach's principle— from Newton's bucket to general relativity* (ed. J. B. Barbour and H. Pfister), pp. 458–73. Birkhäuser, Boston.

Einstein, A. (1916a). Hamiltonsches Prinzip und allgemeine Relativitätstheorie. *Sitzber. kgl.-preuß. Akad. Wiss. Berlin, Sitzung der phys.-math. Klasse*, **XLII**, 1111–6.

Einstein, A. (1916b). Näherungsweise Integration der Feldgleichungen der Gravitation. *Sitzber. kgl.-preuß. Akad. Wiss. Berlin, Sitzung der phys.-math. Klasse*, **XXXII**, 688–96.

Einstein, A. and Fokker, A. D. (1914). Die Nordströmsche Gravitationstheorie vom Standpunkt des absoluten Differentialkalküls. *Ann. Phys. (Leipzig)*, 4th series, **44**, 321–8.

Eppley, K. and Hannah, E. (1977). The necessity of quantizing the gravitational field. *Found. Phys.*, **7**, 51–68.

Everett, H. (1957). 'Relative state' formulation of quantum mechanics. *Rev. Mod. Phys.*, **29**, 454–62. Reprinted in Wheeler, J. A. and Zurek, W. H. (ed.), *Quantum theory and measurement*. Princeton University Press, Princeton (1983).

Faddeev, L. D. and Popov, V. N. (1967). Feynman diagrams for the Yang–Mills field. *Phys. Lett. B*, **25**, 29–30.

Farrugia, Ch. J. and Hajicek, P. (1979). The third law of black hole mechanics: a counterexample. *Commun. Math. Phys.*, **68**, 291–9.

Fels, M. E. and Torre, C. G. (2002). The principle of symmetric criticality in general relativity. *Class. Quantum Grav.*, **19**, 641–76.

Feng, J. L., March-Russell, J., Sethi, S., and Wilczek, F. (2001). Saltatory relaxation of the cosmological constant. *Nucl. Phys. B*, **602**, 307–28.

Fertig, C. and Gibble, K. (2000). Measurement and cancellation of the cold collision frequency in an ^{87}Rb fountain clock. *Phys. Rev. Lett.*, **85**, 1622–5.

Feynman, R. P. (1963). Quantum theory of gravitation. *Acta Physica Polonica*, **XXIV**, 697–722.

Feynman, R. P. and Hibbs, A. R. (1965). *Quantum mechanics and path integrals.* McGraw Hill, Boston.

Fierz, M. and Pauli, W. (1939). On relativistic wave equations for particles of arbitrary spin in an electromagnetic field. *Proc. Roy. Soc. A*, **173**, 211–32.

Fischer, A. E. (1970). The theory of superspace. In *Relativity* (ed. M. Carmeli, S. I. Fickler, and L. Witten), pp. 303–57. Plenum Press, New York.

Fischer, A. E. (1986). Resolving the singularities in the space of Riemannian geometries. *J. Math. Phys.*, **27**, 718–38.

Ford, L. H. (1982). Gravitational radiation by quantum systems. *Ann. Phys. (NY)*, **144**, 238–48.

Fradkin, E. S. and Vasiliev, M. A. (1977). Hamiltonian formalism, quantization and S-matrix for supergravity. *Phys. Lett. B*, **72**, 70–4.

Fré, P., Gorini, V., Magli, G., and Moschella, U. (ed.) (1999). *Classical and quantum black holes.* Institute of Physics Publishing, Bristol.

Friedman, J. L. and Sorkin, R. D. (1980). Spin 1/2 from gravity. *Phys. Rev. Lett.*, **44**, 1100–3.

Frieman, J., Brandenberger, R., Kiefer, C., Müller, V., Mukhanov, V., Sato, K., *et al.* (1997). Are we making progress in relating cosmology and fundamental theories? In *The evolution of the Universe* (ed. G. Börner and S. Gottlöber), pp. 141–56. Wiley, Chichester.

Fritelli, S., Lehner, L., and Rovelli, C. (1996). The complete spectrum of the area from recoupling theory in loop quantum gravity. *Class. Quantum Grav.*, **13**, 2921–32.

Frolov, V. P. and Novikov, I. D. (1998). *Black hole physics.* Kluwer, Dordrecht.

Frolov, V. P. and Vilkovisky, G. A. (1981). Spherical symmetric collapse in quantum gravity. *Phys. Lett. B*, **106**, 307–13.

Fuji, Y. and Maeda, K. (2003). *The scalar-tensor theory of gravitation.* Cambridge University Press, Cambridge.

Fulling, S. A. (1973). Nonuniqueness of canonical field quantization in Riemannian space–time. *Phys. Rev. D*, **7**, 2850–62.

Gasperini, M. and Veneziano, G. (2003). The pre-big bang scenario in string cosmology. *Phys. Rep.*, **373**, 1–212.

Gasser, J. and Leutwyler, H. (1984). Chiral perturbation theory to one loop. *Ann. Phys. (NY)*, **158**, 142–210.

Gerlach, U. H. (1969). Derivation of the ten Einstein field equations from the semiclassical approximation to quantum geometrodynamics. *Phys. Rev.*, **177**, 1929–41.

Geroch, R. and Hartle, J. B. (1986). Computability and physical theories. *Found. Phys.*, **16**, 533–50.

Gibbons, G. W. and Hartle, J. B. (1990). Real tunneling geometries and the large-scale topology of the universe. *Phys. Rev. D*, **42**, 2458–68.

Gibbons, G. W., Hawking, S. W., and Perry, M. J. (1978). Path integrals and the indefiniteness of the gravitational action. *Nucl. Phys. B*, **138**, 141–50.

Giulini, D. (1995*a*). On the configuration space topology in general relativity. *Helv. Phys. Acta*, **68**, 86–111.

Giulini, D. (1995*b*). What is the geometry of superspace? *Phys. Rev. D*, **51**, 5630–5.

Giulini, D. (1999). The generalized thin-sandwich problem and its local solvability. *J. Math. Phys.*, **40**, 2470–82.

Giulini, D. and Kiefer, C. (1994). Wheeler–DeWitt metric and the attractivity of gravity. *Phys. Lett. A*, **193**, 21–4.

Giulini, D. and Kiefer, C. (1995). Consistency of semiclassical gravity. *Class. Quantum Grav.*, **12**, 403–11.

Goroff, M. H. and Sagnotti, A. (1985). Quantum gravity at two loops. *Phys. Lett. B*, **160**, 81–6.

Gousheh, S. S. and Sepangi, H. R. (2000). Wave packets and initial conditions in quantum cosmology. *Phys. Lett. A*, **272**, 304–12.

Green, M. B., Schwarz, J. H., and Witten, E. (1987). *Superstring theory*, 2 vols. Cambridge University Press, Cambridge.

Greene, B. (1997). String theory on Calabi–Yau manifolds. Available on http://arxiv.org/abs/hep-th/9702155.

Grib, A. A., Mamayev, S. G., and Mostepanenko, V. M. (1994). *Vacuum effects in strong fields*. Friedmann Laboratory Publishing, St. Petersburg.

Gross, D. and Periwal, V. (1988). String perturbation theory diverges. *Phys. Rev. Lett.*, **60**, 2105–8.

Grumiller, D., Kummer, W., and Vassilevich, D. V. (2002). Dilaton gravity in two dimensions. *Phys. Rep.*, **369**, 327–430.

Haag, R., Lopuszanski, J. T., and Sohnius, M. (1975). All possible generators of supersymmetries of the S matrix. *Nucl. Phys. B*, **88**, 257–74.

Hájíček, P. (2001). Unitary dynamics of spherical null gravitating shells. *Nucl. Phys. B*, **603**, 555–77.

Hájíček, P. (2003). Quantum theory of gravitational collapse (lecture notes on quantum conchology). In *Quantum gravity: from theory to experimental search* (ed. D. Giulini, C. Kiefer, and C. Lämmerzahl), pp. 255–99. Lecture Notes in Physics 631. Springer, Berlin.

Hájíček, P. and Kiefer, C. (2001*a*). Embedding variables in the canonical theory of gravitating shells. *Nucl. Phys. B*, **603**, 531–54.

Hájíček, P. and Kiefer, C. (2001*b*). Singularity avoidance by collapsing shells in quantum gravity. *Int. J. Mod. Phys. D*, **10**, 775–9.

Hájíček, P. and Kijowski, J. (2000). Covariant gauge fixing and Kuchař decomposition. *Phys. Rev. D*, **61**, 024037.

Halliwell, J. J. (1988). Derivation of the Wheeler–DeWitt equation from a path integral for minisuperspace models. *Phys. Rev. D*, **38**, 2468–81.

Halliwell, J. J. (1989). Decoherence in quantum cosmology. *Phys. Rev. D*, **39**, 2912–23.

Halliwell, J. J. (1991). Introductory lectures on quantum cosmology. In *Quantum cosmology and baby universes* (ed. S. Coleman, J. B. Hartle, T. Piran, and S. Weinberg), pp. 159–243. World Scientific, Singapore.

Halliwell, J. J. (1998). Effective theories of coupled classical and quantum variables from decoherent histories: a new approach to the back reaction problem. *Phys. Rev. D*, **57**, 2337–48.

Halliwell, J. J. and Hartle, J. B. (1991). Wave functions constructed from an invariant sum over histories satisfy constraints. *Phys. Rev. D*, **43**, 1170–94.

Halliwell, J. J. and Hawking, S. W. (1985). The origin of structure in the Universe. *Phys. Rev. D*, **31**, 1777–91.

Halliwell, J. J. and Louko, J. (1991). Steepest descent contours in the path integral approach to quantum cosmology. 3. A general method with applications to anisotropic minisuperspace models. *Phys. Rev. D*, **42**, 3997–4031.

Hanson, A. J., Regge, T., and Teitelboim, C. (1976). *Constrained Hamiltonian systems*. Accademia Nazionale dei Lincei, Roma.

Harrison, E. (2000). *Cosmology*, 2nd edn. Cambridge University Press, Cambridge.

Hartle, J. B. (1987). Prediction in quantum cosmology. In *Gravitation in astrophysics* (ed. B. Carter and J. B. Hartle), pp. 329–60. Plenum Press, New York.

Hartle, J. B. (1997). Quantum cosmology: problems for the 21st century. In *Physics in the 21st century* (ed. K. Kikkawa, H. Kunitomo, and H. Ohtsubo), pp. 179–99. World Scientific, Singapore.

Hartle, J. B. and Hawking, S. W. (1983). Wave function of the Universe. *Phys. Rev. D*, **28**, 2960–75.

Hartle, J. B. and Schleich, K. (1987). The conformal rotation in linearised gravity. In *Quantum field theory and quantum statistics*, Vol II (ed. I. Batalin, C. J. Isham, and G. A. Vilkovisky), pp. 67–87. Adam Hilger, Bristol.

Hawking, S. W. (1975). Particle creation by black holes. *Commun. Math. Phys.*, **43**, 199–220.

Hawking, S. W. (1979). The path-integral approach to quantum gravity. In *General relativity. An Einstein centenary survey* (ed. S. W. Hawking and W. Israel), pp. 746–89. Cambridge University Press, Cambridge.

Hawking, S. W. (1982). The boundary conditions of the universe. *Pontificia Academiae Scientarium Scripta Varia*, **48**, 563–74.

Hawking, S. W. (1984). The quantum state of the universe. *Nucl. Phys. B*, **239**, 257–76.

Hawking, S. W. and Ellis, G. F. R. (1973). *The large scale structure of space–time*. Cambridge University Press, Cambridge.

Hawking, S. W. and Penrose, R. (1996). *The nature of space and time*. Princeton University Press, Princeton.

Hayward, G. (1993). Gravitational action for spacetimes with nonsmooth boundaries. *Phys. Rev. D*, **47**, 3275–80.

Hehl, F. W. (1985). On the kinematics of the torsion of space–time. *Found. Phys.*, **15**, 451–71.

Hehl, F. W., Kiefer, C., and Metzler, R. J. K. (ed.) (1998). *Black holes: theory and observation.* Lecture Notes in Physics 514. Springer, Berlin.

Hehl, F. W., Lemke, J., and Mielke, E.W. (1991). Two lectures on fermions and gravity. In *Geometry and theoretical physics* (ed. J. Debrus and A. C. Hirshfeld), pp. 56–140. Springer, Berlin.

Hehl, F. W., von der Heyde, P., Kerlick, G. D., and Nester, J. M. (1976). General relativity with spin and torsion. *Rev. Mod. Phys.*, **48**, 393–416.

Heisenberg, W. (1979). *Der Teil und das Ganze.* Dtv, München.

Heitler, W. (1984). *The quantum theory of radiation.* Dover Publications, New York.

Helfer, A. D. (1996). The stress-energy operator. *Class. Quantum Grav.*, **13**, L129–34.

Henneaux, M. and Teitelboim, C. (1992). *Quantization of gauge systems.* Princeton University Press, Princeton.

Heusler, M. (1996). *Black hole uniqueness theorems.* Cambridge University Press, Cambridge.

Heusler, M., Kiefer, C., and Straumann, N. (1990). Self-energy of a thin charged shell in general relativity. *Phys. Rev. D*, **42**, 4254–6.

Higgs, P. W. (1958). Integration of secondary constraints in quantized general relativity. *Phys. Rev. Lett.*, **1**, 373–4.

Hod, S. (1998). Bohr's correspondence principle and the area spectrum of quantum black holes. *Phys. Rev. Lett.*, **81**, 4293–6.

Hogan, C. J. (2000). Why the universe is just so. *Rev. Mod. Phys.*, **72**, 1149–61.

Hogan, C. J. (2002). Holographic discreteness of inflationary perturbations. *Phys. Rev. D*, **66**, 023521.

Hojman, S. A., Kuchař, K. V., and Teitelboim, C. (1976). Geometrodynamics regained. *Ann. Phys. (NY)*, **96**, 88–135.

Hornberger, K., Uttenthaler, S., Brezger, B., Hackermueller, L., Arndt, M., and Zeilinger, A. (2003). Collisional decoherence observed in matter wave interferometry. *Phys. Rev. Lett.*, **90**, 160401.

Horowitz, G. T. (1990). String theory as a quantum theory of gravity. In *General relativity and gravitation* (ed. N. Ashby, D. F. Bartlett, and W. Wyss), pp. 419–39. Cambridge University Press, Cambridge.

Horowitz, G. T. (1998). Quantum states of black holes. In *Black holes and relativistic stars* (ed. R. M. Wald), pp. 241–66. The University of Chicago Press, Chicago.

Horowitz, G. T. and Polchinski, J. (1997). Correspondence principle for black holes and strings. *Phys. Rev. D*, **55**, 6189–97.

Huang, K. (1992). *Quarks, leptons & gauge fields*, 2nd edn. World Scientific, Singapore.

Infeld, L. (ed.) (1964). *Relativistic theories of gravitation.* Pergamon Press, Oxford.

Isham, C. J. (1981). Topological θ-sectors in canonically quantized gravity. *Phys. Lett. B*, **106**, 188–92.

Isham, C. J. (1984). Topological and global aspects of quantum theory. In *Relativity, groups and topology II* (ed. B. S. DeWitt and R. Stora), pp. 1059–290. North-Holland, Amsterdam.

Isham, C. J. (1987). Quantum gravity. In *General relativity and gravitation* (ed. M. A. H. Mac Callum), pp. 99–129. Cambridge University Press, Cambridge.

Isham, C. J. (1989). Quantum topology and quantization on the lattice of topologies. *Class. Quantum Grav.*, **6**, 1509–34.

Isham, C. J. (1993). Canonical quantum gravity and the problem of time. In *Integrable systems, quantum groups, and quantum field theory* (ed. L. A. Ibort and M. A. Rodríguez), pp. 157–287. Kluwer, Dordrecht.

Isham, C. J. (1994). Prima facie questions in quantum gravity. In *Canonical gravity: from classical to quantum* (ed. J. Ehlers and H. Friedrich), pp. 1–21. Lecture Notes in Physics 434. Springer, Berlin.

Israel, W. (1986). Third law of black-hole dynamics: a formulation and proof. *Phys. Rev. Lett.*, **57**, 397–99.

Iwasaki, Y. (1971). Quantum theory of gravitation vs. classical theory. *Prog. Theor. Phys.*, **46**, 1587–609.

Jackiw, R. (1995). *Diverse topics in theoretical and mathematical physics*. World Scientific, Singapore.

Jacobson, T. (1995). Thermodynamics of spacetime: the Einstein equation of state. *Phys. Rev. Lett.*, **75**, 1260–3.

Johnstone Stoney, G. (1881). On the physical units of nature. *Phil. Magazine, Ser. 5*, **11**, 381–90.

Joos, E., Zeh, H. D., Kiefer, C., Giulini, D., Kupsch, J., and Stamatescu, I.-O. (2003). *Decoherence and the appearance of a classical world in quantum theory*, 2nd edn. Springer, Berlin.

Kaku, M. (1999). *Introduction to superstrings and M-theory*, 2nd edn. Springer, New York.

Kamenshchik, A. Yu. (1990). Normalizability of the wave function of the universe, particle physics and supersymmetry. *Phys. Lett. B*, **316**, 45–50.

Kastrup, H. A. (1996). The quantum levels of isolated spherically symmetric gravitational systems. *Phys. Lett. B*, **385**, 75–80.

Kaul, R. K. and Majumdar, P. (2000). Logarithmic correction to the Bekenstein–Hawking entropy. *Phys. Rev. Lett.*, **84**, 5255–7.

Kazakov, D. I. (1988). On a generalization of renormalization group equations to quantum field theories of an arbitrary type. *Theor. Math. Phys.*, **75**, 440–2.

Kiefer, C. (1987). Continuous measurement of minisuperspace variables by higher multipoles. *Class. Quantum Grav.*, **4**, 1369–82.

Kiefer, C. (1988). Wave packets in minisuperspace. *Phys. Rev. D*, **38**, 1761–72.

Kiefer, C. (1989). Non-minimally coupled scalar fields and the initial value problem in quantum gravity. *Phys. Lett. B*, **225**, 227–32.

Kiefer, C. (1990). Wave packets in quantum cosmology and the cosmological constant. *Nucl. Phys. B*, **341**, 273–93.

Kiefer, C. (1991). On the meaning of path integrals in quantum cosmology. *Ann. Phys. (NY)*, **207**, 53–70.

Kiefer, C. (1992). Decoherence in quantum electrodynamics and quantum gravity. *Phys. Rev. D*, **46**, 1658–70.

Kiefer, C. (1994). The semiclassical approximation to quantum gravity. In *Canonical gravity: from classical to quantum* (ed. J. Ehlers and H. Friedrich), pp. 170–212. Lecture Notes in Physics 434. Springer, Berlin.

Kiefer, C. (1998). Towards a full quantum theory of black holes. In *Black holes: theory and observation* (ed. F. W. Hehl, C. Kiefer, and R. J. K. Metzler), pp. 416–50. Lecture Notes in Physics 514. Springer, Berlin.

Kiefer, C. (1999). Thermodynamics of black holes and Hawking radiation. In *Classical and quantum black holes* (ed. P. Fré, V. Gorini, G. Magli, and U. Moschella), pp. 17–74. Institute of Physics Publishing, Bristol.

Kiefer, C. (2001). Path integrals in quantum cosmology. In *Fluctuating paths and fields* (ed. W. Janke, A. Pelster, H.-J. Schmidt, and M. Bachmann), pp. 729–40. World Scientific, Singapore.

Kiefer, C. (2003a). Quantum aspects of black holes. In *The galactic black hole* (ed. H. Falcke and F. W. Hehl), pp. 207–25. Institute of Physics Publishing, Bristol.

Kiefer, C. (2003b). On the interpretation of quantum theory—from Copenhagen to the present day. In *Time, quantum and information* (ed. L. Castell and O. Ischebeck), pp. 291–9. Springer, Berlin.

Kiefer, C. (2004a). Is there an information-loss problem for black holes? In *Decoherence and entropy in complex systems* (ed. H.-T. Elze). Lecture Notes in Physics 633. Springer, Berlin.

Kiefer, C. (2004b). Arrow of time from timeless quantum gravity. In *Time and matter* (ed. I. Bigi and M. Fäßler). World Scientific, Singapore, to appear.

Kiefer, C. and Louko, J. (1999). Hamiltonian evolution and quantization for extremal black holes. *Ann. Phys. (Leipzig)*, 8th series, **8**, 67–81.

Kiefer, C. and Singh, T. P. (1991). Quantum gravitational correction terms to the functional Schrödinger equation. *Phys. Rev. D*, **44**, 1067–76.

Kiefer, C. and Wipf, A. (1994). Functional Schrödinger equation for fermions in external gauge fields. *Ann. Phys. (NY)*, **236**, 241–85.

Kiefer, C. and Zeh, H. D. (1995). Arrow of time in a recollapsing quantum universe. *Phys. Rev. D*, **51**, 4145–53.

Kiefer, C., Polarski, D., and Starobinsky, A. A. (1998). Quantum-to-classical transition for fluctuations in the early universe. *Int. J. Mod. Phys. D*, **7**, 455–62.

Kiefer, C., Polarski, D., and Starobinsky, A. A. (2000). Entropy of gravitons produced in the early universe. *Phys. Rev. D*, **62**, 043518.

Kodama, H. (1990). Holomorphic wavefunction of the universe. *Phys. Rev. D*, **42**, 2548–65.

Kokkotas, K. D. and Schmidt, B. G. (1999). Quasi-normal modes of stars and

black holes. *Living reviews in relativity*, available on
http://relativity.livingreviews.org/Articles/lrr-1999-2/.

Kraus, P. and Wilczek, F. (1995). Self-interaction correction to black hole radiance. *Nucl. Phys. B*, **433**, 403–20.

Kretschmann, E. (1917). Über den physikalischen Sinn der Relativitätspostulate A. Einsteins und seine ursprüngliche Relativitätstheorie. *Ann. Phys. (Leipzig)*, 4th series, **53**, 575–614.

Kuchař, K. (1970). Ground state functional of the linearized gravitational field. *J. Math. Phys.*, **11**, 3322–34.

Kuchař, K. (1971). Canonical quantization of cylindrical gravitational waves. *Phys. Rev. D*, **4**, 955–86.

Kuchař, K. (1973). Canonical quantization of gravity. In *Relativity, astrophysics and cosmology* (ed. W. Israel), pp. 237–88. D. Reidel, Dordrecht.

Kuchař, K. (1981). Canonical methods of quantization. In *Quantum Gravity 2. A second Oxford symposium* (ed. C. J. Isham, R. Penrose, and D. W. Sciama), pp. 329–76. Oxford University Press, Oxford.

Kuchař, K. V. (1992). Time and interpretations of quantum gravity. In *Proc. 4th Canadian Conf. General relativity Relativistic Astrophysics* (ed. G. Kunstatter, D. Vincent, and J. Williams), pp. 211–314. World Scientific, Singapore.

Kuchař, K. V. (1993). Canonical quantum gravity. In *General relativity and gravitation 1992* (ed. R. J. Gleiser, C. N. Kozameh, and O. M. Moreschi), pp. 119–50. Institute of Physics Publishing, Bristol.

Kuchař, K. V. (1994). Geometrodynamics of Schwarzschild black holes. *Phys. Rev. D*, **50**, 3961–81.

Kuchař, K. V. and Ryan, M. P. (1989). Is minisuperspace quantization valid?: Taub in mixmaster. *Phys. Rev. D*, **40**, 3982–96.

Kuchař, K. V. and Torre, C. G. (1991). Strings as poor relatives of general relativity. In *Conceptual problems of quantum gravity* (ed. A. Ashtekar and J. Stachel), pp. 326–48. Birkhäuser, Boston.

Kuo, C.-I. and Ford, L. H. (1993). Semiclassical gravity theory and quantum fluctuations. *Phys. Rev. D*, **47**, 4510–19.

Laflamme, R. (1987). Euclidean vacuum: justification from quantum cosmology. *Phys. Lett. B*, **198**, 156–60.

Laflamme, R. and Louko, J. (1991). Reduced density matrices and decoherence in quantum cosmology. *Phys. Rev. D*, **43**, 3317–31.

Lämmerzahl, C. (1995). A Hamilton operator for quantum optics in gravitational fields. *Phys. Lett. A*, **203**, 12–17.

Lämmerzahl, C. (1996). On the equivalence principle in quantum theory. *Gen. Rel. Grav.*, **28**, 1043–70.

Lämmerzahl, C. (1998). Quantum tests of the foundations of general relativity. *Class. Quantum Grav.*, **15**, 13–27.

Lämmerzahl, C. (2003). The Einstein equivalence principle and the search for new physics. In *Quantum gravity: from theory to experimental search* (ed. D. Giulini, C. Kiefer, and C. Lämmerzahl), pp. 367–400. Lecture Notes in

Physics 631. Springer, Berlin.

Lanczos, C. (1986). *The variational principles of mechanics*, 4th edn. Dover Publications, New York.

Landau, L. and Peierls, R. (1931). Erweiterung des Unbestimmtheitsprinzips für die relativistische Quantentheorie. *Z. Phys.*, **69**, 56–69.

Lapchinsky, V. G. and Rubakov, V. A. (1979). Canonical quantization of gravity and quantum field theory in curved space–time. *Acta Physica Polonica*, **10**, 1041–8.

Lee, H. C. (ed.) (1984). *An introduction to Kaluza–Klein theories*. World Scientific, Singapore.

Leinaas, J. M. (2002). Unruh effect in storage rings. In *Quantum aspects of beam physics* (ed. P. Chen), pp. 336–52.

Leutwyler, H. (1964). Gravitational field: equivalence of Feynman quantization and canonical quantization. *Phys. Rev.*, **134**, B1155–82.

Leutwyler, H. (1986). Anomalies. *Helv. Phys. Acta*, **59**, 201–19.

Lewandowski, J., Newman, E. T., and Rovelli, C. (1993). Variations of the parallel propagator and holonomy operator and the Gauss law constraint. *J. Math. Phys.*, **34**, 4646–54.

Lidsey, J. E. (1995). Scale factor duality and hidden supersymmetry in scalar-tensor cosmology. *Phys. Rev. D*, **52**, 5407–11.

Lifshits, E. M. (1946). About gravitational stability of the expanding world (in Russian). *Zh. Eksp. Teor. Phys.*, **16**, 587–602.

Linde, A. D. (1990). *Particle physics and inflationary cosmology*. Harwood, New York.

Loll, R. (2003). A discrete history of the Lorentzian path integral. In *Quantum gravity: from theory to experimental search* (ed. D. Giulini, C. Kiefer, and C. Lämmerzahl), pp. 137–71. Lecture Notes in Physics 631. Springer, Berlin.

Louis-Martinez, D., Gegenberg, J., and Kunstatter, G. (1994). Exact Dirac quantization of all 2D dilaton gravity theories. *Phys. Lett. B*, **321**, 193–8.

Louis-Martinez, D. and Kunstatter, G. (1994). Birkhoff's theorem in two-dimensional dilaton gravity. *Phys. Rev. D*, **49**, 5227–30.

Louko, J. and Matschull, H.-J. (2001). The 2+1 Kepler problem and its quantization. *Class. Quantum Grav.*, **18**, 2731–84.

Louko, J. and Whiting, B. F. (1995). Hamiltonian thermodynamics of the Schwarzschild black hole. *Phys. Rev. D*, **51**, 5583–99.

Louko, J., Whiting, B., and Friedman, J. (1998). Hamiltonian spacetime dynamics with a spherical null-dust shell. *Phys. Rev. D*, **57**, 2279–98.

Louko, J. and Winters-Hilt, S. N. (1996). Hamiltonian thermodynamics of the Reissner–Nordström-anti-de Sitter black hole. *Phys. Rev. D*, **54**, 2647–63.

Lüscher, M., Narayanan, R., Weisz, P., and Wolff, U. (1992). The Schrödinger functional—a renormalizable probe for non-abelian gauge theories. *Nucl. Phys. B*, **384**, 168–228.

Lüst, D. and Theisen, S. (1989). *Lectures on string theory*. Lecture Notes in Physics 346. Springer, Berlin.

MacCallum, M. A. H. (1979). Anisotropic and inhomogeneous relativistic cosmologies. In *General relativity. An Einstein centenary survey* (ed. S. W. Hawking and W. Israel), pp. 533–80. Cambridge University Press, Cambridge.

Major, S. A. (1999). A spin network primer. *Am. J. Phys.*, **67**, 972–80.

Marsden, J. E. and Hughes, T. J. R. (1983). *Mathematical foundations of elasticity.* Prentice-Hall, Englewood Cliffs, New Jersey.

Mashhoon, B. (1995). On the coupling of intrinsic spin with the rotation of the earth. *Phys. Lett A*, **198**, 9–13.

Matschull, H.-J. (1995). Three-dimensional canonical quantum gravity. *Class. Quantum Grav.*, **12**, 2621–703.

Matschull, H.-J. (2001). The phase space structure of multi-particle models in 2+1 gravity. *Class. Quantum Grav.*, **17**, 3497–560.

McAvity, D. M. and Osborn, H. (1993). Quantum field theories on manifolds with curved boundaries: scalar fields. *Nucl. Phys. B*, **394**, 728–88.

Misner, C. W. (1957). Feynman quantization of general relativity. *Rev. Mod. Phys.*, **29**, 497–509.

Misner, C. W. (1972). Minisuperspace. In *Magic without magic: John Archibald Wheeler* (ed. J. R. Klauder), pp. 441–73. Freeman, San Francisco.

Misner, C. W., Thorne, K. S., and Wheeler, J. A. (1973). *Gravitation.* Freeman, San Francisco.

Mohaupt, T. (2001). Black hole entropy, special geometry and strings. *Fortschr. Phys.*, **49**, 3–161.

Mohaupt, T. (2003). Introduction to string theory. In *Quantum gravity: from theory to experimental search* (ed. D. Giulini, C. Kiefer, and C. Lämmerzahl), pp. 173–251. Lecture Notes in Physics 631. Springer, Berlin.

Møller, C. (1962). The energy–momentum complex in general relativity and related problems. In *Les théories relativistes de la gravitation* (ed. A. Lichnerowicz and M. A. Tonnelat), pp. 15–29. Editions du Centre National de la Recherche Scientifique, Paris.

Moniz, P. V. (1996). Supersymmetric quantum cosmology. Shaken, not stirred. *Int. J. Mod. Phys. A*, **11**, 4321–82.

Moniz, P. V. (2003). FRW minisuperspace with local $N = 4$ supersymmetry and self-interacting scalar field. *Ann. Phys. (Leipzig)*, 8th series, **12**, 174–98.

Motl, L. (2003). An analytical computation of asymptotic Schwarzschild quasinormal frequencies. *Adv. Theor. Math. Phys.*, **6**, 1135–62.

Mott, N. F. (1931). On the theory of excitation by collision with heavy particles. *Proc. Cambridge Phil. Soc.*, **27**, 553–60.

Mukhanov, V. F. (1986). Are black holes quantized? *JETP Lett.*, **44**, 63–6.

Mukhanov, V. F. and Chibisov, G. V. (1981). Quantum fluctuations and a nonsingular universe. *JETP Lett.*, **33**, 532–5.

Neitzke, A. (2003). Greybody factors at large imaginary frequencies. Available on http://arxiv.org/abs/hep-th/0304080.

Nesvizhevsky, V. V. *et al.* (2002). Quantum states of neutrons in the Earth's gravitational field. *Nature*, **415**, 297–9.

Neugebauer, G. (1998). Black hole thermodynamics. In *Black holes: theory and observation* (ed. F. W. Hehl, C. Kiefer, and R. J. K. Metzler), pp. 319–38. Lecture Notes in Physics 514. Springer, Berlin.

Núñez, D., Quevedo, H., and Sudarsky, D. (1998). Black hole hair: a review. In *Black holes: theory and observation* (ed. F. W. Hehl, C. Kiefer, and R. J. K. Metzler), pp. 187–98. Lecture Notes in Physics 514. Springer, Berlin.

Okun, L. B. (1992). The fundamental constants of physics. *Sov. Phys. Usp.*, **34**, 818–26.

Osterwalder, K. and Schrader, R. (1975). Axioms for euclidean Green's functions II. *Comm. Math. Phys.*, **42**, 281–305.

Padmanabhan, T. (1985). Physical significance of Planck length. *Ann. Phys. (NY)*, **165**, 38–58.

Page, D. N. (1991). Minisuperspaces with conformally and minimally coupled scalar fields. *J. Math. Phys.*, **32**, 3427–38.

Page, D. N. and Geilker, C. D. (1981). Indirect evidence of quantum gravity. *Phys. Rev. Lett.*, **47**, 979–82.

Palais, R. S. (1979). The principle of symmetric criticality. *Comm. Math. Phys.*, **69**, 19–30.

Parentani, R. (2000). The background field approximation in (quantum) cosmology. *Class. Quantum Grav.*, **17**, 1527–47.

Parikh, M. K. and Wilczek, F. (2000). Hawking radiation as tunneling. *Phys. Rev. Lett.*, **85**, 5042–5.

Paternoga, R. and Graham, R. (1998). The Chern–Simons state for the non-diagonal Bianchi IX model. *Phys. Rev. D*, **58**, 083501.

Paz, J. P. and Sinha, S. (1992). Decoherence and back reaction in quantum cosmology: multidimensional minisuperspace examples. *Phys. Rev. D*, **45**, 2823–42.

Pauli, W. (1955). Schlußwort durch den Präsidenten der Konferenz. *Helv. Phys. Acta Suppl.*, **4**, pp. 261–7.

Pauli, W. (1985). *Scientific correspondence with Bohr, Einstein, Heisenberg a.o. Vol II: 1930–1939.* Springer, Berlin.

Peet, A. W. (1998). The Bekenstein formula and string theory (N-brane theory). *Class. Quantum Grav.*, **15**, 3291–338.

Penrose, R. (1971). Angular momentum: an approach to combinatorial space–time. In *Quantum theory and beyond* (ed. T. Bastin), pp. 151–80. Cambridge University Press, Cambridge.

Penrose, R. (1981). Time-asymmetry and quantum gravity. In *Quantum gravity*, Vol. 2 (ed. C. J. Isham, R. Penrose, and D. W. Sciama), pp. 242–72. Clarendon Press, Oxford.

Penrose, R. (1996). On gravity's role in quantum state reduction. *Gen. Rel. Grav.*, **28**, 581–600.

Percival, I. (1997). Atom interferometry, spacetime and reality. *Physics World*, **10**(3), 43–8.

Peres, A. (1962). On Cauchy's problem in general relativity–II. *Nuovo Cimento*, **XXVI**, 53–62.

Pilati, M. (1977). The canonical formulation of supergravity. *Nucl. Phys. B*, **132**, 138–54.

Pilati, M. (1982). Strong coupling quantum gravity. I. Solution in a particular gauge. *Phys. Rev. D*, **26**, 2645–63.

Pilati, M. (1983). Strong coupling quantum gravity. II. Solution without gauge fixing. *Phys. Rev. D*, **28**, 729–44.

Pinto-Neto, N. and Santini, E. S. (2002). The consistency of causal quantum geometrodynamics and quantum field theory. *Gen. Rel. Grav.*, **34**, 505–32.

Planck, M. (1899). Über irreversible Strahlungsvorgänge. *Sitzber. kgl.-preuß. Akad. Wiss. Berlin, Sitzungen der phys.-math. Klasse*, pp. 440–80.

Poincaré, H. (1970). *La valeur de la science*. Flammarion, Paris.

Polarski, D. and Starobinsky, A. A. (1996). Semiclassicality and decoherence of cosmological fluctuations. *Class. Quantum Grav.*, **13**, 377–92.

Polchinski, J. (1998*a*). *String theory. Vol I. An introduction to the bosonic string*. Cambridge University Press, Cambridge.

Polchinski, J. (1998*b*). *String theory. Vol II. Superstring theory and beyond*. Cambridge University Press, Cambridge.

Pullin, J. (1999). An overview of canonical quantum gravity. *Int. J. Theor. Phys.*, **38**, 1051–61.

Randall, L. and Sundrum, R. (1999). A large mass hierarchy from a small extra dimension. *Phys. Rev. Lett.*, **83**, 3370–3.

Rees, M. (1995). *Perspectives in astrophysical cosmology*. Cambridge University Press, Cambridge.

Regge, T. and Teitelboim, C. (1974). Role of surface integrals in the Hamiltonian formulation of general relativity. *Ann. Phys. (NY)*, **88**, 286–318.

Reuter, M. (1998). Nonperturbative evolution equation for quantum gravity. *Phys. Rev. D*, **57**, 971–85.

Reuter, M. and Saueressig, F. (2002). A class of nonlocal truncations in quantum Einstein gravity and its renormalization group behavior. *Phys. Rev. D*, **66**, 125001.

Rosenfeld, L. (1930). Über die Gravitationswirkungen des Lichtes. *Z. Phys.*, **65**, 589–99.

Rosenfeld, L. (1963). On quantization of fields. *Nucl. Phys.*, **40**, 353–6.

Rovelli, C. (1991*a*). Ashtekar formulation of general relativity and loop-space non-perturbative quantum gravity: a report. *Class. Quantum Grav.*, **8**, 1613–75.

Rovelli, C. (1991*b*). Time in quantum gravity: an hypothesis. *Phys. Rev. D*, **43**, 442–56.

Rovelli, C. (1998). Loop quantum gravity. *Living reviews in relativity*, available on www.livingreviews.org/Articles/Volume1/1998-1rovelli/.

Rovelli, C. (2000). Notes for a brief history of quantum gravity. Presented at 9th Marcel Grossmann Meeting, Rome, Italy, 2-9 Jul 2000. Available on http://arxiv.org/abs/gr-qc/0006061.

Rovelli, C. and Gaul, M. (2000). Loop quantum gravity and the meaning of diffeomorphism invariance. In *Towards quantum gravity* (ed. J. Kowalski-Glikman), pp. 277–324. Lecture Notes in Physics 541. Springer, Berlin.

Rovelli, C. and Smolin, L. (1990). Loop space representation of quantum general relativity. *Nucl. Phys. B*, **331**, 80–152.

Rovelli, C. and Smolin, L. (1995). Discreteness of area and volume in quantum gravity. *Nucl. Phys. B*, **442**, 593–622. Erratum *ibid.*, **456**, 753–4.

Rubakov, V. A. (2001). Large and infinite extra dimensions. *Phys. Usp.*, **44**, 871–93.

Ryan, M. P. (1972). *Hamiltonian cosmology.* Lecture Notes in Physics 13. Springer, Berlin.

Ryan, M. P. and Waller, S. M. (1997). On the Hamiltonian formulation of class B Bianchi cosmological models. http://arxiv.org/abs/gr-qc/9709012.

Scherk, J. and Schwarz, J. H. (1974). Dual models for non-hadrons. *Nucl. Phys. B*, **81**, 118–44.

Schrödinger, E. (1939). The proper vibrations of the expanding universe. *Physica*, **6**, 899–912.

Schwinger, J. (1963). Quantized gravitational field. *Phys. Rev.*, **130**, 1253–8.

Seidel, E. (1998). Numerical approach to black holes. In *Black holes: Theory and observation* (ed. F. W. Hehl, C. Kiefer, and R. J. K. Metzler), pp. 244–65. Lecture Notes in Physics 514. Springer, Berlin.

Sen, A. (1982). Gravity as a spin system. *Phys. Lett. B*, **119**, 89–91.

Sexl, R. U. and Urbantke, H. K. (2001). *Relativity, groups, particles.* Springer, Wien.

Shi, Y. (2000). Early Gedanken experiments revisited. *Ann. Phys. (Leipzig)*, 8th series, **9**, 637–48.

Singh, T. P. and Padmanabhan, T. (1989). Notes on semiclassical gravity. *Ann. Phys. (NY)*, **196**, 296–344.

Smith, G. J. and Bergmann, P. G. (1979). Measurability analysis of the magnetic-type components of the linearized gravitational radiation field. *Gen. Rel. Grav.*, **11**, 133–47.

Sorkin, R. (2003). Causal sets: discrete gravity. Available on http://arxiv.org/gr-qc/0309009.

Starobinsky, A. A. (1979). Spectrum of relict gravitational radiation and the early state of the universe. *JETP Lett.*, **30**, 682–5.

Stelle, K. S. (1977). Renormalization of higher-derivative quantum gravity. *Phys. Rev. D*, **16**, 953–69.

Stelle, K. S. (1978). Classical gravity with higher derivatives. *Gen. Rel. Grav.*, **9**, 353–71.

Straumann, N. (1984). *General relativity and relativistic astrophysics.* Springer, Berlin.

Straumann, N. (2000). Reflections on gravity. Available on http://arxiv.org/abs/astro-ph/0006423.

Strominger, A. and Vafa, C. (1996). Microscopic origin of the Bekenstein–Hawking entropy. *Phys. Lett. B*, **379**, 99–104.

Stueckelberg, E. C. G. (1938). Die Wechselwirkungskräfte in der Elektrodynamik und in der Feldtheorie der Kernkräfte (Teil 1). *Helv. Phys. Acta*, **11**, 225–44.

Sundermeyer, K. (1982). *Constrained dynamics*. Lecture Notes in Physics 169. Springer, Berlin.

Symanzik, K. (1981). Schrödinger representation and Casimir effect in renormalizable quantum field theory. *Nucl. Phys. B*, **190** [FS3], 1–44.

Tabensky, R. and Teitelboim, C. (1977). The square root of general relativity. *Phys. Lett. B*, **69**, 453–6.

Teitelboim, C. (1977). Supergravity and square roots of constraints. *Phys. Rev. Lett.*, **38**, 1106–10.

Teitelboim, C. (1980). The Hamiltonian structure of space-time. In *General relativity and gravitation*, Vol 1 (ed. A. Held), pp. 195–225. Plenum, New York.

Teitelboim, C. (1984). The Hamiltonian structure of two-dimensional space–time. In *Quantum theory of gravity* (ed. S. M. Christensen), pp. 327–44. Adam Hilger, Bristol.

Thiemann, T. (1996). Anomaly-free formulation of non-perturbative, four-dimensional, Lorentzian quantum gravity. *Phys. Lett. B*, **380**, 257–64.

Thiemann, T. (2001). Introduction to modern canonical quantum general relativity. Available on http://arxiv.org/abs/gr-qc/0110034.

Thiemann, T. (2003). Lectures on loop quantum gravity. In *Quantum gravity: From theory to experimental search* (ed. D. Giulini, C. Kiefer, and C. Lämmerzahl), pp. 41–135. Lecture Notes in Physics 631. Springer, Berlin.

Thiemann, T. and Kastrup, H. A. (1993). Canonical quantization of spherically symmetric gravity in Ashtekar's self-dual representation. *Nucl. Phys. B*, **399**, 211–58.

't Hooft, G. and Veltman, M. (1974). One-loop divergencies in the theory of gravitation. *Ann. Inst. Henri Poincaré A*, **20**, 69–94.

Tomboulis, E. (1977). $1/N$ expansion and renormalization in quantum gravity. *Phys. Lett. B*, **70**, 361–4.

Torre, C. G. (1993). Is general relativity an 'already parametrized' theory? *Phys. Rev. D*, **46**, 3231–4.

Torre, C. G. (1999). Midisuperspace models of canonical quantum gravity. *Int. J. Theor. Phys.*, **38**, 1081–102.

Torre, C. G. and Varadarajan, M. (1999). Functional evolution for free quantum fields. *Class. Quantum Grav.*, **16**, 2651–68.

Tsamis, N. C. and Woodard, R. P. (1987). The factor-ordering problem must be regulated. *Phys. Rev. D*, **36**, 3641–50.

Tsamis, N. C. and Woodard, R. P. (1993). Relaxing the cosmological constant. *Phys. Lett. B*, **301**, 351–7.

Unruh, W. G. (1976). Notes on black-hole evaporation. *Phys. Rev. D*, **14**, 870–92.

Unruh, W. G. (1984). Steps towards a quantum theory of gravity. In *Quantum theory of gravity* (ed. S. M. Christensen), pp. 234–42. Adam Hilger, Bristol.

Uzan, J.-P. (2003). The fundamental constants and their variation: observational status and theoretical motivations. *Rev. Mod. Phys.*, **75**, 403–55.

van Dam, H. and Veltman, M. J. (1970). Massive and massless Yang–Mills and gravitational fields. *Nucl. Phys. B*, **22**, 397–411.

van de Ven, A. E. M. (1992). Two-loop quantum gravity. *Nucl. Phys. B*, **378**, 309–66.

van Nieuwenhuizen, P. (1981). Supergravity. *Phys. Rep.*, **68**, 189–398.

Vaz, C., Kiefer, C., Singh, T. P., and Witten, L. (2003). Quantum general relativity and Hawking radiation. *Phys. Rev. D*, **67**, 024014.

Veneziano, G. (1993). Classical and quantum gravity from string theory. In *Classical and quantum gravity* (ed. M. C. Bento, O. Bertolami, J. M. Mourão, and R. F. Picken), pp. 134–80. World Scientific, Singapore.

Vilenkin, A. (1988). Quantum cosmology and the initial state of the Universe. *Phys. Rev. D*, **37**, 888–97.

Vilenkin, A. (1989). Interpretation of the wave function of the Universe. *Phys. Rev. D*, **39**, 1116–22.

Vilenkin, A. (2003). Quantum cosmology and eternal inflation. In *The future of theoretical physics and cosmology* (ed. G. W. Gibbons, E. P. S. Shellard, and S. J. Rankin), pp. 649–66. Cambridge University Press, Cambridge.

Vilkovisky, G. (1984). The Gospel according to DeWitt. In *Quantum theory of gravity* (ed. S. M. Christensen), pp. 169–209. Adam Hilger, Bristol.

von Borzeszkowski, H.-H. and Treder, H.-J. (1988). *The meaning of quantum gravity*. D. Reidel, Dordrecht.

von Neumann, J. (1932). *Mathematische Grundlagen der Quantenmechanik*. Springer, Berlin. For an English translation of parts of this book, see Wheeler, J. A. and Zurek, W. H. (ed.), *Quantum theory and measurement*. Princeton University Press, Princeton (1983).

Wald, R. M. (1984). *General relativity*. The University of Chicago Press, Chicago.

Wald, R. M. (2001). The thermodynamics of black holes. *Living reviews in relativity*, www.livingreviews.org/Articles/Volume4/2001-6wald/.

Weinberg, S. (1964). Photons and gravitons in S-matrix theory: derivation of charge conservation and equality of gravitational and inertial mass. *Phys. Rev.*, **135**, B1049–56.

Weinberg, S. (1972). *Gravitation and cosmology. Principles and applications of the general theory of relativity*. Wiley, New York.

Weinberg, S. (1993). *Dreams of a final theory*. Hutchinson Radius, London.

Weinberg, S. (1995). *The quantum theory of fields, Vol I (Foundations)*. Cambridge University Press, Cambridge.

Weinberg, S. (1996). *The quantum theory of fields, Vol II (Modern applications)*. Cambridge University Press, Cambridge.

Weinberg, S. (2000). *The quantum theory of fields, Vol III (Supersymmetry)*. Cambridge University Press, Cambridge.

Werner, S. A. and Kaiser, H. (1990). Neutron interferometry—macroscopic manifestations of quantum mechanics. In *Quantum mechanics in curved space–time* (ed. J. Audretsch and V. de Sabbata), pp. 1–21. Plenum Press, New York.

Wess, J. and Bagger, J. (1992). *Supersymmetry and supergravity*, 2nd edn. Princeton University Press, Princeton.

Wheeler, J. A. (1968). Superspace and the nature of quantum geometrodynamics. In *Battelle rencontres* (ed. C. M. DeWitt and J. A. Wheeler), pp. 242–307. Benjamin, New York.

Wheeler, J. A. (1990). Information, physics, quantum: the search for links. In *Complexity, entropy, and the physics of information* (ed. W. H. Zurek), pp. 3–28. Addison-Wesley, Redwood City.

Williams, R. (1997). Recent progress in Regge calculus. *Nucl. Phys. B (Proc. Suppl.)*, **57**, 73–81.

Witten, E. (1988). 2+1 dimensional gravity as an exactly soluble system. *Nucl. Phys. B*, **311**, 46–78.

Witten, E. (1995). String theory in various dimensions. *Nucl. Phys. B*, **443**, 85–126.

Woodard, R. P. (1993). Enforcing the Wheeler–DeWitt constraint the easy way. *Class. Quantum Grav.*, **10**, 483–96.

Woodhouse, N. M. J. (1992). *Geometric quantization*, 2nd edn. Clarendon Press, Oxford.

Yoneya, T. (1974). Connection of dual models to electrodynamics and gravidynamics. *Progr. Theor. Phys.*, **51**, 1907–20.

Zeh, H. D. (1970). On the interpretation of measurement in quantum theory. *Found. Phys.*, **1**, 69–76. Reprinted in Wheeler, J. A. and Zurek, W. H. (ed.), *Quantum theory and measurement*. Princeton University Press, Princeton (1983).

Zeh, H. D. (1986). Emergence of classical time from a universal wave function. *Phys. Lett. A*, **116**, 9–12.

Zeh, H. D. (1988). Time in quantum gravity. *Phys. Lett. A*, **126**, 311–7.

Zeh, H. D. (1994). Decoherence and measurements. In *Stochastic evolution of quantum states in open systems and measurement processes* (ed. L. Diósi and B. Lukács). World Scientific, Singapore.

Zeh, H. D. (2001). *The physical basis of the direction of time*, 4th edn. Springer, Berlin.

Zeh, H. D. (2003). There is no 'first' quantization. *Phys. Lett. A*, **309**, 329–34.

Zel'dovich, Ya. B. and Starobinsky, A. A. (1984). Quantum creation of a universe with nontrivial topology. *Sov. Astron. Lett.*, **10**, 135–7.

Zhuk, A. (1992). Integrable multidimensional quantum cosmology. *Class. Quantum Grav.*, **9**, 2029–38.

Zurek, W. H. (2003). Decoherence, einselection, and the quantum origins of the classical. *Rev. Mod. Phys.*, **75**, 715–75.

INDEX

RETURN TO: PHYSICS LIBRARY

351 LeConte Hall   510-642-3122

LOAN PERIOD 1 **1-MONTH**	2	3
4	5	6

ALL BOOKS MAY BE RECALLED AFTER 7 DAYS.
Renewable by telephone.

DUE AS STAMPED BELOW.

This **book** will be held in PHYSICS LIBRARY until APR 1 8 2005	
MAY 0 1 2005 AUG 20 2007	
MAY 3 1 2005 SEP 20 2007	
AUG 1 8 2005 NOV 23 2007	
APR 0 7 2006 DEC 24 2007	
OCT 2 8 2006 APR 2 4 2009	
MAY 1 8 2007 JAN 1 5 2010	
AUG 2 0 2007 APR 2 1 2011	
JUN 2 1 2007 JUN 1 7 2011 AUG 2 2 2012	

FORM NO. DD 22
500   4-03

UNIVERSITY OF CALIFORNIA, BERKELEY
Berkeley, California 94720–6000